The Vaccine Race

The Vaccine Race

Science, Politics, and the Human Costs of Defeating Disease

Meredith Wadman

VIKING

VIKING
An imprint of Penguin Random House LLC
375 Hudson Street
New York, New York 10014
penguin.com

PHOTOGRAPH SOURCES AND CREDITS
INSERT: *p. 1* (top): Norman Cohen; (middle) The Wistar Institute, Wistar Archive Collections, Philadelphia, PA; (bottom) University of Pennsylvania, University Archives and Records Center; *p. 2* (top): University of Pennsylvania, University Archives and Records Center; (bottom left) Leonard Hayflick; (bottom right) Katherine Aird; *p. 3* (top and bottom): Eva Herrström; *p. 4* (top): Margareta Böttiger; (bottom) Ernholm family; *p. 5* (top): Smithsonian Institution, Division of Medicine & Science, National Museum of American History; (middle) U.S. National Library of Medicine, History of Medicine Division; (bottom) American Media, Inc.; *p. 6* (top): James A. Poupard; (bottom) Frank P. Montone/ Special Collections Research Center, Temple University Libraries, Philadelphia, PA: *p. 7* (top): University of Pennsylvania, University Archives and Records Center; (bottom) Edward A. Hubbard/U.S. National Library of Medicine; *p. 8* (top): Smithsonian Institution, Division of Medicine & Science, National Museum of American History; (middle) Royal Prince Alfred Hospital Museum and Archives; (bottom) Frederick A. Murphy, University of Texas Medical Branch, Galveston, Texas; *p. 9* (top): The Wistar Institute, Wistar Archive Collections, Philadelphia, PA; (bottom) Mary and Steve Wenzler; *p. 10* (top): Philadelphia Archdiocesan Historical Research Center, Robert and Theresa Halvey Photograph Collection; (bottom) Merck Archives—Merck, Sharp & Dohme Corp., 2016; *p. 11* (top and bottom): March of Dimes Foundation; *p. 12* (top): Dorothy M. Horstmann papers (MS 1700). Manuscripts and Archives, Yale University Library; (middle) Children's Hospital of Philadelphia; (bottom) Merck Archives—Merck, Sharp & Dohme Corp., 2016; *p. 13* (top): Leonard Hayflick; (bottom) National Institutes of Health; *p. 14* (top): Jerry Huffman; (middle) poster reprinted with the permission of the Helen Keller National Center for Deaf-Blind Youths and Adults; *p. 15* (top): Frederick A. Murphy, University of Texas Medical Branch, Galveston, Texas; (bottom) The Wistar Institute, Wistar Archive Collections, Philadelphia, PA; *p. 16* (top and bottom): Meredith Wadman

LIBRARY OF CONGRESS CATALOGING-IN-PUBLICATION DATA
Names: Wadman, Meredith, author.
Title: The vaccine race : science, politics, and the human costs of defeating disease / Meredith Wadman.
Description: New York, New York : Viking, [2016] | Includes bibliographical references and index.
Identifiers: LCCN 2016044189 (print) | LCCN 2016045456 (ebook) |
ISBN 9780525427537 (hardback)
ISBN 9780698177789 (ebook)
Subjects: | MESH: Measles-Mumps-Rubella Vaccine–history |
Human Experimentation–history | History, 20th Century | United States
Classification: LCC RA644.M5 (print) | LCC RA644.M5 (ebook)
NLM WC 11 AA1 | DDC 614.5/23–dc23
LC record available at https://lccn.loc.gov/2016044189

Printed in the United States of America
10 9 8 7 6 5 4 3 2 1

Set in Garamond Premier Pro / Designed by Francesca Belanger

A sane society whose riches are happy children, men and women, beautiful with peace and creative activity, is not going to be ordained for us. We must make it ourselves.

—Helen Keller

CONTENTS

PART THREE: THE WI-38 WARS

The Vaccine Race

Prologue

The role of the infinitely small in nature is infinitely great.
— Louis Pasteur, nineteenth-century French microbiologist[1]

On October 9, 1964, a baby girl was born at Philadelphia General Hospital. She arrived early, when her mother was about thirty-two weeks pregnant. The baby weighed 3.2 pounds and was noted to be blue, floppy, and not breathing. The only sign of life was her slow heartbeat. Nonetheless, she clung to life, and her seventeen-year-old mother named her.

One month later the baby was still in the hospital, and a doctor leaning close with a stethoscope heard a harsh heart murmur. A chest X-ray showed that she had a massively enlarged heart because a hole in the muscular organ was preventing it from pumping blood efficiently. Doctors also noticed that the baby was staring into space, not fixing her gaze on anything. An ophthalmologist was called in. It emerged that the baby had cataracts blinding both eyes. Later other signs indicated that she was profoundly deaf, although a formal hearing test was never conducted.

In January 1965, after surgery attempting to repair one of the cataracts, the mother took her three-month-old daughter home. Nine days later the baby was back in the hospital with diarrhea. She remained in the hospital, where she suffered from recurring respiratory infections. She had trouble gaining weight, which is a common problem in infants with heart problems like hers. A psychologist who assessed her in July 1965, after a second heart defect was found, judged the nine-month-old to be the size of a two- or three-month-old infant and at about that stage of development; she couldn't sit up or grasp an object placed in her hand.

The baby needed heart surgery if she was going to survive. Just before her first birthday, surgeons cut a seven-inch incision in her chest wall and repaired her heart. After the operation she remained in the hospital. The chronic respiratory infections continued. The baby was sixteen months old and weighed eleven pounds when she died of pneumonia at 3:30 a.m. on February 18, 1966.

She had lived all but nine days of her brief life at Philadelphia General Hospital.

The young mother had told the doctors something when she brought her daughter back to the hospital. When she was one month pregnant, she had had German measles, which is also known as rubella.[2]

The early 1960s marked a coming of age for the study of viruses like the one that causes rubella—tiny infectious agents that invade cells and hijack their machinery in order to reproduce themselves. Biologists, with new tools in hand, were racing to capture viruses in throat swabs or urine or even snippets of organs from infected people and to grow them in lab dishes. Isolating a virus in the lab made it possible to make a vaccine against it. And making antiviral vaccines promised huge inroads against common childhood diseases like measles, mumps, and rubella, along with less-common killers like hepatitis. The principle of vaccination is simple: if a person is injected with, or swallows, a tiny amount of a virus—either a killed virus or a weakened live virus—that person will develop antibodies against the virus. Then, if he or she is exposed in the future to the naturally occurring, disease-causing form of the virus, those antibodies will attack the invader and prevent it from causing disease.

But if the concept is simple, making effective vaccines is anything but. In the early 1960s that reality was all too evident in recent tragedies. In 1942 as many as 330,000 U.S. servicemen were exposed to the hepatitis B virus in a yellow fever vaccine that was contaminated with blood plasma from donors—plasma that was used to stabilize the vaccine. It turned out that some of those donors carried hepatitis B. About 50,000 of the vaccinated servicemen came down with the dangerous liver disease and between 100 and 150 died.[3] In 1955 a California-based company called Cutter Laboratories made polio vaccine with live, disease-causing virus in it, paralyzing 192 people, many of them children, and killing ten.[4] Every senior U.S. government employee involved in overseeing the Cutter process lost his or her job—right up through the director of the National Institutes of Health and the secretary of health, education, and welfare.[5]

Then in the summer of 1961, Americans learned that the monkey kidney cells used to make the famous Salk polio vaccine often harbored a monkey virus called SV40. Tens of millions of American children had already received contaminated injections, and while the jury was still out on the tainted vaccine's long-term health consequences, the unknown risks were weighing on regulators in the United States and elsewhere.

It was against this backdrop that, on a drizzly June morning in 1962, a young scientist went to work in his lab at the Wistar Institute of Anatomy and Biology, an elegant 1890s brownstone tucked in the heart of the University of Pennsylvania campus. Leonard Hayflick had just turned thirty-four years old. A serious, slight, reserved man with close-cropped dark hair, Hayflick was a product of working-class Philadelphia and was hungry to make his name. He was in love with biology and was plenty smart—he had come to believe—but that fact was far from appreciated. His boss, the famous polio-vaccine pioneer Hilary Koprowski, saw him as a mere technician, hired to serve up bottles of lab-grown cells to the institute's impressive cadre of biologists.

This hadn't deterred the ambitious Hayflick. The previous year the junior scientist had published a paper challenging a major piece of scientific dogma: the belief that cells grown in a lab bottle, if properly nurtured, would multiply indefinitely. His findings had been met with skepticism from some outstanding biologists. Let the critics carp, he thought. Time would prove that he was right—that normal cells cultivated in the lab eventually died, just like human beings.

On this drizzly day, however, Hayflick's mind was not on cell death but on cell birth. Today, he hoped, he was going to launch a group of normal human cells that would revolutionize vaccine making. He had been waiting months for this opportunity—waiting for the arrival of the lungs that would be the source of these new cells. Cells were needed to make antiviral vaccines because outside of cells viruses can't multiply. And huge quantities of viruses were needed to produce vaccines. Now, at last, the lungs were here in his bustling second-floor lab, two purplish things floating in clear pink fluid in a glass bottle. They had traveled all the way from Sweden packed on wet ice, courtesy of a Koprowski colleague who was a top virologist at the prestigious Karolinska Institute in Stockholm.

Several days earlier a woman living near Stockholm had had an abortion. Most Swedish physicians frowned upon the procedure, but it was legal, even for not-strictly-medical reasons. The woman was sixteen or seventeen weeks pregnant and had several children already. Her husband, she told her doctors, was an unsupportive alcoholic. The decision was clear. She sought out a sympathetic gynecologist, Eva Ernholm, one of the rare women in Swedish medical ranks, to perform the procedure.

After the abortion the eight-inch-long female fetus was wrapped in a sterile green cloth and delivered to a yellow brick outbuilding on the grounds of the National Bacteriological Laboratory in northwest Stockholm. Here, in what they nicknamed the "monkey house" because it was also home to monkeys used in making polio vaccine, young PhD and medical students were occasionally called on to dissect out the lungs of aborted fetuses for shipping to the Wistar Institute. It wasn't a pleasant task, but when their boss, Sven Gard, the top virologist at the Karolinska Institute, asked them to do it, they obliged, slipping on head covers and changing from white wooden clogs to red or blue ones when they entered the sterile rooms. Other employees, working nearby in a grand building with a spiral staircase, were responsible for packing the lungs on ice and transporting them to Bromma Airport for the transatlantic flight that would eventually bring them to Philadelphia.

Hayflick was convinced that compared with monkey kidney cells, which were often laden with lurking viruses, normal human cells would serve as cleaner, safer vehicles for making antiviral vaccines. And he knew that he was uniquely positioned to produce a long-lasting supply of such cells. He had spent the previous three years perfecting the procedure that would do it.[6]

Hayflick took the bottle with the little lungs floating in it into a tiny room off his lab—what passed for a "sterile" room in 1962. He picked up a pair of tweezers, dipped them in alcohol, and passed them through the flame of a Bunsen burner. He waited for them to cool and then, gently, one at a time, lifted the lungs out of the bottle and laid them on a petri dish. The underdeveloped organs were each no larger than his thumb above the knuckle. He assembled two scalpels, held the blades at right angles to each other, and began carefully slicing the lungs. He didn't stop until he had cut them into innumerable pieces, each about the size of a pinhead.

Hayflick nudged the minute pieces of lung into a wide-mouthed glass flask. The translucent pink fluid inside the flask looked innocent enough, but it was full of digestive enzymes from slaughtered pigs. These biological jackhammers broke up the "mortar" between the lung cells, freeing millions upon millions of them.

Later, he transferred the resulting cells into several flat-sided glass bottles and poured nutritious solution over them. He loaded the bottles onto a tray and walked them into an incubation room beside his lab. Here the temperature was a cozy 96.8 degrees Fahrenheit. He laid the bottles on their sides on a wooden shelf and closed the door carefully behind him.

The cells began to divide.

Hayflick already had a name for them: WI-38.

The WI-38 cells that Hayflick launched on that long-ago summer day were used to make vaccines that have been given to more than 300 million people—half of them U.S. preschoolers. A copycat group of cells, developed using the method that Hayflick pioneered, has been used to make an additional 6 billion vaccines. Together these vaccines have protected people the world over from a whole range of viral illnesses: rubella, rabies, chicken pox, measles, polio, hepatitis A, shingles, and adenovirus—a respiratory infection that flourishes where people live in close quarters. (Every U.S. military recruit—more than 9 million of them since 1971—is vaccinated with an adenovirus vaccine made using WI-38 cells.)[7]

In the United States the rubella vaccine made in WI-38 cells and still given to young children has wiped out homegrown rubella. That vaccine was developed at the Wistar, by Hayflick's colleague Stanley Plotkin, in the midst of a devastating rubella epidemic that swept the country in 1964 and 1965. That rubella outbreak damaged tens of thousands of American babies—including the baby described above who lived most of her short life at Philadelphia General Hospital. This book will tell the story of that epidemic and of the race that followed to develop a rubella vaccine.

How can it be that these WI-38 cells launched so long ago are still in use today? Partly because Hayflick made such a large initial stock of them: some eight hundred tiny, wine bottle–shaped ampules that he froze in the summer of 1962. Partly because the cells, when frozen, stop dividing, but then gamely begin replicating again when they are thawed—even after decades. And partly because of the power of exponential growth. Each petite glass vial that Hayflick froze contained between 1.5 million and 2 million cells. And the cells in those vials had, on average, the capacity to divide about forty more times. Early on, Hayflick did the math and determined that the newly derived cells covering the floor of just one of his small glass lab bottles, if allowed to replicate until they died, would produce 22 million tons of cells. He had created in those eight hundred vials a supply of cells that for practical purposes was almost infinite.

And so, in addition to their use in vaccine making, the WI-38 cells became the first normal, noncancerous cells available in virtually unlimited quantities to scientists probing the mysteries of cell biology. Because they were easily infected with many viruses, they became important to disease detectives tracking viruses in the 1960s, before more sophisticated technology came along. Biologists still reach for WI-38 cells when they need a normal cell to compare against a cancerous one or to bombard with a potential new drug to see if it's toxic. The cells are also a workhorse of aging research, because they so reliably age in lab dishes. They are held in such high regard by scientific historians that original ampules of WI-38, and of polio vaccine made using it, are part of the collection of the National Museum of American History.

In the 1960s and 1970s the cells became the object of a bitter, epochal feud between Hayflick and the U.S. government, first over whether they were safe for vaccine making and then over who owned them. Hayflick's preternaturally proprietary feelings for the cells—he once described them as "like my children"—led him to defiantly decamp from the Wistar to a new job three thousand miles away at Stanford University with the entire stock of WI-38 cells. His escape infuriated the Wistar's director, Koprowski, who had his own money-making designs on the cells.

Hayflick's flight with the cells would eventually make him the target of a career-derailing investigation by the National Institutes of Health, which had funded his work deriving WI-38.[8] Then, just as the tug-of-war over ownership of the WI-38 cells peaked in the second half of the 1970s, profound changes occurred in attitudes and laws governing who could make money from biological inventions. In the space of a very few years, biologists went from being expected to work for their salaries and the greater good—and nothing more—to being encouraged by their universities and the government to commercialize their inventions for the benefit of their institutions, the U.S. economy—and themselves.

Although the WI-38 cells were launched long before these changes took place—and eighteen years before the Supreme Court decreed that a living entity like a WI-38 cell could be patented—that is not to say that money has not been made from them. The huge drug company Merck in particular has made billions of dollars by using the WI-38 cells to make the rubella vaccine that is part of the vaccine schedule for U.S. babies and preschoolers—ensuring more than seven million injections each year, not including those in more than forty other countries where the Merck vaccine is sold. The Wistar Institute too until the late 1980s enjoyed a handsome royalty stream from vaccines made by its scientists using the WI-38 cells—including a much-improved rabies vaccine that replaced sometimes-dangerous injections. Cell banks today charge several hundred dollars for a tiny vial of the cells.

But the tale of the WI-38 cells involves much more than money—and more than the highly unusual story of Hayflick, the iconoclastic scientist who launched them. It involves the silent, faceless Swedish woman whose fetus was used to derive the cells without her consent. It involves the dying patients into whose arms the WI-38 cells were injected with the aim of proving that the cells did not cause cancer. It touches on the ordinary American children who perished from rabies before WI-38 cells were used to make a better vaccine, and on the U.S. military recruits who died from adenovirus infections when the Pentagon stopped giving service members the vaccine against that virus, made in WI-38 cells. It involves the abortion opponents who, now as then, harbor a deep moral abhorrence of any vaccines made using human fetal cells.

It is also about Stanley Plotkin, a young scientist who stubbornly fought powerful competitors by using the WI-38 cells to develop a superior rubella vaccine—and the purely political roadblocks that nearly stopped him. And it is about the one-, two-, and three-year-old orphans on whom Plotkin tested that vaccine, with the blessing of the archbishop of Philadelphia. It involves the irony of the untold millions of miscarriages, stillbirths, and infant deaths that have been prevented by a rubella vaccine made using cells from an aborted fetus.

These pages are full of medical experiments that we find abhorrent today. Young, healthy prisoners are injected with hepatitis-tainted blood serum; premature African American babies with experimental polio vaccine; intellectually disabled children with untried rubella vaccine.

We recoil in horror. It is easy to condemn out of hand the scientists who conducted these experiments on the most voiceless and powerless among us. And their actions were in many cases horrifying and inexcusable. But it is more instructive—and perhaps more likely to prevent similar betrayals in the future—to try to understand why they did what they did.

The experiments began, in large part, during World War II.*

They grew out of the exigencies of the war, when an ends-justify-the-means mentality took over in U.S. medicine in the interest of keeping soldiers healthy at the front, because civilization was at stake.[9] Everyone was expected to do their part for the cause—even institutionalized people with grave disadvantages or disabilities. When the war ended, the mentality didn't. In the two decades following the war and in several cases into the 1970s, medical researchers experimented on people—almost always vulnerable people—making them sick and sometimes killing them.[10]

These scientists were perceived and perceived themselves as part of a heroic quest to defeat disease. They were ambitious, driven, and well funded by the U.S. government. And they got results.

_____

*A striking exception is the Tuskegee Syphilis Study, in which U.S. government researchers purposely left syphilis untreated in 399 poor, illiterate African Americans. That notorious study began in 1932.

During World War II and in the two decades following it, childhood mortality declined strikingly, in large part because of dramatic inroads against infectious diseases. Antibiotics that became available in the 1940s turned often-lethal diseases like typhoid fever and dysentery into less-grim maladies and slashed both the incidence of tuberculosis and its lethality. Vaccines against diphtheria, polio, and whooping cough hammered these childhood killers. Infectious diseases as a cause of death among children were rare by the middle of the 1960s.

The men who conducted unethical human experiments in this era were not medical outliers. They were top physicians and researchers operating with the full backing of the U.S. government, private funders, and esteemed medical schools and hospitals. But in 1966, when a landmark paper in the *New England Journal of Medicine* exposed the harm being done to powerless people in scores of experiments, the government implemented new protections.[11] The surgeon general launched a requirement that people give their informed consent to participate in research studies funded by the U.S. government's health agencies and that researchers win preapproval for their human experiments from an independent committee charged with examining the risks and benefits to participants.[12] Since then, those protections have been strengthened, expanded, and written into U.S. law. Today's system of human-subject protections is not perfect. In fact, it has serious shortcomings and vocal critics.[13] But it is worlds better than the feeble effort that existed half a century ago.

To remove the history of human exploitation from vaccines and medicines that were developed in the postwar era is impossible. The knowledge that allowed their development is woven into them. Should we therefore shun them? Definitely not. Take rubella as a case in point. As I write this in the summer of 2016, 1,700 babies in a dozen countries have been born with abnormally small heads or other brain malformations; their mothers were infected with the Zika virus while pregnant.[14] Zika's emergence is a vivid reminder of what life was like in the United States in 1964. Then, there was no rubella vaccine and tens of thousands of American babies were born gravely damaged by the rubella virus, which selectively harms fetuses in the womb. Like Zika, rubella homes in on the brains of fetuses; it also ravages their eyes, ears, and hearts. But today, thanks to the vaccine that was perfected in experiments on institutionalized orphans and intellectually disabled children, indigenous rubella has been wiped out in the Western Hemisphere. Cases occur only when they are imported from other countries.

We can't turn the clock back. The only way we can partially make it up to these children and untold others is to honor their contributions by making them meaningful—by continuing to vaccinate against rubella and the other diseases that made childhood a perilous journey before vaccines against them existed. We also need to strive constantly to enforce and improve the regulations and laws that protect research subjects so that in the future such abuses never happen again. We might also remember, when judging the men who took advantage of vulnerable human beings in order to advance both human health and their own careers, that they were creatures of their time, just as we are of ours. Rather than training our criticism on them, it might be more useful to ask ourselves this: what are we doing or accepting or averting our eyes from today that will cause our grandchildren to look at us and ask, *How could you have let that happen?*

PART ONE

The Cells

CHAPTER ONE

Beginnings

Philadelphia, 1928–48

Once upon a time there was a boy. He lived in a village that no longer exists, in a house that no longer exists, on the edge of a field that no longer exists, where everything was discovered and everything was possible.

—Nicole Krauss, *The History of Love*[1]

When he was about eight years old, Leonard Hayflick had a scare that caused him to run to his mother in tears and imprinted itself vividly on his memory. Hiking one day with friends in Cobbs Creek Park, near his home in southwest Philadelphia, he slipped when crossing the creek on stepping-stones. One of his sneakers got soaked.

The young boy immediately panicked. Polio is spread through contaminated water, and the terror of the paralyzing, sometimes-fatal disease was acute in the mid-1930s. Hayflick sat down, crying, and took his shoe and sock off, desperately rubbing his foot with the nearest chunk of dirt or grass he could find. He went home to his mother, who tried to comfort him.

His fears were understandable. What had been a rare disease in the nineteenth century had become all too common in the twentieth. Annual summer surges in polio cases had mothers keeping their children out of public swimming pools. Not even the most privileged Americans were safe. Indeed, because the wealthy grew up in cleaner environments, they were less likely to be exposed to polio and to develop protective antibodies as children. President Franklin D. Roosevelt, the man who was running the nation from a wheelchair, had been paralyzed by the dread disease at age thirty-nine.

In fact, infectious diseases of all kinds were a real threat in the 1930s. Children died of scarlet fever; of influenza; of tuberculosis; of measles. There were no reliable vaccines to prevent these common maladies. The first antibiotics wouldn't be prescribed until the late 1930s. Hayflick remembers the orange signs from Philadelphia health authorities that would appear periodically on the front doors of afflicted households, proclaiming in huge black font: **This house is quarantined because of the presence of measles**—or another infectious disease.

But Hayflick had not contracted polio from his brush with the water of Cobbs Creek. He was luckier than thousands of American children in the 1930s.

Hayflick came from humble beginnings. He was born on May 20, 1928, to Nathan and Edna Hayflick, the new, young owners of a narrow brick row house in a working-class neighborhood in southwest Philadelphia. Hayflick's parents were part of a Jewish migration across the Schuylkill River out of the slums of south Philadelphia that began before World War I and continued in the 1920s.[2] The new arrivals launched thriving synagogues and Hebrew schools. The sidewalks were wide and the families young. The schools were less than first-rate, but that did not tamp down the ambitions of many families who were determined to build better lives.

Hayflick's own father, Nathan, when he was eight years old, had been living in a south Philadelphia row house, occupied by thirteen family members, abutting a rough red-light district.[3] In this densely packed neighborhood of dark, cobblestoned streets and alleys that often lacked pavement or sewers, filth and excrement stuck in the cracks, frozen, during the winter and then thawed in the spring.[4] Single outhouses served dozens of people. The Philadelphia authorities, notorious for their indifference and corruption, did virtually nothing to improve conditions. In fact, they paid attention to sanitation in south Philadelphia only when outbreaks of cholera, typhus, or diphtheria blighted the area and threatened to spread.[5] The crowding and filth made the slum a perfect incubator for the devastating influenza pandemic of 1917–18, when hospitals, homes, and morgues were overwhelmed and corpses spilled into the street.[6] The Hayflick clan seemed to emerge unscathed, although Nathan's mother soon died of tuberculosis. Nathan, in his midteens, went to work driving the horse-drawn family milk cart.

Within a few years he had landed a job at the Climax Company, a leading Philadelphia denture-designing firm. He would advance to become a master denture designer, serving clients including Albert Einstein. Lunching one day at a popular diner, he met Edna Silver, a quiet, thoughtful young beauty who was also from south Philadelphia and who, like him, was the child of Eastern European immigrants. The couple married in 1927 and moved across the river.

Their three-bedroom row house was soon full. Leonard Hayflick was born the year after they married, and eighteen months later Edna gave birth to a daughter, Elaine. The date was November 11, 1929. Two weeks earlier the bottom had fallen out of the U.S. stock market, launching the Great Depression.

By 1933 half of the city's banks had failed and just 40 percent of the work-force was fully employed.[7] The Philadelphia County Relief Board began dis-tributing shoes to schoolchildren—seven hundred pairs every day.[8] Nathan Hayflick's wages were cut and the Hayflicks became one of ninety thousand city families to lose their houses.[9] They moved to a nearby rental. Eventually the family's finances recovered enough for Leonard's parents to buy a modest row house in the same neighborhood, where Hayflick spent his teenage years.

Despite the difficulties his family endured during the Depression, Hayflick says he doesn't remember ever being hungry or aware of the family's financial duress.

"I never had a motivation to make money, ever," he says. "The Depression and my mother's and father's experiences played no role in my outlook on life."[10] But it did, he told another interviewer in 2003, impact him in the fol-lowing way: "Being brought up in the Depression has a lot to do with my work ethic, my belief in myself, and [the belief] that I should have confidence in what I think is true and correct as long as it is demonstrably so."[11]

In that interview he also recalled that his parents' broad-mindedness in-stilled in him a bent for challenging convention. "My mother and father were very liberal. . . . I take enjoyment in challenging dogma. If there is anything that I challenge, it is orthodoxy."

Hayflick was exposed to lab life early. On Saturdays he sometimes tagged along with his hardworking father to the lab at the Climax Company. Nathan Hayflick kept his son occupied by sitting the boy in front of a Bunsen burner with plaster molds and easily melted Melotz metal to liquefy and pour into them. Leonard Hayflick learned from his uneducated but talented father to be at home in a lab. He also saw in him the costs of a stunted education. His father was fascinated by all manner of scientific questions but hadn't the tools or the energy to pursue them: upon returning home after his thirteen-hour workdays, he would fall asleep on the couch.

Hayflick's mother taught him not to be afraid to ask questions. Her an-swers were patient and explicit when he asked why street-corner newspaper vendors were shouting—whether their three-inch headlines were reporting the kidnapping of the Lindbergh baby when Hayflick was not yet four years old or Hitler's troops marching into Austria when he was nine.

A chance gift from his favorite uncle ignited a passion when Hayflick was ten or eleven years old. Jacob Silverman, a smart, natty thirtysomething bach-elor, gave his nephew a Gilbert Company chemistry set. It came with test tubes, a test tube holder, tongs for grasping the test tubes, and an alcohol lamp with a ground-glass head on it to cover the wick and keep the alcohol from evaporating.

The young boy was stunned to learn that the universe was composed—as was believed at that time—of a mere ninety-two elements, and that they behaved in such extraordinary ways when they were combined. He was entranced by the color changes, the bursts of flame, and the substances that mysteriously precipitated out in the bottom of his test tubes. At age eighty-six Hayflick would still have the set's manual and alcohol lamp.

It wasn't long before Hayflick had exhausted the chemicals and experiments that came with the set. With a neighborhood friend named Teddy Cooper, he began biking across large swaths of southwest Philadelphia in search of new chemicals and the glassware that would show them off to the best advantage. The duo got to know a kind sales clerk at Dolbey and Company, a chemical and glassware supplier near the University of Pennsylvania. The thin, bespectacled salesman eventually let them look at obsolete supplies in the store's basement, and they came home laden with outmoded retorts and condensers.

Soon Hayflick was busy building his own basement lab; he walled in a corner, put in a workbench, and mounted shelves where he proudly displayed his chemicals in labeled bottles. He and Cooper also asked the friendly salesclerk to sell them some metallic sodium—a volatile element the consistency of a hard stick of butter that's kept under kerosene to mute its explosiveness and that bursts into a flame of burning hydrogen when it's submerged in water. The clerk told Hayflick he couldn't do it without a letter from his mother. Hayflick went home and composed one, which his trusting mother signed in her beautiful cursive. Hayflick and Cooper began riding the neighborhood's back alleys after rainstorms. They deposited chunks of the metallic sodium in the water-filled holes where laundry poles stood in good weather, then sped gleefully away.

By the time he graduated from high school, Hayflick had developed a keen sensitivity to injustice, especially when it involved him. He won the Bausch + Lomb Honorary Science Award for being the best science student at John Bartram High School but marched into the principal's office and returned it indignantly after learning he had placed second to a female classmate in the race for the coveted Philadelphia Mayor's Scholarship. That scholarship would have paid his tuition at any university or college in the country.

"It was a bitter, bitter disappointment," Hayflick said emphatically during a 2012 interview, sounding as if he could still taste the letdown nearly seventy years later.[12] "It was clearly a consolation prize. The first prize was given to a girl 'apple polisher' whose mother sent pies and gifts to her teachers."

Temple University offered him a scholarship, but Hayflick didn't accept it. He had his sights set higher. He wanted to go to the prestigious University of Pennsylvania, known to many as "Penn." The university sat on an ivied hundred-acre campus in west Philadelphia that Hayflick's father passed on the streetcar on his way to work. It was founded by Benjamin Franklin. It had a medical school as old as Harvard's. And its history was peopled with leading medical and scientific figures like the physician William Osler and the anatomist and paleontologist Joseph Leidy. As he prepared to graduate from high school in January 1946, Hayflick applied to Penn and was accepted. His parents scraped together the nearly $250 that would pay his tuition for the spring semester.

Hayflick was a slight seventeen-year-old who stood less than five feet nine inches tall, and he found himself lost in a sea of former servicemen who were flooding the university on the GI Bill. Not only was Hayflick intimidated, but it soon became apparent that his parents couldn't sustain the tuition, especially with his sister preparing to enter college too. So, shortly after his eighteenth birthday in May 1946, Hayflick took a leave of absence from Penn and enlisted in the U.S. Army. The support of the GI Bill would pay his tuition when he returned to Penn and—crucially, it would turn out—would also provide $75 in monthly living expenses.

In the army Hayflick learned to repair antiaircraft guns at the Aberdeen Proving Ground in Maryland and later landed at Fort Benning, Georgia, where he was chosen for a job as a teacher in a program that allowed army re-cruits to finish high school. The classes didn't attract many soldiers, and Hay-flick happily used the time in his small office to read. The position also came with a chauffeured car that he regularly summoned to drive him to the post's library.

When Hayflick, by then almost twenty, reenrolled at the University of Pennsylvania in the spring of 1948, he was unsure what academic route to pur-sue. He took chemistry, math, zoology, and English; he also recalls enrolling in an accounting course at the Wharton School of Business. But he was soon dis-tracted from his studies because his family was in crisis.

In the late 1940s Nathan Hayflick was persuaded by his nephew, Norman Silverman, to leave the Climax Company after three decades to join Silver-man's fledgling denture-designing business, Victory Laboratories, just over the Delaware River in Camden, New Jersey. Silverman was a charming, ambitious young entrepreneur and a relative newcomer to denture design. Nathan Hayflick was by now a master craftsman, but not a businessman. The partnership turned into a disaster as the two men's personalities clashed. Nathan Hayflick fell into a depression that concerned his son so much that Leonard Hayflick began cutting classes, leaving Penn at midday to cross the river and take his father to lunch in order to get him out of the lab for a couple of hours.

Nathan Hayflick didn't have an exit strategy: his old job at the Climax Company had been filled. He was a man in his late forties with a grade-school education and a craft. Leonard Hayflick decided there was only one way out for his father. Nathan Hayflick needed to go into business for himself, bringing his loyal clientele of dentists with him. While a student at Penn, Hayflick set about building his father a laboratory. He had absorbed enough of the denture-designing craft by osmosis to be sure that he could do it. The entire effort would have to be accomplished on a shoestring: the four Hayflicks were now surviving on the $75 in monthly living expenses that the GI Bill provided to Leonard Hayflick.[13]

Through a distant relative, Hayflick located rental space above the Latin Casino, a popular center-city nightclub. With the help of Al Ketler, a close friend and fellow Penn student who was skilled in carpentry and plumbing, Hayflick equipped the lab with a plaster bench, grinding instruments, water lines, and a casting machine that he and Ketler built out of a fifty-gallon steel drum. He and Ketler sneaked into the ancient building's basement, confronted a tangle of wires and pipes, selected a likely-looking gas pipe, and cut it with a metal saw, allowing them to run an extension to supply the lab's Bunsen burners. "We were young and not too smart," Hayflick recalled in a 2014 interview. "Fortunately, we didn't die."[14]

What did suffer were Hayflick's grades. He didn't think he had a choice; it was a matter of his family's economic survival. In the end, the Nathan Hayflick Dental Laboratory opened, the senior Hayflick got back on his feet, and his son turned back to his studies.

CHAPTER TWO

Discovery

Philadelphia and Galveston, 1948–58

Oh, you may be sure that Columbus was happy not when he had discovered America, but when he was discovering it.

—Fyodor Dostoevsky, *The Idiot*

Early during his studies at the University of Pennsylvania, Hayflick enrolled in an introductory course in what was then called bacteriology.

On his first day in the lab, a technician walked in carrying something that changed Hayflick's life. It was a tray of test tubes containing a nutritious, gelatin-like substance called agar, which was a dull chicken-soup color. The test tubes had been tilted so that the agar, initially poured into the tubes in liquid form, solidified on a slant, maximizing the surface area available for bacteria to grow on. The technician had then used a fine needle to inoculate each tube with a different kind of bacterium, dragging the needle in a wave pattern along the yellow brown agar to "streak" it. What the young Hayflick saw was the resulting bacterial growth. The slants were streaked with a rainbow of colors, from yellow to purple, green, white, and pink.

Hayflick was blown away. He decided on the spot to major in bacteriology. (The discipline would soon be relabeled "microbiology," in order to encompass viruses as well as bacteria.)

Hayflick's love affair with microbes began at the dawn of a golden age for the study of viruses. The study of bacteria—bigger organisms that can survive independently outside of cells—was older. Scientists had been growing and examining bacteria in lab dishes since the late 1870s, when a German microbiologist named Robert Koch developed practical methods of growing pure cultures of bacteria in the lab. Koch also decisively laid out the steps that biologists needed to take to prove that a given bacterium was causing a particular disease. Biologists began to link specific bacteria with diseases, to understand how they were transmitted, to track outbreaks, and to launch the first therapeutic salvos against bacterial banes like diphtheria and syphilis.

The ease with which bacteria could be grown in lab dishes was vital not only for studying them but also for the discovery and testing of the antibiotics that were new miracles in the late 1940s: drugs like sulfa and streptomycin. The most famous among these was discovered when a short, slight Scotsman named Alexander Fleming noticed something strange on an agar plate on which he was growing *Staphylococcus* bacteria in his lab at St. Mary's Hospital in London. A mold had accidentally taken root on the petri dish, and the bacteria, which were thriving elsewhere on the dish, wouldn't grow anywhere near the mold. That moldy invader, it emerged, made a substance that Fleming named penicillin.

Virology was a slightly younger and decidedly less well-equipped science. Viruses had been known to exist since the early 1890s, when a young Russian scientist named Dmitry Ivanovsky took sap from the yellowed, stunted leaves of plants with tobacco mosaic disease and passed it through a filter containing pores too tiny for bacteria to slip through. (Bacteria are enormous compared with viruses. Consider that HIV, a typical-sized virus, is a mere golf ball compared with the soccer ball that is *Streptococcus pyogenes*, the diminutive bacterium that causes strep throat today but regularly killed kids when Hayflick was a child.)

The filtered fluid from the diseased tobacco leaves was able to infect other, healthy tobacco plants.[1] Soon a Dutch botanist, Martinus Beijerinck, who had done similar experiments, demonstrated that whatever was causing the tobacco-plant disease could reproduce itself but needed living cells in which to do so.[2] He became convinced that the disease-causing entity was a liquid, and he christened it with a Latin name, "virus," which means "slimy fluid."

The same year, 1898, a pair of German scientists, Friedrich Loeffler and Paul Frosch, found that they were able to pass along a devastating animal affliction, foot-and-mouth disease, by taking fluid from the sores of infected calves, filtering it, and using the filtrate to infect other animals—the first proof of animal infection by these mysterious new entities, which scientists began calling "filterable agents."[3] (It would be some time before the term "virus" came into common use.) The German duo also surmised, correctly, that the infectious agent wasn't a liquid but a particle so small that the filter did not capture it.[4] The pair also developed what was probably the first killed-virus vaccine, taking fluid from the sores of infected animals, heating it to destroy its infectivity, and injecting it into nonimmune cows and sheep, the overwhelming majority of which were then protected from the disease.[5]

In the next half century some dozen viruses that cause human diseases were identified, including the viruses that cause yellow fever, rabies, polio, and influenza. Dozens of animal and plant viruses were also found. In 1927, Thomas Rivers, an eminent American microbiologist, defined viruses as "obligate parasites," meaning that they could reproduce only by invading living cells.[6] In 1928, the year of Hayflick's birth, Rivers published *Filterable Viruses*, a collection of essays of which he was editor, describing the roughly sixty-five viruses that had been identified to that date.[7]

But identifying viruses was hardly tantamount to understanding them, never mind fighting them. And the fact was, for the first half of the twentieth century, scientists investigating human-infecting viruses were hobbled by the difficulty of getting *at* them. That was because, unlike bacteria, which live happily and independently in lab dishes as long as they're nourished with nutritious substances like agar, viruses need living cells in order to survive.

At its most basic, a virus consists of a circular or linear thread of genetic material—DNA or its chemical cousin, RNA—and a protective protein coat. It reproduces itself by invading a cell and forcing the host cell's machinery to make tens, hundreds, or thousands of copies of the virus in one huge burst, sometimes in the space of minutes. These new viruses—each virus is called a virus "particle"—then bud or burst out of the cell and proceed to invade other cells. (Some viruses can also move directly from cell to cell.)

Despite their formidable talent for hijacking, viruses are helpless on their own. They are not independent organisms that propel themselves around, eat, digest, excrete waste, or have sex. Their sole business is to invade living cells so they can reproduce. So while they can survive on nonliving objects and surfaces for hours, days, weeks, and sometimes months, they are merely inert chemicals when they are sitting in, say, a test tube. It is "only in the interior of a living cell [that a virus's] hidden forces are liberated," the Swedish virologist Sven Gard—who will play a role in this story—observed as he presented the Nobel Prize in 1954.

But how then to study human viruses? Sometimes scientists relied on human heroism. Walter Reed, the famous U.S. Army physician who proved that yellow fever was caused by a mosquito-borne virus, did so by enlisting volunteers who were willing to be bitten by mosquitoes that had recently fed on yellow fever–infected patients. One of Reed's medical colleagues, Jesse Lazear, was infected and died in that decisive experiment in 1900.

Sometimes virologists could use living animals to study human diseases, like the group of British investigators who, during an outbreak of influenza in 1932, noticed that some of their furry laboratory ferrets, being used for other purposes, were sneezing. They had caught human influenza from the sick scientists. (Later the reverse also happened.) From then on, the group studied influenza by using a pipette to drop throat garglings from infected people onto ferrets' noses. When the disease was at its height, they would sacrifice the ferrets and study their tissues. But observing the damage to animals—rather than observing the virus itself—was hardly satisfactory. And yet viruses were too small to be seen with the light microscopes that were then available. Besides which, many human viruses did not infect other animals.

In some limited cases scientists succeeded in growing human viruses in lab dishes, using a little-understood art called tissue culture. Tissue culture today is more commonly called cell culture. It means growing living cells in the laboratory, outside of the animals or plants that they came from. (It will play a central role in this book.) Ross Harrison, a brilliant, driven biologist then at Johns Hopkins University, is credited with launching tissue culture in 1907, by growing bits of frog embryo brain in the lab. By nourishing the frog brain cells with fluid from the frogs' lymph glands, he kept the cells alive for weeks.

In the following three decades virologists would manage, with difficulty, to grow several viruses in tissue culture—for instance, in fresh, minced hen's kidney bathed with blood serum (the liquid, noncellular part of blood). But their successes were sporadic and inconsistent. Soon the viruses would die out in their dishes. Over these decades the only practical accomplishment to come from the use of tissue culture in virology was this: in the 1930s Max Theiler, a South African–born virologist at the Rockefeller Institute in New York City, weakened the human yellow fever virus by growing it in minced chicken embryos, developing the yellow fever vaccine that is still used today.

This achievement was the exception when it came to viruses—in stark contrast to the strides being made against their bacterial counterparts. By the middle of the twentieth century, bacterial diseases were being beaten back by vaccines against once-common killers like diphtheria, tuberculosis, and whooping cough—and by antibiotics. These new wonder drugs shut down bacteria, but they didn't target viruses, which, because of the way they co-opt the native machinery of host cells, are harder to take aim at without producing off-target side effects. The first antiviral drugs wouldn't begin to be developed until the 1960s. And so, as Hayflick stood, transfixed, before those rainbow streaks in a bacteriology lab at the University of Pennsylvania, viral illnesses like measles, rubella, and hepatitis remained stubborn, dangerous banes—with polio the most visible and frightening among them.

The discovery that changed everything for virus hunters happened in 1948, just as Hayflick returned to Penn from the army. In the spring of that year, an unassuming, middle-aged scientist was laboring in a small lab at the Boston Children's Hospital. John Enders came from a wealthy New England banking family. He enjoyed the poetry of T. S. Eliot, favored old tweed jackets, and had flown rickety biplanes as a flight instructor during World War I. He had failed as a real estate agent before developing a passion for biology and earning a PhD at Harvard. Enders and his younger colleague Thomas Weller, a pediatrician from a family of physicians, had been trying to improve tissue-culture techniques for a decade—interrupted by World War II, when Weller had served in the Army Medical Corps. They had lately been joined by a third virus hunter, Frederick Robbins, an infectious-disease physician who had served in the army in North Africa and Italy, winning a Bronze Star. The trio was about to break open the world of virology, allowing scientists to grow a plethora of viruses in many tissues in lab dishes. With the application of their techniques, viruses would no longer die out in their dishes but would continue to multiply, allowing for study—and vaccine making.

Apart from sheer, dogged hard work, there were three key developments that led to the Enders team's success. First, tissue culturists were figuring out, slowly, how to improve the nourishing solutions that they used to keep cells alive in lab dishes. Second, the Boston scientists put to good use a roller-tube system invented fifteen years earlier. Something like a Ferris wheel, it slowly rotated cells in test tubes, lying on their sides, eight to ten times every hour. This allowed the cells to be first washed in nutrient fluid, then exposed to air, in an attempt to mimic the conditions in the human body with its ceaseless flow of oxygenated blood to and waste removal from cells. Third, and crucially, as the 1940s progressed, Enders also took advantage of a new tool: antibiotics, which he began religiously applying to his test-tube cultures. He hoped, correctly as it turned out, that they would eliminate contaminating bacteria but leave viruses thriving.

One day in March 1948, Enders made a seemingly off-the-cuff suggestion to his junior partners. It involved polio. Poliovirus had resisted attempts to corral it in culture dishes, with this exception: In the mid-1930s a brilliant, ambitious young scientist named Albert Sabin, working with his senior colleague Peter Olitsky at the Rockefeller Institute in New York City, managed to grow polio in nerve cells from the brains and spinal cords of two aborted human fetuses.[8] (The spinal cord, like the brain, is composed mainly of nerve cells, called neurons.) The duo had obtained the three- to four-month-old fetuses from a physician colleague at Bellevue Hospital, dissected their organs, and stored them in a lab refrigerator. Their experiment was groundbreaking not only for what it reported about polio but also for being one of the first published studies to use human fetal tissue in the lab.

In the 1930s abortion was a crime in every U.S. state in most circumstances.[9] And indeed, most of the estimated 800,000 abortions that were conducted annually during the economically stressed 1930s were illegal.[10] The exceptions were known as therapeutic abortions; they reflected an unwritten understanding between legal authorities and physicians that the latter would be allowed to conduct abortions they deemed medically necessary or advisable. Therapeutic abortions were conducted at abortion clinics or medical offices by licensed physicians. But this was not an age of placard-carrying demonstrators outside clinics. These were procedures done out of public view. In the same way, fetal tissue research was conducted out of sight of the public and at the will of researchers.

No one had tried growing polio in human cells. Since polio attacked the nervous system to cause paralysis, Sabin and Olitsky surmised that it would grow in nerve cells. And so they minced the fetal brains and spinal cords and placed the resulting bits of tissue in wide-bottom flasks. Then they added poliovirus from the ground-up spinal cords of infected monkeys. The virus multiplied in the fetal nerve cells—as proven by the fact that when the scientists took fluid from the cultures bathing the fetal cells and injected it into the brains of monkeys, the animals became paralyzed.

Ironically, the paper that Olitsky and Sabin published actually slowed the hunt for a polio vaccine. Why? Because the Rockefeller duo also reported "complete lack of growth" of polio in cultures of other organs from the fetuses, including kidneys and lungs. The "special affinity of the virus for nervous tissue" disqualified the virus for vaccine-making purposes, because viral vaccines contain tiny bits of the cells in which they are made, and nerve cells, when injected into people, were known to occasionally cause a dangerous and sometimes fatal allergic reaction: an inflammation of the brain and spinal cord called encephalomyelitis.

And so very little progress against polio was made until twelve years later, at the Enders lab in Boston. Early in 1948 Enders, Weller, and Robbins were deep into studies trying to grow mumps, measles, influenza, and chicken pox viruses in culture. Enders had called on a physician colleague at the Boston Lying-In Hospital, to provide aborted embryos and fetuses.

He received several from abortions conducted at 2.5 to 4.5 months of pregnancy, as well as a stillborn infant delivered at seven months of pregnancy. Weller minced the skin, muscle, and connective tissue from the fetal arms and legs and distributed it in flasks. The plan was to inoculate the flasks with chicken pox virus from the throat of a sick child. But as they were preparing to do so, Enders casually suggested that Weller and Robbins also inoculate an equal number of flasks with poliovirus from infected mouse brains that were also on hand in the lab. The experiment was timely because virologists, Enders among them, were beginning to doubt the handed-down wisdom that polio would only grow in nervous tissue. For one thing, the virus was being found in quantity in the feces of polio-infected people. Enders was skeptical that a virus that resided strictly in nerve cells could turn up in such profusion in the intestinal tract.

The chicken pox cultures grew nothing. But the poliovirus grew spectacularly, and when the scientists injected fluid from the polio flasks into the brains of mice and monkeys, they became paralyzed. The Enders lab had made an enormous leap. It turned out that Sabin and Olitsky had failed in their experiments thirteen years earlier because they were using a particular strain of polio that would grow only in nerve tissue. Other strains were far less choosy.

The momentous discovery was described in a short article buried in the back of the journal *Science* in January 1949.[11] When polio virologists saw the report, "it was like hearing a cannon go off," Rivers, the dean of U.S. microbiologists, recalled later.[12] Not only had they cornered polio, but Enders and his colleagues had delivered the methods that would allow scientists to grow, without limit, many kinds of viruses in many kinds of tissues.

The Enders lab's breakthrough soon made possible the isolation of scores of new viruses—including viruses that infected only humans and grew only in human cells. And scientists could now readily study the effects of those viruses on cells in the lab, rather than in living animals. Of most immediate import for the public, the discovery also made possible within a few years the industrial-scale growth of poliovirus in nonnervous tissue in lab dishes, allowing the development of polio vaccines.

In 1954 the Nobel Committee in Sweden honored Enders, Weller, and Robbins with that year's prize in physiology or medicine. Their discovery had thrown open the doors to virology. It also made tissue culture a vital part of coming advances. He didn't know it yet, but Leonard Hayflick would land squarely in the middle of the new push forward.

Hayflick graduated from the University of Pennsylvania in the spring of 1951 with a BA in arts and sciences and a double major in microbiology and chemistry. He went to work as a research assistant at a drug company called Sharp & Dohme in Glenolden, a Philadelphia suburb. There he helped make a product to dissolve clotted blood and pus in infected surgical wounds. Soon Sharp & Dohme merged with the big drug company Merck, which had built a brand-new research facility twenty-seven miles northwest of Philadelphia in West Point.

Sharp & Dohme had been something of a scientific backwater. At the state-of-the-art Merck labs Hayflick was exposed to new and exciting things. He began to learn about viruses like bacteriophages, which attack and invade bacteria. He saw firsthand the excitement of the hunt for new antibiotics at a time when the drugs were transforming medical practice. He also saw for the first time highly educated commercial scientists in action. A revolutionary ambition began to take hold in Hayflick's head and heart. He had never let himself consider getting a PhD.[13, 14]

Hayflick applied and was admitted to the doctoral program in medical microbiology at the University of Pennsylvania. He enrolled in the fall of 1952. He had saved enough money to pay the tuition and to get by if he continued to live at home. After his first year in the program, he would receive university scholarships and a fellowship that supported him through the rest of his PhD studies.

Just before Hayflick left Merck, Sharp & Dohme in June 1952, he met a talented young artist who worked preparing slides for scientists at the West Point facility. The pair discovered that they both planned to travel in Europe that summer and that their separate itineraries had them crossing paths in Paris.

Ruth Louise Heckler was a slim, self-assured twenty-six-year-old with a broad smile and a quiet demeanor that comported well with Hayflick's own. She was from a churchgoing Pennsylvania Dutch family in Lansdale, a railroad town not far north of Philadelphia, where her father worked as an accountant for the Lehigh Coal & Navigation Company. She had studied life drawing and book illustration at the Philadelphia Museum School of Industrial Art before coming to Merck.

Heckler was drawn to Hayflick's mind—his ability to analyze problems clearly and quickly.[15] He was drawn to her quiet self-assurance, her intelligence, and her questioning of religious authority; she had rejected the Lutheranism of her childhood and begun attending Quaker meetings. He also found her beautiful: one day as they walked in Paris, he put his arm around her waist. On October 2, 1955, with Hayflick in the final year of his PhD, the couple were married in a simple, intimate service at the 150-year-old Arch Street Friends Meeting House in Philadelphia. There was a dry reception in the hall next door, and then the newlyweds walked to a nearby Reform synagogue on Broad Street, where a rabbi blessed their union.

For his PhD thesis Hayflick studied a mysterious group of microbes then called pleuropneumonia-like organisms, or PPLOs. (They have long since been renamed *Mycoplasma*.) These microbes had been known for two centuries to cause a highly contagious pneumonia in cows in Europe, but they were still poorly understood. Too large to be viruses and yet smaller than bacteria, they defied categorization and their links to other animal and human diseases were murky.

Hayflick was intrigued by PPLOs and grew to be equally taken with the newly exciting art of tissue culture. His graduate mentor—assistant professor Warren Stinebring, a soft-spoken, stocky former college football player—was full of energy about a course he had just taken in tissue culture, one of the first of its kind. He wanted to train Hayflick. Hayflick didn't want to be distracted from his PPLOs but agreed to a compromise: for his thesis project he would grow PPLOs in tissue culture. Hayflick was working in primitive conditions by today's standards. He grew his PPLOs in a chicken incubator bought for less than $40 from a Sears Roebuck catalog.

Early in his graduate studies Hayflick began to spend time at a nearby institute that would have a huge and lasting impact on his professional life. He recalls being asked to investigate an outbreak of a middle-ear infection in a famous colony of pink-eyed, snow-white research rats. The albino rats were known to all as Wistar rats, because they were developed and resided at the Wistar Institute of Anatomy and Biology. They were an important laboratory tool, but the infections had upset their balance and left them spinning in purposeless circles. PPLOs were possibly the culprit.

The Wistar, as people called it, was a gracious, *V*-shaped, three-story building of light brown brick located on prime real estate in the heart of the University of Pennsylvania campus. A stone's throw from the iconic statue of Benjamin Franklin on Penn's main quad, the institute was the oldest freestanding biological research organization in the country. It was completely independent of Penn, having been founded in 1892 by a wealthy, eminent Philadelphia family. The Wistars included Caspar Wistar, an eighteenth-and-nineteenth-century physician and anatomist who wrote the first U.S. textbook of anatomy and in the process amassed and preserved a huge number of anatomical specimens. Caspar Wistar's great-nephew, Isaac Wistar, a Civil War brigadier general and a prominent Philadelphia attorney, established and endowed the institute to preserve and display his great-uncle's impressive collection.[16]

The brain of Isaac Wistar—at his request—was preserved in a big glass jar in the basement of the institute, along with his right arm, shriveled from a Civil War wound. His ashes were, and still are, in an urn that overlooks the atrium. (If officials of the newly founded institute had had their way in the 1890s, they would also have displayed the gray matter of the psychopath Henry Holmes. The Wistar tried without success to obtain his brain for study after the hanging of the serial killer who haunted the 1893 Chicago World's Fair.)[17]

In the mid-1950s the Wistar Institute was a strange mix of faded elegance and creepiness. It boasted terra-cotta detail on its facade, an airy atrium surrounding a broad wrought-iron staircase, and a public museum on the first floor that was the stuff of horror movies. There were reptiles from Borneo and the bladder stones of Chief Justice John Marshall (removed without benefit of anesthesia). There were human bones gathered on the field after the 1815 Battle of Waterloo and a wide selection of human skulls used to teach medical and dental students. There were seven wax-injected human hearts. There was an intact skeleton of what had been Siamese twins. And floating in formalin, in patented display cases, there was the largest collection of embryos and fetuses in the country, many of them with abnormalities like clubfoot and cleft palate.[18]

But despite the crowds of schoolchildren who regularly trooped through the locally famous museum, the Wistar Institute in the mid-1950s was slowly dying from decades of neglect. Its wiring and plumbing were failing. Its senior staff comprised exactly three scientists, two of them in their eighties. And since 1940 its inertia-ridden board of managers had left the institute to be run by a less-than-ideal acting director. Perceiving the lack of leadership, junior scientists came and went very quickly.[19]

This acting director—a short, quick, domineering man named Edmond Farris—was a middling PhD scientist and not a physician at all, but he had made himself indispensable to certain Philadelphia couples by launching an infertility clinic that he ran out of the Wistar, fueled by the sperm donations of University of Pennsylvania students.[20] In addition to artificial insemination, Farris's services included microscopic examination of the male partner's sperm for deficiencies. He also ran pregnancy tests by injecting a woman's urine into a prepubescent female rat from the Wistar colony. If the rat went into heat despite its immaturity, that indicated the presence in the woman's urine of a female hormone made only during pregnancy. (Early in 1956, two decades before home pregnancy tests were available, Hayflick and his wife took advantage of the in-house services. The couple's first child, Joel, arrived later that year.)

The lab that Hayflick chose for pursuing his rat assignment, on the otherwise-empty second floor, had antique Bunsen burners and wrought-iron filigree. Hanging outside the door, suspended from the high ceiling of the atrium, was the skeleton of a seventy-foot finback whale sold to the institute in 1897 by the renowned paleontologist Edward Drinker Cope.[21] Far from putting him off, the eerie, empty environs fascinated Hayflick, who enjoyed working alone in the lab on the second floor or thumbing through the collection of ancient scientific books in the eighteen-thousand-volume library. Occasionally he encountered one of Edmond Farris's happy customers climbing the wrought-iron staircase with a new baby in her arms.[22]

In the spring of 1956 Hayflick received his PhD. He had, indeed, shown that PPLOs could be grown in tissue culture.[23] (He also confirmed that a PPLO had sickened the Wistar rats.) He was no longer an uncertain undergraduate. And he had new, outside affirmation of his abilities. He had won a postdoctoral fellowship endowed by A. C. McLauglin, a Colorado oil tycoon. It would take him to Galveston, Texas, to the lab of Charles Pomerat, the man who was arguably the best tissue culturist in the world. The fellowship paid a considerable sum in Hayflick's world: $5,500, tax free. He and Ruth moved to Galveston in August 1956.

The charismatic Pomerat, a bald, portly man who wore a butcher's apron and favored white duck trousers, ran a big lab in the basement of the psychiatry building at the University of Texas Medical Branch in Galveston. It was a place that hummed with activity, its tone set by its chatty leader, who was not only a pioneer cell culturist but also an outstanding chef and an accomplished artist. Pomerat had pioneered a new tool: time-lapse microscopic photography of cells in action, with exposures made every thirty or forty seconds and rendered on reel-to-reel films.

At any given time Pomerat would have several cameras peering down long tubes running down to the microscopes, where they focused on cells in a minuscule chamber. The lab was full of a constant clicking of shutters and attendant flashes of light emanating from the tops of the microscopes.

Hayflick used the cameras to study adenoviruses, a class of viruses that had recently been discovered in human tonsils and in the adenoid tissue after which they were named: glandular tissue in the back of the throat. He was able to observe the effects, hour by hour, as adenoviruses destroyed cells. Holes would appear in the cells' cytoplasm; the cells would sprout abnormal, armlike extensions. Finally, they would break apart. Hayflick did not make any grand discoveries in Galveston, nor could he publish in journals the reel-to-reel films he produced. But he became increasingly expert in cell culture, and he rubbed shoulders with and learned from first-rate scientists like Morris Pollard, an eminent virologist. Hayflick also met a colleague of his own age who would play an important part in his career. Paul Moorhead was a blue-eyed Arkansan with adamantly liberal politics and a passion for chromosomes—the long, stringy bundles of DNA that are housed in a cell's nucleus and contain its genetic material.

Ruth gave birth to Joel in November 1956. The Hayflicks' second child, Deborah, was born thirteen months later, while Hayflick was still in the Pomerat lab. Once or twice a night he would wake, give a bottle to a baby or two, then drive to the lab to adjust the microscope, which would inevitably slide out of focus after a few hours.

Early in his second year in Galveston, Hayflick began looking toward his next step. He heard that the Wistar Institute, after nearly two decades under an acting director, had finally hired a permanent chief. He was a polio vaccine pioneer named Hilary Koprowski, and he was looking for a cell culturist. Hayflick applied and received an offer. It was "scut work," providing cell cultures to Wistar scientists, and not the pure research position he would have preferred.[24] Still, it could lead to bigger things, and he was sure he could squeeze in his own research on the side. It would also take him and Ruth back to their families and friends in Philadelphia. He began work at the Wistar, his old stomping ground, in April 1958, one month shy of his thirtieth birthday.

CHAPTER THREE

The Wistar Reborn

Philadelphia, April–December 1958

I told this guy, I said, "You know, Hilary Koprowski is Wistar. . . . Hilary built the Wistar. No matter what you think about him, it's great because of Hilary."
—Maurice Hilleman, former Merck vaccine chief, in a 2004 interview[1]

Hilary Koprowski was a brilliant, erudite, Polish-born virologist who was equal parts disarming charm and ruthless ambition. He was a short, stocky man with prominent cheekbones and light, piercing eyes. He spoke with a marked accent. A polymath who quoted the poets Arthur Rimbaud and Ezra Pound as fluently as he discussed viruses and antibodies, Koprowski was a graduate of the Warsaw Conservatory of Music who had considered a career as a concert pianist at the same time as he pursued a degree at the Warsaw University Medical School. A magnetic combination of old-world romantic, ambitious, forward-thinking scientist, and gregarious *bonhomme*, Koprowski was equally at home playing cards on the floor of a young technician's apartment or wining and dining Europe's finest biologists. Wherever he was, he made a lasting impression—as he did on the lab technician Barbara Cohen, who was twenty-one years old when Koprowski hired her as the first member of his polio research group at the Wistar. "He would just fix you with these ice blue eyes, just, like, *freeze* you," Cohen recalled in a 2014 interview. "He would just be intensely in your presence."[2]

"When he wished to charm, he was gay and confidential, his eyes flashed warmly, he used your Christian name like a caress, he would take you by the arm as if you and he were the only two people who mattered, two superior beings in a rather ridiculous, muddleheaded world," the author John Rowan Wilson observed.[3] A consummate networker, Koprowski was nonetheless not above sneaking out the back door of his lab into an adjoining lab's sterile room when a boring visitor arrived at his receptionist's desk. His eyes would sparkle and his voice dance with delight so that "you could not be cross with him" as he dished up choice remarks about the person he was avoiding, recalled Ursula Roth, a Koprowski technician in the mid-1960s.[4] Nor did Koprowski shy from foisting an eminent but dull British scientist on his junior colleagues for the day, before taking that scientist home to an elegant dinner. He was hoping for a return invitation, because the colorless colleague owned a piano on which Beethoven had composed a sonata and on which Koprowski longed to lay his strong, stout fingers.[5]

A bon vivant, Koprowski obtained a state alcohol permit that made the Wistar an island of merriment on the dry University of Pennsylvania campus and his office a hub of happy hours where he mixed Bloody Marys using Gdansk vodka—a liqueur speckled with flakes of twenty-two-karat gold.[6] "The Wistar operated in Philadelphia like a small independent European municipality. I think Hilary saw himself as Cosimo de' Medici," recalled Peter Doherty, a Nobel Prize–winning immunologist whom Koprowski recruited to the institute in 1975.[7]

If he recognized his faults, Koprowski had a hard time owning them. Once, when a young Wistar scientist named Michael Katz failed to win NIH funding for a grant, Koprowski, who wanted Katz doing other work, celebrated the failure to the young man's face: "Congratulations! Now you can devote yourself to real science." Stung, Katz told Koprowski to go do something that was anatomically impossible. He was summoned to Koprowski's office at 5:00 p.m. that same day. The young man walked in fully expecting to be fired. Koprowski was sitting behind his desk with a twinkle in his eye and a pitcher of ice-cold martinis in front of him. "Olive or lemon peel?" he inquired.[8]

Koprowski could be imperious, domineering, and brutal in his dealings with those he perceived to have been disloyal. "He could run over them with a bulldozer," his son Christopher recalled.[9]

"Hilary had a penchant for roughing up people," Edwin Lennette, a mentor of the young Koprowski, told Koprowski's biographer, Roger Vaughan.[10]

He could also be generous. The virologist Robert Gallo recalls Koprowski quietly upgrading Gallo's airline ticket so that the younger man could join him in first class as the two men flew to meetings in Europe and Asia.[11]

Koprowski disliked confrontation, and often sent his loyal lieutenant, a genteel, chain-smoking New Englander named Tom Norton, to send packing whatever scientist had fallen out of his good graces at the Wistar. Stormy partings with his former acolytes occurred regularly. "When you decided to leave Wistar, Koprowski considered it betrayal no matter how you did it. When I told him I was leaving, he tried to trap me, personally and emotionally," the cancer scientist Vittorio Defendi recalled.[12]

Koprowski's enemies—and there were many—were driven to distraction by his irrepressibility, his manipulativeness, and his happy refusal to acknowledge, never mind attend to, their criticisms.

"There can be no doubt that he is a splendid scientist," I. S. Ravdin, the powerful surgeon in chief at the Hospital of the University of Pennsylvania and a member of the Wistar's board of managers, wrote to the outgoing university provost Jonathan Rhoads in 1960, slightly more than two years after Koprowski's arrival at the Wistar. "But there is also no doubt that he can cause more trouble than any man with whom I have ever been associated."[13]

Ravdin and other top University of Pennsylvania officials soon failed in an attempt to oust Koprowski, who blithely ignored their demands for itemized expense reports, wouldn't disclose whether he was consulting on the side, and was being accused by his former bosses at the drug company Lederle Laboratories of having stolen the polio vaccine that he developed at the company, bringing it to the Wistar.[14] Koprowski outmaneuvered them, helped by a revolt by loyal Wistar scientists.[15] The institute would remain his domain for another thirty-one years.

Koprowski was born in 1916 in Warsaw. His mother was a dentist and his father a textile manufacturer. Hilary Koprowski fled Nazi-occupied Poland in 1940 with his dominating mother and his young, very-pregnant wife, Irena Koprowska.[16]

The family eventually landed in Brazil, after a hair-raising flight during which Koprowska, separated from her husband, defied Nazi orders to continue working as a physician and fled France with her newborn baby in her arms; and Koprowski escaped from Italy on the very day that Mussolini closed the borders to men capable of bearing arms. In Rio de Janeiro Koprowski was forced to earn a living by teaching piano to ungrateful students while his wife worked as a pathologist conducting autopsies in the green marble morgue of the city's biggest hospital.

Several months later, through a chance meeting with an old high school friend on a Rio sidewalk, Koprowski was hired by the Rockefeller Foundation, which maintained a lab in the city. There the New York City–based research powerhouse was busy refining the young yellow fever vaccine. Koprowski spent three years doing work with yellow fever and several other viruses before the family received a U.S. visa that Koprowski had applied for years earlier.

En route to the United States in 1944, the aging merchant marine vessel that the Koprowskis were aboard laid over in Trinidad. With typical enterprise, Koprowski sought out a leading Trinidadian scientist, J. L. Pawan, who had made the groundbreaking discovery that rabies was transmitted by bats and that humans could be infected by vampire-bat bites. Koprowski was already fascinated by vampire literature. And he had read widely on rabies after he observed a rabid vampire bat—its brain was later dissected to look for the microscopic hallmarks of rabies—when he was working for the Rockefeller Foundation in Rio. He turned up unannounced at the door of Pawan's lab. The senior scientist spoke with Koprowski at length about rabies. Koprowski's resulting, keen interest in the disease and in making an improved vaccine against it would last his whole life.[17]

In the United States, Koprowski presented himself, without a job, at the Rockefeller Institute for Medical Research and succeeded in impressing the senior virologist, Peter Olitsky—the same man who had worked with the young virologist Albert Sabin to grow polio in brain tissue from human fetuses a decade earlier. Through Olitsky's influence, in January 1945 Koprowski landed a research job in the virology department of Lederle Laboratories in Pearl River, New York, twenty-six miles northwest of Manhattan.[18]

Lederle, the pharmaceutical arm of the chemical giant American Cyanamid, was a superbly outfitted research complex, thanks to the backing of Cyanamid's science-loving chief executive, William Graham Bell. While there, Koprowski became locked in an all-out race with the man who became his archrival, Albert Sabin. Sabin was ten years older than Koprowski. Driven and extremely bright, he was a refugee from pogroms in eastern Poland who had landed with his family on U.S. shores at the age of fifteen, speaking no English. By 1931 he had worked his way through medical school at New York University, and by the time Koprowski arrived at Lederle, Sabin had established himself as a top polio expert, first at the Rockefeller Institute and then at the University of Cincinnati.[19]

Both Koprowski and Sabin were bent on inventing the first live polio vaccine. A live vaccine consists of a naturally occurring virus that has been weakened so as to produce a low-grade infection that generates antibodies in a vaccinee without making that person sick. By contrast, a killed vaccine uses viruses that have been killed by chemical or physical processes to prompt an immune response.

However, Jonas Salk, a New York City native who was the son of unschooled Russian immigrants, would beat both Koprowski and Sabin to the glory of inventing the first polio vaccine. Salk's vaccine contained naturally occurring polio virus that had been killed by the chemical formaldehyde. The injected vaccine tricked the immune system into recognizing and responding as if the virus were alive, generating antibodies in the blood. Salk's vaccine was licensed by U.S. regulators in 1955, turning him into an instant public hero.

But Koprowski, Sabin, and many other virologists were convinced from the beginning that a vaccine that contained live, weakened virus would be more effective than a killed vaccine. Salk's vaccine required several injections and later booster shots, and even these did not seem to prevent immunity from declining with time. What was more, Salk's vaccine, which was injected into muscle, would not generate robust levels of antibodies in the throat or the walls of the digestive tract, which are the body's main ports of entry for the polio virus.

Naturally occurring polio virus enters the body through the mouth via infected water or food. It then multiplies in the digestive tract and is excreted in the feces. In most people this infection is mild, and many, many people were infected without even being aware of it in the prevaccine era. It is when polio invades the blood from the digestive tract and travels to the spinal cord and brain that it causes paralysis and even death. Sabin, Koprowski, and many other immunologists believed that a live vaccine, given in a drink or on a sugar cube, would mimic the natural route of human infection and in so doing would confer lifelong immunity, producing robust levels of antibodies both in the walls of the digestive tract and in the blood. An oral vaccine would also be cheaper and easier to administer; it wouldn't require injections or highly trained health personnel. And it would be shed in recipients' feces and, in environments with poor sanitation and unclean water, passed on to other, unvaccinated people, provoking a protective immune response in some of them too—so-called passive immunization. The corresponding danger was that a live vaccine virus, shed in the feces, could mutate over time, reverting to an infective form and spreading the disease, rather than protecting people.

Making a live vaccine meant striking a delicate balance. A scientist needed to weaken the virus enough to stop it from causing disease, but not so much that it failed to cause a mild infection that provoked a protective immune response. Koprowski spared no effort to win the race to develop the first, and best, live vaccine. On a wintry evening in 1948, he swallowed his own experimental polio vaccine: a gray, fatty, viscous glop of cotton-rat brain and spinal cord that he had pulverized in a Waring blender.[20] It was infected with a naturally occurring polio virus that he had weakened—he hoped.

He had done so by injecting blood serum and cerebrospinal fluid, which bathes the brain and spinal cord, from a twenty-nine-year-old man with polio directly into the brains of mice. When a mouse came down with polio, he injected that mouse's ground-up brain and spinal cord into a new group of mice, and so on through many groups of mice. Then he took the virus from the mice and sequentially injected the brains of several groups of cotton rats, furry rodents that resemble mice and were common lab animals at the time, again using the infected brains and spinal cords of sickened animals to inject each new group.[21] The idea behind these "passages" of the virus through multiple rodent generations is that, as the virus adapts to causing disease in a different species, it becomes less good at causing it in human beings; this weakening process was fundamental to the development of a live vaccine.

Koprowski next fed his live vaccine to chimpanzees. They developed antibodies against polio and did not get sick when they were exposed to naturally occurring virus.[22, 23] But this wasn't proof of safety: chimpanzees did not contract polio naturally, as human beings did.

While it was a time-honored tradition among scientists to be the first to receive their own vaccines, swallowing the cotton-rat vaccine from the Waring blender didn't put Koprowski at risk for polio; like so many people at the midpoint of the twentieth century, he already had antibodies to the virus from being exposed to polio in the course of daily living.[24] To know if his vaccine worked, Koprowski needed human "volunteers" who, unlike himself, weren't already immune.

In 1950 Koprowski tested his vaccine on intellectually disabled children at Letchworth Village in Thiells, New York, an institution where "naked residents, unkempt and dirty, huddled in sterile dayrooms."[25] His use of people with mental disorders was not without precedent. During the war, under the sponsorship of the U.S. government, leading researchers had infected psychotic residents at an Illinois state hospital with malaria to test the effectiveness of experimental drugs.[26] They had also tested trial influenza vaccines by requiring intellectually disabled people to breathe in influenza virus through aviation masks or to inhale a nebulized spray into their nostrils for four minutes; both vaccinated people and unvaccinated controls were forced to breathe in the virus.[27] One of the leaders of these experiments was the young Jonas Salk.[28]

Koprowski told the story of the Letchworth Village trial in a lecture three decades later. He said that George Jervis, a friend and colleague who was the laboratory director at Letchworth Village, sought him out in the late 1940s and asked him to test the children living there to see if they had protective polio antibodies circulating in their blood. Jervis, Koprowski said, feared a polio epidemic: hand-to-mouth transmission by polio-infected feces was a special concern among the mentally ill children living at Letchworth. Koprowski drew blood from the children and found no antipolio antibodies in about 60 percent of them. Jervis then asked him to test his experimental vaccine at Letchworth, according to Koprowski. "I realized we would never get official permission from the state of New York. Therefore, we asked permission from the parents of these children."[29]

Over the course of thirteen months beginning in February 1950, Koprowski fed the vaccine—the same gray glop that he himself had ingested but now disguised in chocolate milk or corn syrup—to twenty children at Letchworth Village. Aside from Koprowski and his lab manager, Tom Norton, the children were the first experimental recipients of live polio vaccine anywhere. Two of the children, including the first child fed the vaccine, were so disabled that they had to receive it through feeding tubes inserted into their stomachs.[30]

The earlier wartime experiments notwithstanding, Koprowski infuriated his peers when he presented the results of his Letchworth Village trial at a meeting of polio-vaccine scientists in 1951. It was no longer wartime, and polio was a dangerous, sometimes lethal virus. Sabin demanded to know how Koprowski had dared put a live poliovirus vaccine in children.[31] Joseph Stokes, the esteemed physician in chief at the Children's Hospital of Philadelphia, asked if he had thought about the fact that the Society for the Prevention of Cruelty to Children could sue him.[32] Others simply sat in stony silence.[33] The scientists couldn't argue with Koprowski's results, however: all of the intellectually disabled "volunteer" children—that is how Koprowski described them in the resulting paper—who lacked antipolio antibodies before they consumed the vaccine promptly developed them. None showed any signs of illness.[34]

It wasn't the last time that Koprowski would give his polio vaccine to institutionalized children. They were easily controlled and readily accessible, and they couldn't talk back. They were certainly no match for Koprowski, who, when he wanted something, used every tool in his considerable arsenal of charm, persuasive power, and underhandedness to make sure that he got it.

The southern spring was in full flush as Leonard and Ruth Hayflick drove the 1,600 miles from Galveston to Philadelphia early in 1958. Their family was twice the size it had been when they left Philadelphia. Joel was now a toddler, and Deborah, aged five months, rode hanging in a drawer suspended from the dashboard of the family's four-door sedan. Soon the Hayflick clan would be still larger: another child was on the way, due that November. Also looming into view ahead was a job that, even if it wasn't a plum research position, was planting him on the ground floor of what sounded like an exciting new phase in the life of the Wistar Institute.

In 1956 Norman Topping, the energetic vice president of medical affairs at the University of Pennsylvania and a new, active member of the Wistar's board of managers, took matters in hand at the decrepit, dying Wistar. It presented "a particular problem," with its ancient whale skeleton and equally ancient labs. The Napoleonic acting director, Farris, with his not quite licit infertility clinic, was also on Topping's radar screen, and not in a good way. "His reputation was not of the best," Topping recalled in his memoir. Topping initiated a search for a permanent director. He knew of Koprowski and his polio work because of his own connections at Lederle. Topping had met the man and was impressed.[35]

Koprowski himself was on the lookout for a new position when he heard from Topping. At Lederle management had changed multiple times, and the situation for scientists had deteriorated badly from the company's heyday in the 1940s. So Koprowski was all ears when Topping offered him the Wistar's directorship, along with space for his research on the third floor of the institute.[36]

The forty-year-old Koprowski saw opportunity where others might have seen a quick route to obscurity. He told Topping that the third floor wasn't enough. He wanted the whole institute, with its 68,000 square feet of floor space. He also wanted—demanded—two full professorships at the University of Pennsylvania in the School of Arts and Sciences and in Research Medicine.[37]

Topping, eager to shed his Wistar worries, agreed, and in late January 1957 Koprowski accepted the Wistar job. It came with a $17,000 annual salary ($145,000 in 2016 dollars), a $2,400 expense account, and a start date of May 1, 1957. It also came with the written promise, in the letter offering the job, that his salary would be reviewed every year "so as to be commensurate to that of the best-paid professor of the University of Pennsylvania." He would also have "the same rights and privileges as to tenure" as a full Penn professor.[38]

"I liked [the Wistar] because it was dead," Koprowski told his biographer, Vaughan.[39] It was an empty shell that he, Hilary Koprowski, was going to remake as a mecca of unconstrained, imagination-fired biological research.

"I wanted no time sheets . . . no specified vacations. No departments, no walls behind which petty jealousies, self-interest and pockets of power could flourish," Koprowski said. "I wanted to attract mature scientists who were talented, experienced—the kind of self-starting, disciplined people who could make it on Wall Street or anywhere else."[40] Koprowski was arguably the best-suited virologist in the world to lure top minds to his new Wistar Institute. He had a near-visionary sense of who was doing original and important work and an astute eye for talent. And he was a networker par excellence. (He once invited the poet T. S. Eliot to tea when he discovered that they were aboard the same transatlantic ship. The poet obliged.)[41]

Koprowski had carefully cultivated relationships with top scientists from Germany to Sweden to the United Kingdom. But he began by bringing with him to the Wistar several close colleagues from Lederle. When they arrived, they may have questioned their decisions. The building was decrepit. Its hallways were lit by bare hundred-watt bulbs powered by a generator; only small portions of the building connected to the city's power grid. The new arrivals blew circuits when they plugged in their equipment. David Kritchevsky, a cholesterol expert and Koprowski confidant, later recalled that when he turned on the water in one lab, it ran rust-colored for twenty minutes before a pipe burst and water flooded the floor.[42]

Koprowski instituted an unprecedented housecleaning. He steam-cleaned the brownstone facade and consolidated the mausoleum-like museum in one wing of the first floor. He sent many anatomical specimens to other museums but relegated the bottle containing founder Isaac Wistar's brain to a basement storeroom.[43] He jettisoned the whale skeleton, sold the famous albino rat colony, and purged the nineteenth-century library, although he ended up keeping its librarian, Bill Purcell, a slight, long-haired freethinker with an extensive collection of erotica.

Koprowski then spent a lot of money—$567,000, or $4.8 million in 2016 dollars—on renovations. Nearly half of it came from the Wistar Institute's cash coffers. Most of the rest came from the U.S. government. He rewired the building for AC current, repaired pipes, and put in central air-conditioning. He created conference rooms and administrative offices. He purchased autoclaves and centrifuges and an electron microscope. In the basement he installed a big room for washing lab glassware, complete with a fourteen-foot Better-Built washing machine. But above all, every possible bit of floor space was dedicated to a score of state-of-the-art labs for virology and biochemistry and pathology—and for tissue culture. For Koprowski's cadre of outstanding scientists, nothing but the best facilities would do.[44]

Despite his domineering personality, Koprowski was not a micromanager. His goal was to set his scientists free, in an atmosphere where they were protected from financial worry and teaching obligations, to pursue the scientific questions that obsessed them. The resulting discoveries, he hoped, would vault his Wistar Institute into the pantheon of U.S. research giants, beside the likes of the Rockefeller Institute and Johns Hopkins University. He was making a big gamble by guaranteeing the salaries of his new, top-notch recruits on the assumption that he would find the government and private funding to support them. That project would become his bugbear. But at the outset his inspiration and his promise lured the brilliant minds he was looking for—people like Rupert Billingham, an eminent British transplantation expert, and Eberhard Wecker, a virologist who left his job at the famed Max Planck Institute in Tübingen, Germany, to join Koprowski's new project.

Koprowski wasn't thinking of Hayflick as a member of this elite group when he hired him. In his view virologists were the stars of the biological drama. Cell culturists like Hayflick were mere supporting actors, providing bottles of cells for the virologists to use in their groundbreaking experiments. What's more, the composition of Koprowski's inner circle—all of them born and, usually, educated abroad—reflected his generally low estimation of American scientists. The slight, serious Hayflick, with his Philadelphia accent and his lack of panache, didn't fit in with those worldly Europeans. He never would.

Still, Hayflick was given prime real estate: a big, brand-new lab that a visitor could find by climbing the grand wrought-iron staircase in the spacious atrium, stopping at the second floor, turning left, and entering the second door on her right. Inside she would find glistening, glass-fronted cabinets stocked with flasks and pipettes, shiny black countertops, and big windows on the far wall looking west along Spruce Street in the heart of the university campus. Here, under the windows, Hayflick positioned an "inverted" chemist's microscope that he adapted to cell-culture work soon after he landed at the Wistar. It allowed him, instead of peering downward through a large vessel, to peer upward through the glass bottoms of his culture bottles, focusing at close range on the cells growing there in single layers. (The innovation spread and is ubiquitous in cell-culture labs today. Hayflick's original inverted microscope was acquired by the National Museum of American History in 2006.)

One wall of the lab was lined with a pair of tiny "sterile" rooms containing not much but black counters with knee space underneath them. The larger of the two could just accommodate two people. Here, in what were the most microbe-free conditions available in the late 1950s, Hayflick and his technicians would work with cultures, flipping on an ultraviolet light in the ceiling when they left at night in the hope that any bacteria lounging on the counters would be dead by morning. (In the morning, avoiding any glance at the UV light, they would crack the door just enough to reach in and flip the switch off.

If a worker forgot to do this, as happened occasionally, he or she would later emerge with singed eyelashes and a sunburn.)

Biologists in labs today, where special filters in highly sophisticated hoods suck microorganisms from the surrounding air, might be amazed at the crude additional measures that Hayflick was reduced to in his ongoing battle against contamination. Every few days, before he left for the day, Hayflick would plant petri dishes full of agar on the countertops and leave the UV light off for the night. In the morning he would collect the petri dishes and incubate them. If after a couple of days they had grown one or two bacterial colonies, that was okay. If there were more than that, he would dispatch a technician with a bucket of antibacterial solution to wash down the rooms.

Next door to Hayflick's lab there was a walk-in room lined with wooden shelves and equipped with a heavy, pea-green metal door. It was an incubation room, where cell cultures could be grown at carefully regulated warm temperatures—often near 98.6 degrees Fahrenheit, which is body temperature. Soon Hayflick was hard at work providing a steady supply of cells to the Wistar's growing cadre of scientists. Some requests from the staff scientists were easy to fill, like those for the hardy, hugely popular HeLa cells.*

Other requests were tougher to accommodate, such as those for freshly harvested cells from particular rodent organs. These required Hayflick to obtain the animals, sacrifice them, and then coax the cells of the organ in demand to grow in a dish. Hayflick was also kept busy tweaking methods for growing the monkey kidney cells that Koprowski's group was using in its race to beat Sabin to a live polio vaccine.[45]

No matter the demands of his job, the hungry Hayflick was not going to be reduced to the role of cell supplier for supposedly greater minds. He began to launch his own experiments. One of the first that occurred to him might sound strange, but to Hayflick it was perfectly natural and obvious, especially because he knew that at least one other group had already succeeded with the same project. Cell biologist Jørgen Fogh and his colleagues Elsa Zitcer and Thelma Dunnebacke at the University of California at Berkeley had developed two continuously replicating cell cultures—known as "cell lines" because they would divide indefinitely in the lab—from a new source. They had used

---

*These cells, obtained without her knowledge in 1951 from the lethal cervical cancer of a thirty-one-year-old African American woman named Henrietta Lacks, have since been made famous by Rebecca Skloot's 2010 book, *The Immortal Life of Henrietta Lacks*.

human amnion, the tough membrane that envelops the fluid in which the growing human fetus floats in the womb.[46]

Ruth Hayflick was expecting the couple's third child in November 1958. On a foggy, rainy day, in the obstetrics department at the Hospital of the University of Pennsylvania—directly across the street from the Wistar Institute—she delivered a baby girl that her parents named Susan. Leonard Hayflick was standing ready near the delivery room, holding a big stainless-steel pan. (Like most fathers in that era, he was not at his wife's side as she gave birth.) When Susan was safely delivered, Hayflick collected the heavy, bloody, purple placenta.

He walked with it for five minutes: out of the delivery suite, through the hospital, and across Spruce Street to the Wistar. It was the Wednesday evening before Thanksgiving and he had the lab to himself. He liked it that way. Working in one of the sterile rooms, he dissected away from the placenta the strong, semiopaque amniotic membrane. He placed it in a solution of trypsin, a digestive enzyme collected from the pancreas of slaughtered pigs. The trypsin readily broke the amnion into its component cells.

That December, as the Hayflicks navigated Hanukkah and Christmas with a two-year-old, a one-year-old, and a newborn baby, the cells from Susan's amnion incubated in small flasks in the warm room beside Hayflick's lab, bathed in a nutritious medium of salts, amino acids, vitamins, and calf serum—the fluid component of calf's blood. Hayflick had already named the cells: WISH, for "Wistar Institute, Susan Hayflick."

On New Year's Eve Day that year, Hayflick peered into his microscope at the dividing WISH cells and discovered that they didn't look at all normal. They showed typical signs of having become cancer cells. In the paper he published about the cells, Hayflick used the cells' volume and multiplying time to calculate that they must have become cancerous not while Susan was in the womb but when one cell mutated on or about December 21, 1958. That mutation, he proposed, had made that single cell far better able to proliferate in lab conditions than the normal cells, which were soon overrun entirely by the furiously multiplying cancer cells.[47] (Hayflick did not commence to worry about Susan's health; by now it was clear that cells in lab bottles often developed abnormalities, presumably related to their lives in the lab.)

Several years later a geneticist named Stanley Gartler at the University of Washington revealed that WISH, along with many other cell lines used in the 1950s and 1960s, had been contaminated with cancerous HeLa cells, which were so vigorous and hardy that they easily infected other cultures where they didn't belong.[48] As Wistar's resident cell culturist, Hayflick supplied HeLa cells to Wistar scientists, including Koprowski's polio researchers, who used them to measure the levels of polio virus in vaccines.[49] So it is easy to imagine how cross-contamination might have occurred.

Hayflick is adamant to this day that his WISH cells were not invaded by HeLa. Someone else, in some other lab, must have obtained them and contaminated them, he said in a 2014 interview, and then sent them into the cell banks that today advertise them to biologists marked with the prominent warning that they are actually HeLa cells. Gartler, in 2016, begged to differ. "It is clear from Hayflick's paper [first describing the derivation of the WISH cells] that they were already contaminated," he wrote in an e-mail.

True to her beginnings, Susan Hayflick grew up to become a medical doctor and an expert geneticist. Today she is a professor of molecular and medical genetics, pediatrics, and neurology at the Oregon Health & Science University in Portland.

CHAPTER FOUR

Abnormal Chromosomes and Abortions

Philadelphia, 1959

The practice of abortion in American hospitals is inequitable, inconsistent, and largely illegal. The basic reason for this is that this aspect of twentieth-century medicine is being governed by nineteenth-century laws.

—Robert E. Hall, an obstetrician and gynecologist at
Columbia University in New York, 1967[1]

O n an exceptionally lovely weekend in April 1959, with the Wistar Institute's impressive renovations recently completed, Koprowski threw his born-again institute a coming-out party. In typical Koprowski fashion, it was a big, bold, first-class affair, kicked off by a VIP dinner under the vaulted ceiling and among the hieroglyphic-inscribed pillars of the Egyptian Room of the university museum.

That weekend the institute's new labs were formally opened, and five hundred biologists packed a two-day symposium entitled "The Structure of Science," where they were treated to a star-studded list of speakers. These included dignitaries like the mayor of Philadelphia, the U.S. surgeon general, and the president of the National Academy of Sciences, as well as top scientists like Peter Medawar, the British transplantation expert who would win a Nobel Prize the following year—and whose younger partner, Rupert Billingham, Koprowski had already recruited to the Wistar as part of his A-team of scientists. But the most buzz may have been around the presence of Francis Crick, who, with James Watson, had described the structure of DNA only six years earlier. (Barbara Cohen, the young lab technician who was working for Koprowski at the time, asked to recall the event fifty-five years later, remembered only being dazzled by Crick's presence.)

Hayflick also launched something new that April.[2] He began pursuing a question that was swirling in the scientific air and that had begun to intrigue him too. Could viruses cause cancer in humans? The notion that viruses might be implicated in the dread disease was not a new one. As early as 1842, Domenico Rigoni-Stern, an observant surgeon in Padua, Italy, noted that nuns, shut away as they were from the world's temptations, were afflicted with cervical cancer far more rarely than other women.[3] There were no tools available to test the implication of his observation—that a sexually transmitted agent might cause the disease. The first hard evidence for a viral role in cancer didn't come until 1908, when two scientists at the University of Copenhagen, Vilhelm Ellerman and Olaf Bang, showed that healthy chickens infected with fluid from chickens with leukemia—fluid filtered to remove cells and bacteria— contracted leukemia.[4] Their finding might have garnered more notice if it had been more widely recognized at that time that leukemia was a cancer.

But three years later a young American pathologist named Peyton Rous, working at the Rockefeller Institute, showed that a virus caused chicken sarcoma—a malignant tumor of connective tissue. Rous had taken fluid from a sarcoma in one chicken, filtered it—again, the ultrafine filter caught bacteria but not viruses—and injected it into other chickens, which then grew the same cancer. The young scientist was greeted mostly with indifference to his discovery that cancer could be transmitted between the chickens "by an agent separable from the tumor cells."[5] It wasn't thought that a finding about chicken cancer could be relevant to human beings. Rous's "agent" would later be named Rous sarcoma virus and would play a huge role in the study of cancer causation.

Conventional wisdom among biologists in the first half of the twentieth century held that cancer was caused by environmental factors, like smoking or chimney soot or, according to others, by gene mutations. As late as the mid-1950s, the influential Australian biologist Frank Macfarlane Burnet, who in 1960 would win a Nobel Prize for his work in immunology, pointedly dismissed the notion that viruses could cause cancer.[6] However, by 1959, when Hayflick tackled the question, many biologists were pushing back, even against the likes of Burnet. They had ammunition in the fact that, over the decades since Rous's discovery, more than a dozen viruses had been discovered to cause either benign or malignant tumors in a variety of animals, including rats, mice, and cats—if not in humans. Then, in 1958, an Irish surgeon named Denis Burkitt threw a tantalizing new piece of information into the mix. He discovered an aggressive childhood lymphoma in sub-Saharan Africa. Its distribution, in malaria-ridden areas, suggested an infectious cause. The same year, the National Institutes of Health (NIH) launched a well-funded effort to track down human "cancer viruses."

Koprowski began organizing, with other leading scientists including Rous, an American Cancer Society conference on the subject of viruses and cancer causation. In July 1959 *Time* magazine ran a cover story featuring two researchers from the NIH, Bernice Eddy and Sarah Stewart, who had discovered a mouse virus that caused tumors in hamsters, rabbits, and rats. "The hottest thing in cancer is research on viruses as possible causes," John Heller, the chief of the NIH's cancer institute told *Time*.[7]

Hayflick recognized that he had a skill set that equipped him well to tackle the topic: he knew a good deal about microbiology, and he knew more than most scientists about growing cells in the lab. He went at the question with the traits that defined his style: thoroughness, patience, determination, and an outsized tolerance for seemingly mundane, repetitive work. (Some former colleagues call his style unimaginative, dogged, and plodding.)

With the help of a surgeon named Robert Ravdin, across the street at the Hospital of the University of Pennsylvania, he got hold of 300 human tumor samples and coaxed 225 of them to grow in lab dishes bathed in nutritious medium.[8]

Hayflick's next step was based on a reasoned assumption: if any of these tumors were caused by a virus, that virus was likely still lurking within them, replicating itself inside the cells and then bursting or budding out, flooding the surrounding fluid medium with millions of individual virus particles. Hayflick began collecting samples of that lab-dish fluid and freezing it.

When he was ready, his next step would be to thaw that fluid and pour it over *non*cancerous cells in culture. If cancer-causing viruses were in the fluid, then it stood to reason that they might infect some of the normal cells, causing them to become cancerous. But where to get certifiably noncancerous cells that could survive and multiply in the lab?

Fortunately for Hayflick, there was now a new benchmark for normalcy in cells. In a classic 1956 paper, "The Chromosome Number of Man," Albert Levan and Joe Hin Tjio, two scientists working in Lund, Sweden, had used a new microscopic technique to pin down the normal number of human chromosomes.[9] Biologists now knew beyond doubt that normal human cells had, residing in their nuclei, forty-six chromosomes: twenty-three inherited from each parent. A cell that carried this normal complement of chromosomes was called a "diploid" cell. (The only normal human cells that don't carry forty-six chromosomes are sperm and eggs, which, because they carry half as many chromosomes, twenty-three, are called "haploid.")

During the 1950s scientists had launched dozens of cell lines from apparently normal human tissue. But with time—sometimes with very little time— these cells began to behave strangely. They displayed bizarre, disorganized shapes and sizes and bloated nuclei. And they developed abnormal numbers of chromosomes. There were lab-grown knee-joint cells with 133 chromosomes, liver cells with anywhere between 65 and 90 chromosomes, and foreskin cells with 72 chromosomes—the last from a four-day-old baby.[10] To Hayflick and his contemporaries, such changes signaled one thing: cancer.*

Abnormal chromosomes had first been associated with cancer seventy years earlier, in 1890, when a young German pathologist named David Paul von Hansemann first recognized chromosomes in abnormal configurations in dividing cancer cells.[11] Von Hansemann, peering through a microscope, saw in cancer cells split, frayed, broken chromosomes and chromosomes that hadn't doubled in number, as they normally should before cell division, but rather tripled or quadrupled. Not long after this, another German scientist, Theodor Boveri, after studying aberrantly fertilized sea urchin eggs, proposed that abnormal numbers of chromosomes resulted when the stringy DNA packages didn't segregate themselves properly during cell division. The resulting cells, he suggested, might ultimately tilt into uncontrolled growth.[12] But with the tools on hand at the time, he had no way of proving his hunch.

It would be 1960 before two researchers, working near Hayflick in Philadelphia, discovered the first link between a chromosomal aberration and a cancer. Peter Nowell, a tumor biologist at the University of Pennsylvania School of Medicine, and David Hungerford, a graduate student at the Fox Chase Cancer Center in Philadelphia, examined the bone marrow cells of adults with chronic myelogenous leukemia, a blood-cell cancer. In almost all of the patients, chromosome 22 was abnormally short. They christened this chromosome, with its lopped-off head, "the Philadelphia Chromosome."[13] Their discovery confirmed what many scientists had long suspected: cancer was, at least in part, a disorder of genes gone awry.

For Hayflick in 1959, even before the discovery of the Philadelphia Chromosome, the aberrant chromosome numbers in these dozens of lab-launched cell lines presented evidence enough that the cells were not normal and would not do for his experiment. Yet as he confronted a paucity of normal cells that had been launched in lab dishes, he did have some indication that the feat was not impossible. Tjio, the codiscoverer of the normal number of human chromosomes, had since moved from Sweden to the University of Colorado, to the lab of another leading scientist there named Theodore Puck. Together the pair had

---

*Some of these dozens of cell lines were among those later shown by geneticist Stanley Gartler at the University of Washington to be contaminated with cancerous HeLa cells; others were also likely HeLa contaminated.

grown apparently normal cells from patients, or their leftover surgical tissues. They had created cell lines from the uterus, the testis, and the prepuce, a fold of skin surrounding the clitoris. They reported that the cells from each line still had forty-six chromosomes and that those chromosomes looked normal under the microscope, even after five months of vigorous dividing in lab dishes.[14]

Even so, Hayflick had reservations about using leftover surgical samples, or even skin samples from volunteers, to try to grow normal cells in lab bottles. As a microbiologist he was well aware that cells from any human being who has been on the planet for any length of time are potentially contaminated with disease-causing viruses. He had seen firsthand during his time in Galveston the adenoviruses that lurked in people's tonsils and adenoids. Herpes simplex virus was known to lie latent in nerve cells. Hepatitis viruses were assumed to simmer quietly in livers, and scientists were discovering all manner of rhinoviruses—a major cause of the common cold—inhabiting human noses and throats. It would be pointless to expose cells obtained from adults to the fluid that had bathed his cancer cells in the lab. If the ostensibly normal cells became cancerous, he wouldn't know if this was due to a virus from the fluid or to some hidden virus already residing in the cells. However, there was one obvious source of tissue that, while not absolutely guaranteed to be virus free, was far more likely to be clean.

A fetus is protected in the womb. Tucked away in its mother's body, it isn't exposed to the raft of illness-inducing microbes that babies and toddlers meet on diaper-changing tables, in preschool classrooms, and on kitchen floors. What's more, when its pregnant mother is exposed to unwelcome bacteria and viruses, the fetus is protected from most of them by its mother's germ-attacking antibodies and her invader-targeting immune cells. Those malevolent microbes that aren't dispensed with in the mother's throat, digestive tract, and blood can still be attacked in the fetus, because some maternal antibodies cross the placenta. There are exceptions: a handful of disease-causing viruses can escape immune defenses and infect the growing fetus. But compared with the viral exposures of adults, the odds of any one of these affecting a given fetus are remote. In particular, Hayflick in 1959 did not have to worry about one such virus that is notorious today: HIV. So as Hayflick cast about for the cleanest tissue he could find, he kept circling back to a conclusion that seemed inevitable: growing cells from aborted fetuses was his best bet for developing normal human cells.

As the work of Albert Sabin and John Enders and their colleagues in the 1930s and 1940s makes clear, Hayflick was not the first scientist to turn to aborted fetuses to probe a biological question. He himself, while working as a graduate student at the Wistar in the mid-1950s, had seen fetuses waiting to be taken to the incinerator in the Wistar courtyard after being dissected for experiments using cells from fetal pituitary glands. As he completed his graduate studies, scientists in Stockholm were using cells from aborted human fetuses in what became a failed effort to make the first human cell–based polio vaccine. And Levan and Tjio, also working in Sweden, had examined lung cells from four aborted fetuses to pin down the normal number of human chromosomes.

As he looked for a source of aborted fetuses, Hayflick was operating in one of two parallel universes that existed in the United States in 1959. According to the law, abortion was a criminal offense in every U.S. state. The 1939 statute that was on the books in Pennsylvania, unlike those in the other forty-nine states, didn't even make an exception if the woman's life would be endangered by carrying a pregnancy to term.[15] It read:

> Whoever, with intent to procure the miscarriage of any woman, unlawfully administers to her any poison, drug or substance, or unlawfully uses any instrument, or other means, with the like intent, is guilty of felony, and upon conviction thereof, shall be sentenced to pay a fine not exceeding three thousand dollars ($3,000), or undergo imprisonment by separate or solitary confinement at labor not exceeding five (5) years, or both.[16]

If the fetus died—in other words, if the abortion was successful—the penalty for the person who performed the abortion was doubled to $6,000 and ten years in solitary confinement at labor. The harsher penalty also applied if the mother died during the procedure.[17]

The law conspicuously failed to define an "unlawful" abortion; the very word suggested that if there were "unlawful" abortions there might also be "lawful" procedures. That ambiguity, however, did not stop Pennsylvania authorities from taking enforcement seriously. They prosecuted both unqualified, back-alley operators and physicians who operated as solo providers: people like Lamar T. Zimmerman in Montgomery County (which includes Philadelphia's upscale northwestern suburbs), a physician who was convicted in 1967 of performing an illegal abortion; and Benjamin King, MD, of Allegheny County (which encompasses Pittsburgh), who was sentenced to two to five years in prison in 1968.[18]

At the same time in a different setting—the Hospital of the University of Pennsylvania and other major hospitals that comprised the other, parallel universe—legal authorities tolerated abortion. They were carrying on a tradition that had evolved over decades, beginning as early as 1867. Then, an Illinois law declared that abortion was criminal "unless done for *bona fide* medical or surgical purposes." It didn't define those purposes but left it to the medical profession to do so.[19]

The term "therapeutic abortion" came to be used to describe those abortions that were understood—at least by supportive physicians and legal authorities—not to be criminal. A therapeutic abortion was performed by a qualified doctor who judged it necessary, even if the reasons for that necessity floated in a legal gray zone that would not finally be dispelled until the 1973 Supreme Court decision in *Roe v. Wade*. In that landmark ruling, the high court struck down state criminal laws and said that, except to protect the mother's health, states could not restrict abortions before fetuses became capable of meaningful life outside the womb.

In the early decades of the twentieth century, so-called therapeutic abortions were performed in doctors' private offices and in homes. During the 1930s they migrated increasingly to clinics and hospitals, and their numbers grew as women responded to the crushing economics of the Depression.[20] In Philadelphia, however, access was clearly limited: a survey presented to the Obstetrical Society of Philadelphia reported that 329 women died in the city from self-induced or nonphysician-induced abortions between 1931 and 1940, or about 10 women for every 1,000 babies born.[21]

The growing numbers of physician-assisted abortions led states to enact tighter abortion laws, like the 1939 statute in Pennsylvania.[22] In turn, doctors and hospitals moved to try to protect themselves legally. In the 1940s and 1950s they set up what they called "therapeutic abortion committees" in hospitals. The committees were made up of small groups of doctors appointed to officially receive and evaluate abortion applications.

The Hospital of the University of Pennsylvania—HUP for short—was a huge, imposing institution whose sheer, ten-story brick facade towered over the south side of the Ivy League university's campus in west Philadelphia and, at certain times of the day, literally cast its shadow on the Wistar Institute. The oldest university-owned teaching hospital in the country, HUP was established in 1874. By 1959 it had accrued all the power that came with being the most prestigious hospital in Philadelphia.

Like many major hospitals, HUP had put a therapeutic abortion committee in place by 1955, and possibly earlier.[23] One doctor who performed abortions as an obstetrician/gynecologist at HUP in the late 1950s through 1962 recalled the committee as a casual group in a 2014 interview. "Frankly, if a patient wanted an elective abortion, you simply called a colleague or two and said, 'Would you approve this?' The committee never even really met. . . . If it ever came to a legal issue . . . we could say we talked on the phone" and approved it.[24]

But by 1963 that relaxed approach had changed, for reasons that aren't clear. Written rules approved by the hospital's medical board that year required that the committee receive written applications from physicians proposing to perform an abortion. Each of three anonymous obstetrician/gynecologists serving on the committee was required to issue a written opinion on each application. In cases where it was appropriate, a different kind of specialist, often a psychiatrist, could be enlisted to pass judgment on an application. If a member of the committee wished, he—and it was almost always a "he"—could interview and examine the patient.[25]

Women who saw private doctors—that is, women who were, in general, wealthier, whiter, and better connected—were far more likely to obtain abortions through the committee than so-called clinic patients—poorer, often black patients who were seen at the subsidized, hospital-based outpatient clinic that HUP operated. Still, by the late 1950s it was increasingly difficult for women of any color to get therapeutic abortions, whose numbers dwindled in the conservative climate of the 1950s.[26] That did not change in the 1960s. One educated estimate published in 1967 put the ratio of illegal abortions to hospital abortions at one hundred to one.[27]

As Hayflick remembers it, he was able to begin obtaining fetuses in 1959 because of Hilary Koprowski's connection to Isidor Schwaner Ravdin, HUP's surgeon in chief and the vice president of medical affairs at the university.

If the Hospital of the University of Pennsylvania was the preeminent hospital in the city, I. S. Ravdin—"Rav," as he was known to close colleagues—was the preeminent power in the hospital. He had been chair of HUP's Department of Surgical Research since 1935—though he may have been prouder of his stint building and running a jungle hospital in Burma during World War II. A short, mustachioed man with dark, receding hair and preternatural energy—his secretary at the time, Betsy Meredith, remembers him as "like a pea on a hot griddle"—Ravdin terrified medical students and residents alike and was used to his orders being obeyed immediately, if not sooner.[28] His influence extended well beyond HUP to the halls of power in Washington. In 1956 he was photographed on a podium at the Republican National Convention, triumphantly raising the hand of President Dwight D. Eisenhower.[29] Two months earlier, Ravdin had been summoned to Washington to operate on Eisenhower, who had developed a life-threatening bowel obstruction. By 1959 Ravdin was busy hatching a grand, new 374-bed building extending HUP, named after himself.

Supplicants coming to see Ravdin would watch him pull out a Dictaphone and fire off a letter that would result in whatever string they wanted pulled in the vast machinery of the hospital getting pulled—or not. One of these requests, according to Hayflick, came from Koprowski, who asked for the ferrying of aborted fetuses to an obscure junior scientist at the Wistar Institute.

Ravdin was a consummate political player in matters of sexuality, contraception, and abortion. He had to be. In heavily Catholic Philadelphia the church was a hugely powerful presence, always hovering in the background of hospital and university politics. When the medical school set up a Division of Family Studies in 1952, the division's affiliation with the liberal Marriage Council of Philadelphia caused the medical school dean to write to Ravdin, flagging concern "that this plan might be mis-interpreted by the Catholic Church as being concerned with birth control."[30]

Still, by 1960 the church and society at large were confronting a wave of change. At HUP it appeared one morning in the form of bright orange flyers blanketing the hospital's huge Gates Pavilion. THE ROMAN CATHOLIC CHURCH AND THE HOSPITAL OF THE UNIVERSITY OF PENNSYLVANIA, ran the title on the eight-by-eleven-inch sheets, which complained vociferously that even Jewish and Protestant patients seen by hospital doctors couldn't get fitted for diaphragms owing to pressure from the church.[31]

The flyers provoked a flurry of letters between hospital higher-ups—not refuting their truth but trying to figure out who had used hospital paper and mimeograph machines to make them. (It appears that the culprits were never found out.) "It is, as you can see, an inflammatory statement and might very easily get the University into several embarrassing situations," Ravdin wrote to Franklin Payne, the chairman of the Department of Obstetrics and Gynecology. "The paper which was used . . . has also been used in your Department, with a mimeograph machine similar to that which your Department has."[32]

Privately Ravdin very likely agreed with the flyers. But his job was to protect his institution, and in the effort he steered a careful path between women's-rights advocates on one side and, on the other, the church. He was photographed beaming at Pope Pius XII in Rome in 1958, in a moment that he recalled as a "wonderful occasion."[33] Still, when he was invited to a Planned Parenthood luncheon a few years later, he wrote to the organizers that he would be "very happy" to attend.[34]

Koprowski's legendary charm had fallen flat with Ravdin, who in mid-1959 was rapidly coming to consider the ebullient Pole manipulative and untrustworthy; one year later Ravdin would lead the failed attempt by a faction of the Wistar Board of Managers to oust Koprowski.[35] But Ravdin almost certainly agreed to Koprowski's request that the hospital ferry fetuses to Hayflick for a simple reason: he believed in scientific progress and was not willing to let personal politics get in the way of it.

Ravdin had been helping Wistar scientists from his earliest days as surgeon in chief. "We can assist you in providing any amount of tumor tissue for your work. Just let me know what you want and when you want it," he wrote to the elderly cancer scientist Margaret Lewis after she asked for samples from malignant tumors in 1947.[36] A decade later William McLimans, a Wistar virologist, wrote to Ravdin that he had found interesting results in tissue that Ravdin had provided from the breast cancers of twelve HUP patients. He wondered if he could look at their medical records. "This letter will give you access to our Record Room in the Hospital," Ravdin wrote back.[37]

Hayflick himself had likely already benefited from I. S. Ravdin's support: when Hayflick needed tumor samples to begin his investigation of whether viruses might cause human cancers, it was I. S. Ravdin's son, Robert Ravdin, then a young general surgeon at HUP with an office four doors down from his father's, who provided them.[38]

In mid-1959 Hayflick began working with the first of a series of fetuses that would arrive, whole, in his lab after he received a phone call from HUP informing him that an abortion had been done and a fetus was available. The gynecological surgeons supplied fetuses from pregnancies of three to four months, so that their major organs were developed enough to be dissected and removed for Hayflick's purposes.[39]

Hayflick's receipt of the fetuses was not subject to paperwork or permissions beyond the go-ahead from I. S. Ravdin that had set the process in motion. The transfers were informal, as virtually all such arrangements were in those days. And while the physical movement of the fetuses was not hidden as one would hide an illicit transaction, it was not paraded openly either, legal realities and moral sensibilities being what they were.

At eighteen weeks of gestation—in medical parlance, this means eighteen weeks since the mother's last menstrual period and therefore roughly sixteen weeks after sperm and egg meet at conception—a human fetus is roughly 5.5 inches long, from the top of its head to its rump. (Its legs are tucked up in the fetal position.) It has arms and legs, fully formed fingers and toes, a nose and mouth and lips and ears and fingernails. It can blink, grasp, sleep, move its mouth, and kick. Its skin is so new and translucent that the underlying vessels look like vivid red highways on a complicated road map. True, its eyes are still closed, and its head is huge compared with the rest of its body, giving it a slightly alien appearance. The nerve pathways in its brain that will lead to consciousness are just beginning to sprout.[40] It cannot survive outside the womb. But there is no mistaking it for anything but an incipient human being.

"I remember receiving whole fetuses at three, four months' gestation. A baby this big," Hayflick said during a 2012 interview, holding his hands about six inches apart. "I remember distinctly not being disturbed by that—don't ask me why—to dissect that. I can't explain it. But my constitution was such that it didn't affect me.

"It was definitely going to end up in an incinerator," he added. "If it was used for research purposes, some good possibly could come out of it for people."[41]

The first fetus that arrived in his lab was a male. After several hours' work dissecting the lungs, breaking up the lung tissue with the enzyme trypsin, and spinning the resulting cells in a centrifuge, Hayflick planted the cells in four rectangular Pyrex flasks known as Blake bottles.[42] He poured in nourishing medium, carefully plugged each glass bottle with an amber-colored silicone stopper, put the bottles on a tray, and walked them into the 96.8-degree heat behind the heavy door of the incubation room beside his lab. He placed the bottles on their flat sides on a wooden shelf. This allowed the maximum amount of "floor space" for the cells, which didn't float in the fluid medium but sank to the bottom. They would multiply until they covered the floor in a single layer.

It was about three days later that Hayflick spotted signs of growth in the bottles of cells from that first fetus: a near-transparent haze on the bottom of the bottle where the cells, once attached to the glass floor of the bottle, had begun dividing.[43] He poured off the growth medium in each bottle, replaced it with a fresh batch, closed the door, and waited. By ten days later the cells covered the bottom of each bottle in a single confluent layer. (Normal, noncancerous cells stop dividing at this point, rather than continuing to multiply and pile up on top of one another. This property is called "contact inhibition.")

Now Hayflick delegated to his technician, Fred Jacks—a young military veteran who would eventually earn his MD and become the first African

American resident at Abington Memorial, a suburban Philadelphia hospital—
the tedious job of "splitting" the bottles so that the cells coating the bottom of
one bottle were halved and half of them were placed in a new Blake bottle. This
involved a long series of steps that required Jacks first to use trypsin, the "jackhammer" digestive enzyme, to loosen the cells from the side of the bottle,
where they were firmly stuck. Later he would bathe the loosened cells in growth
medium and redistribute half of them to a new bottle by sucking them up into
and then shooting them out of a glass pipette. His mouth was protected only
by a wad of cotton placed in the upper end of the pipette. Labs all over the
country used this technique, occasionally leading to unsavory accidents. Today
most pipettes are operated by thumb-controlled pistons, much like syringes.

Not long after the initial split, the lung cells seemed to kick into much higher
gear, multiplying fast enough that Hayflick soon instituted a strict schedule of
splitting the bottles every third or fourth day. (The frugal Hayflick would take
the spent culture medium home in lab bottles and use it to fertilize his roses,
daffodils, and tulips.)

As two bottles became four, and four became eight, and eight became sixteen, and sixteen became thirty-two, it became clear to Hayflick that the bottles would soon take over the communal incubation room next door to his lab.
So he set Jacks to freezing the cells, transferring them first into tiny wine
bottle–shaped glass ampules not quite two inches high, which he would seal by
melting closed the neck of the ampule with a quick pass through a Bunsen
burner. Each ampule contained a fraction of an ounce of growth medium. And
floating in each were between three million and four million cells.[44]

Living human lung contains many types of cells, even just sixteen weeks
after conception. In a fetus of this age, there are cubelike endothelial cells lining blood vessels; taller, columnlike epithelial cells lining the thousands of tiny
airways called bronchioles; and underlying smooth-muscle cells giving them
support, to name a few. But holding the lung together is connective tissue, and
its components are made by fibroblasts: long, spindly cells with tapered ends,
shaped something like a compass needle. Fibroblasts, it turned out, were the
fittest when it came to survival in the lab. They were the only cells still living in
Hayflick's bottles after a few weeks of life in the lab.

In the summer of 1960 Hilary Koprowski published a sleek, eight-by-eleven-
inch booklet that featured write-ups on the work of all the labs at his rejuve-
nated institute. The Wistar Institute's biennial report for 1958 and 1959
featured just one photo on its cover. It showed Hayflick's fibroblasts. Inside, a

description of the photo read: "These cells are from a normal fetal human lung with a [normal] number of chromosomes."[45]

Hayflick's excitement comes through between the lines of his written report explaining the significance of the cover photo. It was possible, he wrote, to grow from human fetuses cells that had not—not yet, anyway—turned cancerous in lab bottles. The fetal cell line pictured on the cover had been growing in such bottles in his incubator for six continuous months. It was named WIHL, for "Wistar Institute Human Lung," and while it had repeatedly outgrown its bottles, so that it had had to be halved and transferred to new bottles dozens of times, "it still retains the diploid number of chromosomes," Hayflick wrote, referring to the normal number of human chromosomes: forty-six.[46]

What was more, he went on, these WIHL cells didn't have any of the telltale microscopic hallmarks of cancer cells: disorganization, irregular sizes, and bloated nuclei. In fact, they looked no different under a microscope from how they had when freshly harvested from the fetus months earlier. They were classical fibroblasts: elongated cells with slightly thicker middles and tapered ends. The WIHL cell strain "is an extremely important tool in this investigation and has very important implications in other fields as well," Hayflick reported. "It is, from all indications, a normal human cell." That, he added, made it an ideal cell for use in his quest to discover cancer-causing viruses.

But Hayflick's research into whether viruses caused cancer was about to get left behind, supplanted by a discovery that he didn't anticipate and that would stand scientific wisdom on its head.

CHAPTER FIVE

Dying Cells and Dogma

Philadelphia, 1960–61

Science, my boy, is made up of mistakes; but they are mistakes which it is useful to make, because they lead little by little to the truth.

—Jules Verne, *Journey to the Centre of the Earth*[1]

B iologists had been growing cells in culture since 1907, when Ross Harrison, the workaholic embryologist at Johns Hopkins, first coaxed those bits of frog brain to grow in a lab dish. For virtually all of that time, they had operated under this simple piece of received wisdom: cells grown in lab dishes, properly treated, should live indefinitely. If they died, the fault lay not with the cells but with the scientist. His glassware wasn't clean, or her medium didn't contain just the right mix of nutrients, or a sloppy technician had sneezed on a plate of cells, launching a fatal infection. This faith in the open-ended life of cells in the lab grew out of the work of a bald, bespectacled, publicity-seeking French scientist named Alexis Carrel. In 1912 Carrel was awarded the Nobel Prize for inventing a much-needed method for surgically joining the two ends of a severed artery. But for Hayflick and other cell culturists, Carrel's half century of influence resulted from an entirely different experiment.

In the early years of the twentieth century, the charismatic Carrel worked at the Rockefeller Institute in New York City, then the pinnacle of American medical science. One mid-January day in 1912, Carrel took a snippet of tissue from the heart of an eighteen-day-old chick fetus and put it in a lab dish, where he began nourishing it with diluted chicken blood plasma. ("Plasma" is another word for blood serum.) Two months later, having grown so "abundantly" that it had been split and planted in new vessels eighteen times, the heart tissue was still alive and even, he reported, beating in the lab dish. Carrel published a paper reporting on his findings, entitling it "On the Permanent Life of Tissues Outside of the Organism."[2]

When, later that same year, he won the Nobel Prize for his artery-joining advance, many scientists and journalists mistakenly thought that Carrel was being honored for the launch of the undying chicken heart. The miraculous beating tissue "was the leading topic of discussion by medical men the world round," the *New York Times* reported in an article on Carrel's Nobel Prize that consumed the entire front page of the broadsheet's Sunday magazine section.[3]

Carrel soon handed off the work of maintaining the beating bit of chicken heart to a laboratory colleague, Albert Ebeling, who looked after it for the next thirty-four years, feeding it with medium, dividing the cells when they outgrew their space, and dispensing with most of them while keeping a residual piece of heart always going in a dish. The undying chicken heart became a favorite with the popular press. "Isolated Tissue Holds Life 12 Years in Test . . . Growth Continues as in Body," the *New York Tribune* proclaimed in 1924, as the chicken heart approached its twelfth "birthday."[4] So did Carrel himself, who at one point landed on the cover of *Time* magazine—although he likely hastened his retirement from the Rockefeller Institute when in 1935 he published a book, *Man, the Unknown*, propounding the use of gas to euthanize criminals, both sane and insane.[5] The heart-in-a-dish outlived him; Carrel died two years before Ebeling finally disposed of the culture in 1946.

The lesson from Carrel's chicken heart was not lost on the biologists who struggled through the first half of the twentieth century both to repeat his chicken-heart experiment—no one could—and to keep other kinds of cells alive indefinitely in culture. When they failed, as they did repeatedly, it was surely their incompetence that was to blame. That belief was only strengthened when, in 1943, a round-faced, bespectacled cell culturist at the National Cancer Institute coaxed a single cancerous mouse cell into unending lab-dish life.[6] Wilton Earle's mouse cells became the first demonstrably immortal cell line—one that scientists could obtain from Earle and observe for themselves replicating endlessly in the lab.

It was 1951—the year that Hayflick earned his undergraduate degree—before the same feat was repeated with human cells. That year an innovative and determined cancer researcher and cell culturist named George Gey, working at Johns Hopkins University, launched the first human cells to survive indefinitely in the lab, by taking the cells of an extremely aggressive cervical cancer from the womb of a dying woman named Henrietta Lacks. Within a few years the HeLa cell line was being studied and used in biology labs all over the world.

With the perspective of hindsight, it's easy to say that Earle's mouse-tumor cells, Gey's HeLa cells, and other undying cell lines that soon followed in the 1950s were able to live endlessly in the lab only because they were cancerous. After all, it is the very definition of cancer that cells escape the normal constraints on growth and divide uncontrollably and endlessly. But for Hay-flick, working nearly sixty years ago, things were not so clear-cut. Indeed, there was no reason to believe that normal, noncancerous fetal cells—just like Carrel's normal chicken heart cells—shouldn't also live open-endedly in lab dishes.

So when one day in the winter of 1960 Hayflick noticed something amiss in the incubation room, he started looking for his own mistakes. By now there were several other fetal cell lines in addition to WIHL—the lung cells from the first fetus he had received—multiplying in glass bottles on the wooden shelves.

During late 1959 he had continued to receive fetuses at random intervals from the Hospital of the University of Pennsylvania. With them he had developed cell lines from several fetal organs: skin and muscle; thymus and thyroid glands; kidneys and heart. He found that when the bottles were first planted with cells, it took about ten days for the cells to establish themselves, growing to cover the floor of a bottle in a semiopaque sheen that was visible to a practiced eye. (Hayflick labeled this initial stage of the cells' growth, when they grew to first cover the bottom of the bottle, phase I.)

Once he had split the contiguous cells, putting half of them into a new bottle, all of the cells began dividing much more rapidly and needed to be split again every three to four days. (He called this period of rapid division, which went on for months, phase II.) He found that no matter how large the variety of cell types in the organ he started with, all of the cultures ended up consisting of only one cell type: the long, tapered fibroblasts that spin the connective tissue that holds organs and cells together. For whatever reasons, other kinds of cells did not thrive in his bottles.

As he planted these cell lines, Hayflick became systematic about naming them. Since he soon developed a second cell line from fetal lungs, he needed to rebrand the first line something other than Wistar Institute Human Lung. He decided to keep it simple and name the lines in numerical order. WIHL he renamed WI-1. The next line he named WI-2, and so on. By September 1960 he would launch WI-25, the last in the series. It came from the lungs of fetus number 19, a female. (There were fewer fetuses than cell lines because Hayflick derived more than one cell line from several of the fetuses, using different organs. For instance, the second line he developed, WI-2, was grown from the skin and muscle of fetus number 1.)[7]

As was normal in science, there had been mishaps. WI-2 had been lost to bacterial contamination after growing for several months. And cells from some organs, he was learning, grew better in the lab than others. Cell lines from the heart were sluggish: his WI-6 cells had given up and died after not even three months. But the kidney cells seemed to be doing well, and his WI-1 lung cells were star growers. He had begun to suspect that lung fibroblasts were the best suited to life in the lab.

But what was wrong with the WI-3 cells on this day? The WI-3 line had been grown from the lung cells of fetus number 2 and, if his hunch was right, these cells should have been dividing and thriving, just like the lung cells of WI-1. But he had noticed during the past several weeks that the WI-3 cells were taking longer than they had in the past to grow to confluence on the bottom of their bottles. What was more, their culture medium wasn't turning from pink to yellow as quickly as it once had. The change in the fluid's color resulted from acid production and meant that the cells were metabolizing actively. A slower progression from pink to yellow meant that the cells were slowing down in their activities of daily living—acting, in oversimplified terms, a lot more like seventy-year-olds than twentysomethings. And now, five months after the launch of the third line, WI-3, there was a new, worrisome sign. The culture medium was normally clear. But today it was cloudy with what he feared was the debris of dying cells.

Hayflick took the bottle of WI-3 cells next door to the lab and peered at the cells through his inverted microscope. He saw what he had expected he might see: there were grainlike bits of debris scattered around. What was more, when he scrutinized the tangled, dense, dark chromosomes that are visible at high magnification in actively dividing cells, they were very few and far between. The cells were slowing, perhaps stopping, their division.

Over the coming months the WI-3 cells would completely degenerate. Healthy fibroblasts tend to line up like soldiers in parallel formations, with each cell immediately next to its neighbors, their finely tapered ends pointing in the same direction. Instead, the WI-3 cells would become spread out in no discernible pattern, pointing in random directions. Black bits of debris, the detritus of dying cells, would litter the white spaces between the cells on Hayflick's microscope field and cling to the cells' surfaces like so much washed-up driftwood.

The changes would come on slowly, but at this first sign of the cells' deterioration, Hayflick began trying to figure out where he was going wrong with WI-3. He adjusted the components of the culture medium—perhaps it was deficient in some essential nutrient—and poured the rejiggered fluid over the lagging cells. But even as he did so, he felt that it wouldn't make a difference. The medium he used was made by technicians in the basement media room, in large batches that lasted for weeks. If WI-3 was struggling, then the other cells that he had bathed with precisely the same medium should be struggling too. Yet the rest of the cell lines were thriving. His suspicion was confirmed when the readjusted medium failed to resuscitate the WI-3 cells, whose bottles continued to grow cloudier with debris.

Perhaps he was screwing up in some other way. Could dirty glassware be the culprit? But the lab had been using the same glassware, scrubbed and sterilized by the same people working in the same glassware-washing facility in the bowels of the building. Perhaps, occasionally, dirty bottles slipped through due to some oversight, but what were the odds of those random dirty bottles always being used for WI-3 and no other cell lines? Virtually nil.

The only really plausible explanation was that WI-3 had become infected. It was easy enough to rule out bacterial contamination: he planted some sluggish WI-3 cells on plates of agar and left the plates incubating. No bacteria grew. There was no bacterial contamination stunting these cells.

Other microbes were trickier to detect. One key group of suspects was the PPLOs—those nuisance organisms that were smaller than bacteria but larger than viruses and that Hayflick had studied as a graduate student. They were a bane of cell culturists: always popping up like weeds where they weren't wanted, even after a scientist thought they'd been eliminated with antibiotics. Fortunately for Hayflick, he had made himself into an expert on identifying them microscopically. Colonies of PPLOs had a "fried egg" appearance, but it took a practiced eye to pick this up. He checked the ailing cultures and found no signs of PPLOs.

There was still the possibility that a virus was dooming the WI-3 cells. Hayflick examined them under his microscope for telltale signs of viral invasion. Viruses produce typical microscopic signs when they sicken cells. Cells become strangely shaped. They bloat. They may detach from the glass bottles that they normally adhere to. They can develop "inclusion bodies," which are clumps of abnormal protein in the nucleus or cytoplasm. Hayflick saw none of these signs in the WI-3 cells. Of course, no matter how hard he looked, he couldn't prove a negative—that is, he couldn't prove that an undiscovered, undetectable virus wasn't lurking in the cells. This was a reality that would eventually come to dog him.

Hayflick did all he could to revive WI-3. He kept splitting the cells into new bottles. There they continued to languish. Then he tried crowding several bottles' worth of degenerating cells into one bottle. Nothing changed.

Had Hayflick not been so industrious and launched twenty-five cell lines from the organs of nineteen fetuses over the course of months and months—rather than, say, deriving just two or three cell lines—he might at this point have concluded, like so many scientists had before him, that he was simply falling short of the high standard set by the illustrious, Nobel Prize–winning Carrel. He might have concluded that his own ignorance was the culprit and the WI-3 cells were its victim.

However, as WI-3 languished in front of him, Hayflick was watching over his other cell lines. They were growing well until one day a few weeks later, when he again visited his incubation room and noticed that some other Blake bottles were becoming cloudy with debris. They were the bottles bearing WI-4, a cell line launched from the kidneys of fetus number 3 that had been doing very well—until recently, when the cells had begun multiplying more slowly. Now, like the WI-3 cells before them, they were grinding to a halt, succumbing to something. What was more, his bottles of WI-5, derived from the muscle of fetus number 3, were beginning to take longer to grow to confluence. By the time these muscle cells stopped dividing, the next group of cells, WI-7, from the thymus and thyroid glands of fetus number 4, had begun to grow sluggishly.

As Hayflick watched his newer cell lines flourish in their bottles while the cells from the first fetuses languished and finally died around them, he took his perplexity to several Wistar colleagues. One of them was Lionel Manson, a portly, avuncular immunologist with a dry, self-deprecating sense of humor and a razor-sharp intellect.

"I was telling Lionel what I found and I said, 'I'm weighing several different explanations,'" Hayflick recalled in a 2014 interview. "And he just said cavalierly: 'Have you thought about it having to do with aging?' And I said: 'No. But,' I said, 'aging is a wastebasket'—at that time it was; it still is to some extent—'a trash basket into which you put everything you can't explain.'"[8]

Still, Hayflick took Manson's flippant suggestion back to his lab and pondered it. And the more he thought about it, the more convinced he became of its merit. It was the only theory that was supported by his now-voluminous data and his months of observations. And it led him inescapably to one conclusion: there was nothing wrong in his methods. What was wrong was the scientific faith in the immortality of cells in lab dishes that dated back nearly fifty years, to Carrel and his never-dying chicken heart. True, cancer cells like Wilton Earle's mouse cells and Gey's HeLa cells had been living in labs for years now, and it seemed they would go on doing so. But normal, noncancerous cells in a lab bottle were not immortal. They aged and died, just like human beings. The truth of it was staring him in the face.

Hayflick rehearsed the objections that he could already hear a chorus of critics raising. True, he couldn't explain how it was that Carrel's chicken heart tissue had continued to live for decades. But wasn't the standard of scientific credibility the repeatability of an experiment? No one had been able to repeat Carrel's experiment. The reason, he was now increasingly certain, was because it wasn't repeatable.

There was also the matter of WI-1, the first cell line he had launched, back in September of 1959. It was still going strong, dividing energetically, six months later. But in this waning winter of 1960, he was now almost certain that this could be explained: WI-1 was a particularly hardy, long-lived line, but not an immortal one. He would bet money that sooner or later it would die too.

There are two subtle but important points worth making at this juncture about nomenclature; both will bear on this story as it moves forward. The first arises from what is actually happening in a bottle of fetal fibroblasts lying on its side incubating. If one takes a bottle, the flat side of which is covered with cells that have grown to confluence, and splits the cells in half, planting half of them in a new bottle of the same size, it might seem to make sense to conclude that, once the bottoms of the two bottles are covered with cells, every cell in that initial bottle (the original bottle is commonly called the "mother" and the new one the "daughter") has divided once. One would be wrong.

Cells, like people, vary in their vigorousness. Some cells divide more sluggishly, while some are eager, rapid replicators. So over a given period of time, some cells will replicate fewer times than others, some perhaps not at all. Which means that the only conclusion that can be drawn when the floors of the two bottles are eventually covered with cells is that the initial *population* in the mother bottle has doubled in size. For it now covers twice the area that it did.

This is why biologists don't speak about individual cells doubling or say that "these cells have now doubled in number five times." Instead they refer to "*population* doubling levels," which they describe with the acronym "PDL." Given the inherent variability of individual cells, it's the only accurate term to use.

There is a second issue with terminology that can be confusing. It is this: Whenever a biologist splits a confluent culture of cells and puts some of them in one or more new bottles, this is called a "passage," because cells have "passed" from one bottle into another. However, scientists will often use the term "passage" as a synonym for a cell population's doubling. This is accurate if, and only if, as the cells move through a sequence of passages, half of them are placed into just one new daughter bottle of the same size as the mother bottle. This is called a 2:1 split.

However, cells can be placed into any number of new bottles. For instance, if three quarters of the cells are removed from a confluent mother bottle and placed in equal portions into three additional daughter bottles of the same size, and then the cells in all four bottles are allowed to grow to confluence again, the original cell population will clearly have quadrupled in size. Yet it will have been through only one "passage," in that the cells will have been "passed" into new bottles just once. The terms "passage" and "PDL" will be used interchangeably in this book—only because, in the experiments involved, the cells were routinely put through 2:1 splits.

As the winter of 1960 turned into spring, the U.S. Food and Drug Administration approved the world's first officially sanctioned oral contraceptive pill. The University of Pennsylvania prepared to open its new Women's Residence Hall. And in a speech to newspaper editors in Washington, DC, the Roman Catholic Democratic presidential candidate John F. Kennedy announced that "my religion is hardly, in this critical year of 1960, the dominant issue of our time."[9]

The Hayflick family had recently moved into a modest, three-bedroom brick house in the leafy suburb of Penn Wynne, just northwest of the city on the edge of Philadelphia's fashionable Main Line. Across the street was an expansive, hilly park where the Hayflick children spent hours playing hide-and-seek, hunting for frogs in a stream, and swinging on a rope swing with a wooden seat. The couple would soon add a fourth bedroom to accommodate their growing family: daughters Rachel and Annie were born in 1963 and 1965 respectively.

That summer and fall Hayflick's fetal cell lines continued to age and die. He dubbed the stage in which their dividing slowed and stopped and they degenerated and finally died "phase III." And he wrestled with the question of how to prove that it was something intrinsic to the cells—some inherent property and not anything in their environment—that was the cause of their mortality. Hayflick's former colleague and friend from Galveston, the chromosome expert Paul Moorhead, had since moved to the Wistar. In a 2012 interview Moorhead recalled that it was he who proposed the simple, elegant experiment that did the trick.[10]

Moorhead's lab at the Wistar reflected his lowly designation as a postdoctoral fellow: it was a tiny cupboard of a place up on the third floor. But the chromosome aficionado from Arkansas had there what counted most: a Zeiss Jena made by the Reichert Company—in his opinion, the best microscope that money could buy. Leaning over the microscope's sturdy black base, he could paste both eyes to the eyepieces and peer at chromosomes at eight hundred magnifications.

About the time that U.S. voters went to the polls and elected the forty-three-year-old Kennedy to replace the seventy-year-old Dwight D. Eisenhower, Hayflick took his oldest and his youngest fetal cell lines and mixed them. The first were the now-elderly WI-1 cells, which, as he had expected, had stopped dividing in late summer, after eleven months. They were now in phase III—still alive, still metabolizing, but ever so slowly. They had been split in their bottles forty-nine times and might perhaps divide once or twice more, if they could screw up the energy. But basically, they were reaching the end.

To these WI-1 cells Hayflick added youthful WI-25 cells that he had launched just weeks earlier and that were vigorously replicating. They would do so, he expected, for months to come, for the WI-25 bottles had been split a mere thirteen times. Hayflick then left the mix of young and old cells, both in the same bottle, being nourished by the same medium, to incubate.[11]

Moorhead's stratagem relied on the fact that, apart from their ages, there was a singular difference between the two groups of cells. The WI-1 cells had come from a male fetus and thus bore one X chromosome and one Y chromosome. The WI-25 cells were from a female fetus, so they bore two X chromosomes. Both kinds of chromosome would be visible under the microscope to Moorhead's expert eye.

After the cells had been incubating for about two months and had been split into new bottles seventeen times, Hayflick handed the mixture to Moorhead. The Arkansan expert carefully prepared the cells, using techniques that stained and spread the chromosomes so that they were individually visible, rather than messed together in a tangled pile. Then he studied them at hundreds of magnifications. He saw virtually only X chromosomes. The younger, female cells, now having undergone about thirty total divisions, were thriving. As for Y chromosomes, Moorhead spotted vanishingly few. The male cells, which had already been elderly at the start of the mixing experiment, were gone. Dead.

The mixing experiment had clinched the case. If the cell-killing factor had been in the glassware or in the medium, or if it had been some other technical slip, all of the cells would have been exposed to the problem and all of them would have been dead. What was more, as Hayflick and Moorhead wrote archly in a paper that would be named a "Citation Classic" because other papers referred to it so frequently, "If a latent virus had been responsible . . . it seems unlikely that it would have been able to discriminate between male and female cells."[12]

Hayflick could now confidently assert that something intrinsic to the cells was behind what he was seeing over and over again with his own eyes. Something inside them was causing them to die. Admittedly, WI-25, the last cell line he had launched, was still dividing, as were several other more recently derived lines. But he was now sure that these too would eventually slow their replicating and then stop. Then they would degenerate and die. Not one of the twenty-five cell lines he had launched was immortal.

Hayflick was thirty-two years old and a virtual unknown in the rarefied universe of top biologists that he would have loved to inhabit. He was faced with the prospect of making an audacious claim. A claim that would challenge fifty years of received wisdom, along with the reputation of the Nobel Prize–winning Carrel. A claim that normal cells aged and, finally, died in their lab dishes. He was nervous. And his confidence wasn't helped by a warning from one of the most respected cell biologists of the time. Gey, the talented developer of the HeLa cells, was visiting the Wistar Institute one day when Hayflick confided in him his new findings. In a 2012 interview Hayflick recalled Gey's response: "Be careful, Lenny. You're going to ruin your career."[13]

Gey couldn't have been more mistaken. Hayflick's finding would one day make his name and distinguish him as the man who opened the door to a whole new realm—the study of cellular aging. It is an area of huge relevance to two of our top health preoccupations today: aging and cancer. But Hayflick had a long road to travel before recognition came.

And how to explain Alexis Carrel's normal chicken-heart cells, replicating faithfully in the lab from their launch in 1912 until Carrel's colleague Ebeling finally dispensed with them in 1946? Years later, in the 1960s, a woman who had worked as a technician for Carrel in the 1930s approached Hayflick after he gave a talk at a scientific meeting. In his lecture Hayflick had speculated that Carrel's method had a fatal flaw related to the fluid that he and his technicians extracted from chick embryos and used to "glue" the chicken-heart cells to the floor of a new culture vessel whenever they overgrew their current dish and needed to be divided. This fluid extract from chick embryos was also used daily to feed the cells. In preparation for this, it was spun in an antiquated centrifuge, a process that was supposed to remove cells, leaving only nutritious fluid. Hayflick proposed that, with or without Carrel's knowledge, the fluid extract actually contained errant fresh cells from the chick embryos; that the culture stayed alive for decades because it was frequently replenished with these new young cells. The woman told Hayflick that he was on the mark; that in the 1930s she had raised questions with Carrel's chief technician indicating that she thought this might be happening. She had been told to forget what she was seeing or risk losing her job. In the midst of the Depression, that was not something she was eager to do.[14]

It is a measure of Hayflick's productivity, energy, and ambition that in the autumn of 1960, as he and Moorhead conducted the male-and-female-cell-mixing experiment, he was also using his new human fetal cells to develop a first-of-its-kind polio vaccine—more on this will follow—as well as running several studies that he knew would be crucial to the successful reception of the landmark paper that he and Moorhead were now putting together.

(Hayflick also, at about this time, codiscovered the cause of walking pneumonia. The culprit was a species of *Mycoplasma*, the tiny microbes that he had studied as a graduate student. *Mycoplasma pneumoniae* was the first of these microbes discovered to cause disease in humans, and the *New York Times* splashed the discovery at the top of its front page.)[15]

That paper would boldly hypothesize that the cells' deaths in their bottles were the outcome of "[aging] at the cellular level."[16] And it would define what later came to be known as "the Hayflick limit"—the number of divisions that a normal cell in culture can undergo before it ceases to divide. Based on the data from his twenty-five fetal cell lines, Hayflick estimated this number at fifty divisions, plus or minus ten. Importantly, freezing the cells didn't affect the Hayflick limit. For instance, when Hayflick froze some WI-1 cells that had divided just nine times, and then thawed them months later and put them in the incubation room, they began dividing again and, seeming to "remember" their age, went through forty-one more divisions over five months before dying in their bottles. What was more, this fact—that the cells would commence dividing again after being frozen and thawed—meant that freezing appropriate numbers of them at young ages could ensure an all-but-endless supply of cells into the future.

But for what practical purpose would such an endless supply of cells ever be needed? It's in the answer to this question that the enormous and lasting public-health impact of Hayflick's work rests. For in his groundbreaking paper with

Moorhead, Hayflick went beyond an iconoclastic assault on the immortality of normal cells. He also suggested that the new cells could make a big contribution to vaccine making.[17] He had conducted the experiments to prove it.

In 1960 the making of new viral vaccines was a top priority for virologists and a goal that was eminently within reach, thanks to the technical breakthroughs of the previous two decades. The fight against polio, fresh in everyone's minds, had shown as much. It had been a terrifying bane—the disease stirred something like the fear that Ebola does today. But in the space of five years, since the launch of Jonas Salk's killed vaccine in 1955, it had been reduced to a preventable disease. What was more, it was becoming clear that a stronger, longer-lasting live polio vaccine was within a year or two of approval by U.S. regulators.

Now the prospect of developing vaccines against other viral diseases beckoned to scientists. Measles, mumps, and German measles, also called rubella, were regular childhood afflictions. Hepatitis was rarer, and gravely serious. Chicken pox was a particular bane for children with weakened immune systems. For some diseases, like rubella and hepatitis, vaccines couldn't yet be made, because the viruses hadn't yet been captured in lab dishes. Virologists set out to hunt them down. For others, like measles and mumps, the viruses had been isolated, and scientists were hurrying to develop vaccines. The U.S. government soon provided the money and muscle to make sure that the new vaccines were used. In 1962 President John F. Kennedy signed the Vaccination Assistance Act, allowing the Communicable Disease Center (CDC) to support mass immunization campaigns and ongoing maintenance programs and to funnel vaccine money and resources directly to state and local health departments.

As well as the challenge of creating new vaccines, there were vexing problems with existing ones. One of two existing rabies vaccines, made in dried animal brains, could produce fatal allergic reactions; the other, made in duck embryos, was not as effective as the animal brain–produced version. And a silent monkey virus had been discovered in monkey kidney cells used to make the Salk polio vaccine. As he prepared his paper in the autumn of 1960, Hayflick, unlike the public, was keenly aware of the monkey-virus problem.

It wasn't lost on Hayflick that his human fetal cells might provide clean, safe alternative microfactories in which to produce viral vaccines, if—and it was a big "if"—they could be infected with disease-causing viruses. There was precisely one way to find out if that was the case. Hayflick began infecting bottles of the WI cells with different viruses. They turned out to have a huge range of virus susceptibility: thirty-one viruses invaded and damaged the cells. These viruses included measles, rabies, herpes simplex, adenovirus, influenza, and polio. The cells even succumbed to invasion by varicella, the virus that causes both chicken pox and shingles but is extremely choosy about which cells it will grow in.[18]

The implications were exciting, at least in Hayflick's eyes. Public and scientific interest in antiviral vaccine making was surging. At this juncture it could be a big advance to introduce a new, safe, plentiful supply of cells for vaccine making. But Hayflick also knew that it was going to be an uphill battle to convince vaccine regulators that his cells were safe. After all, the hot area in cancer research was the purported potential of as-yet-unidentified hidden viruses to cause cancer. Vaccine-approving agencies were going to want assurances that no such unidentified viruses lurked in his human fetal cells. Nothing would put them off the cells more quickly than the possibility of such a virus getting into a vaccine and later causing cancer in vaccinated people.

Hayflick examined his two dozen fetal cell lines under the microscope repeatedly. He found no telltale signs of viral infection. He took the fluid bathing the cells and injected it into cultures of other kinds of cells, and into animals. Neither the cells nor the animals showed any signs of infection.[19]

Then he turned again to Moorhead to scrutinize the cells' chromosomes. Virtually all cancers had abnormal-looking chromosomes and abnormal numbers of chromosomes. If any of the WI cell lines showed such deviations, that too would mean a no-go from regulators. Their thinking would run like this: If the fetal cells harbored abnormal chromosomes, they were either cancerous or would soon become so. If this was the case, then there was every chance that a hidden virus in the cells had caused the malignant changes. And if a hidden, cancer-causing virus was in the cells and they were used to make a vaccine— well, there was an epidemic of cancer just waiting to happen.

Hayflick waited while Moorhead stared at sample after sample of the fetal cells, counting chromosomes painstakingly, hour after hour, week after week. He looked at young cells, which had divided just nine times. And he was careful to look at old ones too: the oldest cells he examined had been through forty replications. Since chromosomes were particularly vulnerable to developing cancer-associated abnormalities during cell division, the older cells, having divided more times, were most at risk of showing anomalies.

When Moorhead finally looked up at the end of weeks of work, in the autumn of 1960, he had good news for Hayflick: these cells, young and old alike, were normal cells, not aberrant ones. They were diploid cells, with twenty-three pairs of chromosomes, for a reassuring, normal total of forty-six. And so Hayflick dubbed the fetal cells "human diploid cell strains." The term "cell strain" was a very deliberate substitution for "cell line" on Hayflick's part. In Hayflick's nomenclature a "line" is a group of cells that will go on dividing endlessly, whereas a "strain" denotes a group of cells that is mortal—that will reach the Hayflick limit and then expire.

Hayflick was now armed with reassuring chromosome data, which Moorhead documented with striking photos showing the chromosomes of several of the WI lines neatly laid out in twenty-three numbered pairs. But Hayflick wasn't finished amassing evidence for the safety of the cells. He enlisted a scientist friend of his, Anthony Girardi, who worked at Merck in nearby West Point, to conduct another key experiment, this time on hamsters—an additional study aimed at persuading regulators that these cells were not cancers-in-waiting.

Hamsters' big cheek pouches don't have the normal immunological defenses that attack foreign invaders, including cancer cells. That makes them useful to biologists, who, as Hayflick worked, had already shown that cancer cells formed ever-enlarging tumors when they were injected into hamster cheek pouches. Hayflick's friend Girardi injected the animals' cheek pouches with vigorously growing WI-25 cells. If the cells had any propensity to turn into cancers, several weeks growing in hamsters' cheeks would give them plenty of opportunity to do so. For comparison, Girardi injected five additional, control hamsters with aggressively cancerous HeLa cells.[20]

The HeLa-injected animals sprouted tumors in their cheek pouches, and those tumors were still growing after three weeks. None of the WI-25–injected animals developed cancer.[21] These findings were reassuring, but for Hayflick they still weren't enough to counter the fears he anticipated among vaccine regulators. Hamsters were not human beings. The cells, he determined, had to be put into people.

Hayflick believed that the risks of such an experiment would be minimal. He was aware that a few years earlier, a high-profile cancer researcher named Chester Southam had taken microscopically normal-looking cells—fibroblasts, the same cell type as Hayflick's fetal cells—from a human embryo and injected them under the skin of three dying cancer patients whom he described as "volunteers." Southam, who was chief of virology at the prestigious Sloan-Kettering Institute for Cancer Research in New York City, reported that the embryonic cells had not grown in the dying patients, unlike the cancer cells that he had simultaneously injected in the same patients.[22]

Southam's name has come to be notorious among medical ethicists, and his studies, which also included injecting aggressive cancer cells into healthy prisoners at the Ohio State Penitentiary, would ultimately lead to a public outcry in 1963. That year three doctors at the Jewish Chronic Disease Hospital in Brooklyn, New York, would resign in protest rather than inject unsuspecting patients with cancer cells for Southam.[23] Their action launched a lawsuit, a state attorney general's investigation, and a media storm that both accelerated and reflected changing public perceptions. Southam ended up being put on probation by New York's medical licensing authority for one year. Many in the medical establishment were not convinced that he had done anything wrong, and soon after his probation ended he was elected president of the American Association for Cancer Research.[24] But by the mid-1960s, ordinary people were becoming less willing to give scientists carte blanche to tinker with human beings on a "Trust me, I know what's best for you" basis.

That was not the case in 1957, which is the year that the Southam study injecting the normal-looking embryonic cells, along with cancer cells, in dying cancer patients was published in *Science*, the leading American science journal. The notion that dying patients, who had nothing to gain by participating and who were enlisted in a study by their doctors, who had everything to gain by their participation, could be called "volunteers" was accepted then. So was the practice of putting such patients at risk. Southam implied as much when he wrote in *Science* that they "had advanced incurable cancer and a very short life expectancy."[25] And so, as they planned for the injection of their own human fetal cells into dying cancer patients, Hayflick and Moorhead felt not only that they were doing nothing wrong, but also that they were following in the footsteps of an eminent man of science. Indeed, their paper would state that Southam's experiment had laid the groundwork for their own.[26]

It made sense for them to turn to Robert Ravdin, the son of the powerful HUP surgeon in chief I. S. Ravdin, for help injecting the diploid cells into dying cancer patients; Ravdin specialized in cancer surgery. A comment that he made in this era suggests that he likely had no compunction about the injections. In 1964, when Chester Southam was on the public hot seat, Robert Ravdin, defending Southam, would tell a reporter that if every subject in a human trial had to be fully informed, everyone would need a PhD.[27]

Being a hard-charging surgeon—a prince of the hospital, if not the king—the junior Ravdin did what surgeons do. He delegated the task downward, to a second-year surgical resident named William Elkins. The twenty-eight-year-old Elkins had an impeccable pedigree: he had attended Princeton as an undergraduate, then gone on to Harvard Medical School. But as a surgical resident at HUP, he was becoming convinced that he was not cut out for the punishing, testosterone-driven world of the operating room. What he wanted to do was science. The next year he would move to the Wistar and begin a long research career in transplantation immunology.[28] As Thanksgiving 1960 approached, he was still slogging it out on the surgical wards when his boss, the junior Ravdin, asked Elkins to do a menial chore: inject some fetal cells belonging to Wistar scientists into a few dying patients.

It's not clear if, how, or when these patients were informed about the experiment or what they understood of the process and its purposes. Both Hayflick and Elkins, recalling these events more than fifty years later, conjectured that Ravdin approached the patients and explained the experiment. However, it would emerge in the mid-1960s that in dozens of studies in this era patients were not informed that they were experimental subjects. It seems at least possible that that was the case here.

Hayflick chose WI-1 cells for the injections. These were not young cells, and he wanted it that way. If the WI cells were going to morph into cancer cells, these older cells were the likeliest to do so. Conversely, if these aged cells *weren't* cancerous, his case that they were normal would be all the stronger for the fact that he had used older cells.

Hayflick gave Elkins two lots of WI-1 cells. He had grown both of them, then frozen them for months, and then thawed them and grown them some more. One group had divided thirty-seven times. The other was still older, having replicated forty-five times.[29] If any one of the cells was cancerous, he would expect it to form a tumor at the injection site. A biopsy of the tumor would reveal abundantly growing cells with the microscopic hallmarks of cancer.

One day late in 1960, Elkins put a small syringe fitted with a fine needle on a tray alongside a test tube containing a sterile solution of salts and water. In each ounce of that saline solution there were some 177 million living WI-1 cells. Then he headed to the surgical wards.

Visiting a dying cancer patient, Elkins turned one of the patient's forearms over, revealing its softer underside. He swabbed the skin with disinfectant and then turned to his tray. He sucked half a milliliter—less than two hundredths of an ounce but containing about three million WI-1 cells—into the syringe, eased the needle under the patient's skin, and pushed the plunger. Then, using a marker, he drew a circle around the injection site. He would be back to check the result, he told the patient.[30]

This faceless patient and five others are identified only by their initials in

Hayflick and Moorhead's landmark paper. In it there is a table describing the injections and their results.[31]

The first patient had cancer that had spread throughout his or her abdomen. The doctors were at the point of simply treating symptoms, not trying to stop the disease. After the patient was injected, a nodule, or bump, developed at the injection site but disappeared on the sixth day. A biopsy of the nodule, to look at its cells under the microscope, was not done.

Three of the patients had breast cancer that had spread. All three were on drugs that suppressed their immune systems, making them less able to fight off WI-1 if it was in fact cancerous. One of them developed a slight fibrous hardening at the injection site on the seventh day; she died on the eighth day. A biopsy of her injection site was negative for WI-1 cells or anything else abnormal. A second developed a nodule on the sixth day that disappeared by the tenth, four days before she died. A biopsy wasn't done. The third breast cancer patient didn't develop a nodule. Nonetheless, her injection site was biopsied on the seventh day, with negative results.

The other two patients had lung cancer and colon cancer respectively. Their diseases had spread through their bodies and they were on chemotherapy. One developed a small nodule on the sixth day. A biopsy showed no WI-1 cells, normal or cancerous. As for the other patient, the paper simply says that nothing had happened at the injection site after nine days, and that no biopsy was taken.

The sum of the results was reassuring: of the nodules developed by four of the patients, three of them melted away, and a biopsy of the fourth showed no cells of any kind. And no nodules had appeared at the injection sites of the other two patients.

By the summer of 1962, another eleven dying cancer patients would be injected with Hayflick's cells, bringing the total to seventeen. A report by Hayflick and others to the World Health Organization in July of that year states that the cells had been "implanted in 17 patients dying of cancer." Again the nodules that developed in some patients disappeared within ten days, and biopsies didn't reveal any living cells.[32] The results from these seventeen anonymous patients would be a key basis on which Hayflick would testify innumerable times over decades that his human diploid cells were normal and didn't cause cancer.

For Hayflick the injections by Elkins were the final experiment in what was now eighteen months of painstaking trial-and-error work—work that had tested not only his scientific thinking and resourcefulness but his, and

Moorhead's, capacity for repetitive, monotonous tasks and meticulous observation. He had established what were, to his knowledge, the first cell lines that had been proven to be, and to remain, certifiably normal when grown in the lab. They could be grown for months and months and yet they still reliably retained the diploid number of forty-six chromosomes. And those chromosomes, scrutinized by Moorhead under the microscope, looked just as they ought to: his diploid cells did not harbor the broken, disjointed, frayed, and otherwise strangely constructed chromosomes that signified cancer. Nor did they appear to cause cancer in hamsters or human beings.

And just as reliably as they harbored forty-six chromosomes, the diploid cells aged and finally died in culture; they could replicate only through fifty or so divisions. They were as mortal as all flesh. The two sets of observations—the cells' normalcy and their inevitable deaths—he now knew, went hand in hand.

His findings could be summarized like this:

Normal (diploid) chromosome number and appearance = finite growth in the lab

A few years later he would take this thinking further:

Normal (diploid) chromosome number and appearance = finite growth in the lab ↔ corresponds to normal cells in human beings

Abnormal (heteroploid) chromosome number and appearance = indefinite growth in the lab ↔ corresponds to cancer in human beings[33]

Normal cells could escape from aging only by acquiring the properties of cancer cells—whether in the lab or in a living human being, he would write.*[34]

Hayflick had also created, he believed, a promising tool to deploy against infectious diseases. For he had demonstrated that his diploid cells could be infected with dozens of disease-causing viruses, making them near-perfect factories for making viral vaccines. They appeared to be clean—free of hidden, lurking, unknown viruses that would scare off regulators—and they could be

---

*Fifty years of study have since made it clear that the process through which a cell becomes cancerous is complicated and involves many steps; it isn't like flipping an on/off switch. Nor are all cells clearly in one camp or the other. For instance, some cells have aberrant numbers of chromosomes and multiply indefinitely in the lab but have not tipped into the uncontrolled growth that is cancer.

produced and then frozen, thawed, and expanded into near-infinite quantities of cells for just such a purpose.

For one junior scientist and his chromosome-expert colleague, it seemed that there was a tremendous amount to be proud of in this paper, which, as Hayflick wrote it up, stretched to thirty-five pages. He knew precisely where he wanted to place it.

In late 1960 the *Journal of Experimental Medicine* was an enviable place for a young, ambitious scientist to be published. Already sixty-five years old, which was a considerable age in the young world of U.S. research, the journal had been founded by William Welch, a giant of American medical research. It had published many of the great biologists. It had also published Carrel's 1912 paper establishing, via the beating chicken heart, that cells could live indefinitely in a culture bottle if they were just treated properly. The journal was put out by what was arguably the pinnacle U.S. research institution: the Rockefeller Institute in New York City. There, none other than Peyton Rous presided over it as editor.

Rous was the biologist who, fifty years earlier, had discovered that he could cause a sarcoma, a malignant connective-tissue tumor, in chickens by infecting them with finely filtered fluid from chickens that already had such tumors.[35] He had posited that an "agent" separate from the tumor cells caused the sarcoma; since then this "agent" had been named the Rous sarcoma virus. With the resurgence in the 1950s of the idea that viruses might cause cancer, Rous's fame and eminence had grown. In 1966 he would be recognized with a Nobel Prize, fifty-five years after publishing his groundbreaking paper.

Before Hayflick put the bulky manuscript and its glossy accompanying photos in the mail, he did two things. According to Moorhead, he suggested to Moorhead that the two men flip a coin to determine whose name should go first on the manuscript. Moorhead remembers declining the offer, saying that Hayflick had done the lion's share of the work and deserved to be in the prestigious place of first author.[36]

Hayflick also took his findings to Koprowski. He briefed his boss on the many disease-causing viruses that infected the diploid cells, and their consequent potential for vaccine making. He told him about their normal chromosomes and their lack of cancerous transformation. And he laid out the audacious claim that the paper would make—that cells aged and eventually died in lab dishes—throwing into question Carrel's "immortal" chicken heart and upending decades of received wisdom. "The next thing you're going to tell me is that these lung cells of yours are breathing," Hayflick remembers Koprowski saying at one point, Hayflick recalled in a 2014 interview.[37]

Nonetheless, the junior scientist invited his boss to join him and Moorhead as an author on the paper. It was a common practice, then as now, for top bosses to share authorship of papers, no matter how little of the actual work they had done. Koprowski demurred. His message was clear: Hayflick and Moorhead could own the paper, and the consequences of their temerity, all by themselves. He did agree, though, to do them the favor of drafting and signing a cover letter to Rous.

Hayflick sent off the paper with its cover letter and waited for what seemed like an eternity for a reply. While he was waiting, he fought back doubts. He was an unknown, a greenhorn, and no matter how good his data, some were going to dismiss him, to say that he had gone off the rails. Gey's cautionary comment came back to haunt him. Was he, in fact, about to ruin his career?

Then an opportunity to buy some peace of mind presented itself. The huge Federation of American Societies for Experimental Biology was holding its 1961 annual meeting in nearby Atlantic City, New Jersey. In mid-April twelve thousand biologists would take over Gilded Age hotels like the Shelburne and the Marlborough-Blenheim and attend events like "The Thyroid Smoker" on the Skyline Terrace at the Traymore. Among the nearly three thousand talks to be given at the seaside resort, one of the most listened-to would be by Theodore "Ted" Puck.

Puck was a powerful, influential figure in cell culture. He was the Colorado-based scientist who had reported growing normal human cells from the skin samples of adult volunteers and from leftover surgical tissues. In the 1958 paper where he had reported this, however, Puck and his coauthor, Tjio, had not noted anything about the cells grinding to a halt. They had reported that the cells were still dividing heartily months after being launched, after dozens of replications.[38] The implication was that they kept dividing indefinitely.

Puck, a Nobel-caliber biologist, was a man whom people turned to with their questions about how to best nurture lab-dish cells. He was known to be meticulous with his culture procedures and to independently test the ingredients of his medium to rule out contamination. (Hayflick had bought his ingredients off the shelf, and used them as is.) What was more, Puck was established as a brilliant man of science and a preeminent cell culturist. How could he have missed something as basic as cells dying, repeatedly, under his nose?

Hayflick wanted badly to put a single question to Puck, and the Atlantic City conference would give him his chance to do that. He just needed to screw up his courage to ask that question. In a 2012 interview Hayflick recalled Puck giving his talk in a big, packed hall. He thought that Puck was pompous, and at the same time, he felt intimidated. He forced himself to raise his hand. Puck called on him.[39]

Hayflick asked Puck if his normal human cells had ever ceased dividing and died, in spite of his assiduous attention to getting every component of the medium just right. Oh, sure, Puck replied. It happened from time to time. But it wasn't a big problem. He would just go back to the freezer and thaw another ampule. That was all Hayflick needed to hear. He knew in that moment that Puck had looked at cells aging and dying in culture. But he hadn't *seen* them.

Hayflick returned to Philadelphia full of confidence. It was a good thing, because ten days later, on April 24, 1961, Peyton Rous at last responded to the Hayflick-Moorhead paper, in a letter to Koprowski. His words were devastating. After apologizing for keeping the young authors waiting because of his busy spring meeting schedule, Rous reported that he and three other editors had read the manuscript and discussed it. Not only was it "too specialized" for the *Journal of Experimental Medicine*, but it was poorly structured and all over the map, bogged down with extraneous observations like the fact that the fetal diploid cells were susceptible to infection by many different viruses.

As for the authors' proposal that cells aged and died in lab bottles, it "seems notably rash," Rous wrote. "The largest fact to have come out from tissue culture in the last fifty years is that cells inherently capable of multiplying will do so indefinitely" if they're properly cared for.

"What a broadside!" Rous observed of his own skewering of Hayflick and Moorhead. "Yet I write it with complete good will."[40]

The man who in 1911 had been ignored and sometimes disparaged when he dared to suggest that viruses might cause cancer was, fifty years later, quite ready to poke holes in Hayflick's audacious challenge to conventional wisdom. Yet Hayflick, with every stubborn, determined bone in his body, wasn't for a moment going to be deterred. Three weeks after receiving Rous's rejection letter, he sent the paper off to the less-venerable but still-respected journal *Experimental Cell Research*. That journal immediately accepted it for publication.

CHAPTER SIX

The Swedish Source

Philadelphia and Stockholm, 1961–62

I believe that life is chaotic, a jumble of accidents, ambitions, misconceptions, bold intentions, lazy happenstances, and unintended consequences, yet I also believe that there are connections that illuminate our world, revealing its endless mystery and wonder.

—David Maraniss, American author and journalist[1]

Sometime around the middle of 1961, an electrical freezer failed at the Wistar Institute. In it, according to Hayflick, were his frozen stocks of all twenty-five WI cell lines. The accident was discovered too late, and he lost all of the cells. One year earlier this would have been a major setback. By this point, however, it was mainly an annoyance. The cells had already yielded up their data; Moorhead had captured their chromosomes, in all their glorious normality, on film; and the paper had been written up, submitted, and accepted. It would appear in print in *Experimental Cell Research* in December 1961.[2] And Hayflick, if he knew anything by now, knew just how to make replacements. So this wasn't the end of the world. It was more like the beginning.

Hayflick had become firmly convinced of the potential of normal human fetal fibroblasts to improve vaccine making. In his view they were far superior to the expensive, sometimes-infected monkey kidney cells then being used to make polio vaccine—the only vaccine against a common viral illness that was then available. What was more, the human diploid cells could be infected with viruses for which vaccines didn't yet exist or were in early development: common viruses including measles, chicken pox, and adenovirus, which frequently caused respiratory infections. Why not use the diploid cells to develop these new vaccines? Perhaps too his fetal fibroblasts could be used to improve on the two rabies vaccines that were then in use—one made from the pulverized brains and spinal cords of rabies-infected rabbits and the other from duck embryos. The first could have serious or fatal side effects. The second was sometimes ineffective.*

---

*The vaccine against smallpox, then a routine childhood vaccination, was also crudely made, by harvesting virus from the scarred skin of calves, sheep, or buffalo that had been

Hayflick determined that he would make a *human* cell strain that, if all went well, might become a gold standard for vaccine manufacturers. Not a strain launched, as he had launched the first twenty-five WI strains, with an exploring scientist's tentativeness and curiosity, but one created with an eye to the future. A human diploid cell strain that vaccine manufacturers would embrace because of its pedigree: it would be free of viruses, free of cancer, and available in such quantities that running out of it would never be a problem.

To his mind such a gold-standard cell strain would have a huge advantage over, for instance, monkey kidney cells—which came from endless new pairs of monkey kidneys, each with its own risk of harboring some unwanted virus. This was because the monkey kidney cells used to grow the polio vaccine virus were used just once, when they were freshly harvested from the kidneys of a newly slaughtered monkey. They were not allowed to divide repeatedly into ever-expanding quantities in lab bottles because regulators feared that the cells might turn cancerous with repeated divisions. As a result, tens of thousands of animals were imported and slaughtered each year to make polio vaccine.[3]

By contrast, a human cell strain derived from just one pair of fetal lungs and then allowed to replicate could be established as clean and safe at the beginning and then used without worry going forward. Not to mention that it would be far less expensive to provide: one pair of fetal lungs would be all that was needed. The costly obtaining and sacrificing of unending numbers of monkeys could stop.

Hayflick didn't relish the prospect of going back to the gynecological surgeons at the Hospital of the University of Pennsylvania to ask for a twentieth fetus. For one thing, they were surgeons and had a limited interest in the esoteric projects of a junior biologist from the research institute across the street. For another, Hayflick knew that he was now going to need more than a fetus. He was going to need a family history of the parents of that fetus. It would have to be a clean history that would reassure vaccine makers there were no infectious diseases or cancer lurking in either parent—conditions that would "scare the hell" out of them, as Hayflick recalled in a 2012 interview.[4] For this kind of cooperation he would need, on the upstream end of the abortion, a

---

infected with a related virus called vaccinia. Vaccinia was one of the many viruses that could invade Hayflick's human diploid cells. But interest in smallpox vaccination was on the wane: the last case of smallpox in the United States had been reported in 1949, and routine childhood vaccination against smallpox would be dropped in the United States in 1971. The last naturally occurring case of the disease anywhere was reported in 1977.

scientist who understood vaccine making and who therefore understood the importance of his project.

Sven Gard was a tall, solemn, soft-spoken Swede who had spent eight months on sabbatical at the Wistar Institute beginning in January 1959. He had occupied a lab across the hall from Hayflick's and one door down. There he worked with the lights half dimmed—perhaps, Hayflick thought, because the low light made a man used to long northern winters feel at home.

The gray-eyed Gard was brilliant and renowned. Fifty-three years old, he was a father figure in virology who inspired both fear and adoration in his students. He was a power player of the sort that Koprowski regularly recruited to the institute. As chairman of virology at the famous Karolinska Institute in Stockholm, which awards the Nobel Prize in physiology or medicine, he was regularly a member of the committee that decided the winners of the coveted award. He had been instrumental in seeing that the prize went to Enders and his colleagues in 1954, for their discovery that polio virus would grow in many kinds of human cells and not only nerve cells, which opened up the quest for a polio vaccine.[5]

Gard himself had played a major role in his country's intense drive to develop its own polio vaccine—an urgent goal because Sweden had a far-flung population and corresponding reservoirs of people who had never been exposed to the virus and had therefore never developed antibodies. That made the country particularly vulnerable to polio epidemics. In the 1950s Sweden had the highest number of cases per capita anywhere.[6]

Gard was a big believer in making human vaccines in human cells. He had been inspired by the idea during a 1952 visit to the Enders lab in Boston, where he saw the polio virus being grown in human cells. On his return to Sweden, where abortion had been legal since 1938, he began using cells from human fetuses obtained from hospitals in Stockholm to develop a Swedish polio vaccine.[7] Unlike Hayflick six years later, Gard's team of virologists didn't try to coax the human fetal cells to replicate over and over again in lab dishes. Instead they used the fetal skin and muscle cells just once, infecting them with polio and then harvesting the virus-laden fluid that bathed the cells.

The skin and muscle were the only organs sufficiently big to give them enough cells to work with. But nonetheless they found that their efforts yielded ten times less vaccine virus than other scientists were producing using monkey kidney cells. While the Gard team did run one human trial, vaccinating two thousand children with the fetal cell–grown polio vaccine, it became clear that they could not generate enough of it to vaccinate a population of seven million people.[8] Gard eventually bowed to the inevitable and used monkey kidney cells to make the Swedish vaccine.

Why didn't Gard and his team simply grow huge numbers of the human fetal cells, coaxing them to replicate in lab dishes? "We never thought of it," says Erik Lycke, who was a twenty-seven-year-old MD/PhD student working for Gard in 1953. "I don't think anyone but Hayflick did."[9]

The Swedes launched their monkey cell–produced polio vaccine in 1957. By 1964 the disease would be virtually eliminated in Sweden. In January 1959 Gard, who would later say he had been "breathing" polio research night and day, at last found time to respond to Koprowski's invitation to take a sabbatical at the Wistar Institute. There he met, among others, Hayflick.

Hayflick recalled that Gard, who died in 1998, overheard him griping about getting fetuses from the gynecologic surgeons at HUP—the hassle of it and their lack of understanding of his purposes—and volunteered that if Hayflick should need fetuses in the future, it was easy for him to get hold of them back in Sweden.

Possibly as early as 1959, Hayflick began to take advantage of Gard's offer, receiving occasional fetuses or fetal organs from Sweden. The time required in transit was not an obstacle; Hayflick had discovered soon after he began working with fetuses that living fetal tissue could be kept for five days at room temperature without dying. Minced tissue floating in growth medium lasted even longer: up to three weeks.[10] Now, in 1961, confronted with the demise of his first twenty-five fetal cell lines, Hayflick turned again to the solemn Swede.

Eva Herrström had been working with Sven Gard since 1952 in the Karolinska Institute's virology department, which was housed on the expansive grounds of Sweden's National Bacteriological Laboratory. She had practically grown up there, where her father, Josua Tillgren, a prominent Stockholm physician, had hired her as an assistant lab technician in the summer of 1943, when she was just seventeen. She never left, and by the mid-1950s she had risen to become Gard's top technician, a position she still held in 1961.

Gard's lab wasn't in the elegant main building designed by the famous modernist architect Gunnar Asplund but in a small, two-story yellow-brick outbuilding where it occupied the main wing of the second floor. The temporary-looking building was nicknamed the Monkey House because the other wing housed African green monkeys used to safety-test polio vaccine. The communal freezer was just outside the windowed door into the monkey wing; whenever someone deposited a sample in it, the monkeys shook their cages, raising a ruckus.

Humble digs or not, Herrström liked her work as Gard's chief lab technician. He took an interest in all of his staff, inviting her and the other technicians when he presented lectures on scientific advances and crediting everybody for his or her contribution. Still, there was no mistaking who was in charge. There was little or no frivolity on the job, no happy hours, no practical jokes or undue familiarity. She knew that she would, until her dying day, call Gard "Professor."

On this particular morning—April 24, 1961—Herrström climbed the metal stairs that ran up the outside of the building to the second floor to learn that a fetus would be arriving and that she needed to prepare its lungs for shipment to the United States.

When she had first worked for Gard almost a decade earlier, he had been trying to make polio vaccine with human cells. Then, Herrström had worked plenty with human fetuses, even learning how to expertly drain amniotic fluid that was used in cell-nurturing medium from the intact pregnant uteruses of cows that arrived regularly from the Stockholm slaughterhouse. But once Sweden moved to using monkey cells to develop the vaccine, that work with human fetuses had gone away—until lately, when Gard had begun asking her to prepare tissue for shipment to an American institute in Philadelphia.

Later, after she had donned a white gown and a car had delivered a tiny bundle wrapped in green surgical cloth, Herrström headed for one of the sterile rooms in the middle of the floor. She washed her hands in disinfectant at the sink under the window, laid out her instruments, and sat down at the shiny linoleum table that was the lone piece of furniture in the room. She unwrapped the bundle.

It really was incredibly beautiful, this little fetus, with everything already in place. With this task she was being given a privileged glimpse into the creation of life. It helped to remember this as she picked up a scalpel. And it helped that she came from a family of physicians. You got used to it. You turned the tragedy around. You said to yourself that at least in this case, something life-giving might emerge from death.[11] What shape that particular good might take in this instance Herrström didn't know. But if Professor Gard said that scientists in Philadelphia needed fetal lungs, her job was to make sure they got them.

At home that night Herrström made a dinner of sliced reindeer meat and repaired her winter gloves. Before turning in at 9:30 p.m., she wrote in her diary: "Work 8.30-[5 pm]. Sent tissue to USA. Stressful."[12]

The lungs that Herrström had earlier that day placed in medium in a small test tube, and the tube that she had then placed in a thermos, and the thermos that had then been packed in a box, on wet ice, were well on their way by then to Philadelphia.

It's probable, although not certain, that the lungs that Herrström removed from that fetus on that spring day went on to become the next cell strain that Hayflick created. (By this point Hayflick was working only with lungs because they were readily dissected and their fibroblasts seemed particularly hardy in culture.) It's also possible that those lungs didn't work out. Perhaps delays caused them to die in transit, or perhaps they were inadvertently infected. Because scarcely one week later Herrström recorded in her diary that she had shipped more tissue, from a new fetus, to Philadelphia: "Tissue to the USA at 12 (noon). Took all morning to prepare."[13]

What is certain is that by the autumn of 1961 Hayflick had launched his twenty-sixth diploid cell strain, from the lungs of a male fetus aborted sixteen weeks into pregnancy and sent to Philadelphia from Gard's lab in Sweden. By that October he had taken time-lapse photos of the new WI-26 cells being attacked by polio virus in a lab dish.[14]

Hayflick's timing in launching the new fetal cell strain was near perfect. He produced and froze what he thought would be plenty of the fibroblasts—about two hundred ampules of them, each containing up to two million cells—shortly before the December publication of "The Serial Cultivation of Human Diploid Cell Strains," the paper that announced to the world his discovery of the Hayflick limit and his cultivation of those first twenty-five normal fetal cell strains. Readers, of course, didn't know that those cells had since died in the freezer failure.

Demand for his human fetal cells soared when the landmark paper was published. With the dozens of viruses that Hayflick had demonstrated to infect the cells, virologists were keen to get hold of them for experiments on the nature of these viral diseases. Companies hoping to launch viral vaccines wanted them. Basic biologists too sought them for all manner of study on the workings and behavior of normal cells in lab dishes. With the new WI-26 cells on hand, Hayflick was ready.

Soon he was handing out ampules of WI-26 left and right to biologists, virologists, and companies aiming to create vaccines against measles, adenovirus, and polio. But before he knew it, to his chagrin, Hayflick began running out of ampules—an embarrassing fact, since he had been determined to make a plentiful supply. What Hayflick hadn't reckoned on was the demand for the cells. Had he seen it coming, he might have been more conservative, thawing one ampule at a time and expanding its contents through several generations in the incubation room, then sending out the much-more-numerous resulting cells. Giving away the ampules themselves was like giving away his seed corn.

"As a consequence of the unanticipated and unprecedented demand" for WI-26 cells, he would write in the Wistar's next biennial report, "depletion" of his stocks had occurred.[15]

He needed to begin again—again. He needed to create a human fetal cell strain that would outlast the current, seemingly bottomless demand, for this new kind of normal, diploid cell. Fortunately, Gard continued to be good on his word. This time Hayflick asked Gard for the lungs of a female fetus, to ensure that if any of the male WI-26 cells he had launched into labs around the world became mistakenly mixed with the new cells during an experiment, the problem could be made quickly apparent by looking at the cells on a microscope slide: in a female cell strain, there should be no Y chromosomes present.

It wasn't easy for Gard to deliver just the right fetus quickly. It had to be female. It had to be large enough—from a pregnancy three to four months along—to have readily dissected lungs that would yield enough tissue. And it had to be from a woman without health problems in her present or past. And so Hayflick waited. At last the lungs of a female fetus arrived from Sweden. Continuing with his numerical order, Hayflick named this next female lung-cell strain WI-27. Shortly after launching it, he handed a sample of the cells to Moorhead. He examined the cells and reported back that there was an abnormality in the WI-27 chromosomes. It was probably inconsequential, but "probably" was nowhere near good enough for vaccine makers, wary as they were of using cells that might be cancers in waiting. The strain was useless for vaccine-making purposes. Hayflick would have to go back to Gard once again.

At this point another scientist might have thrown up his hands and moved on to something else. The groundbreaking work deriving these normal human fetal cell strains was done. Hayflick's methods were now published for all to see. What was more, he had launched a goodly supply of WI-26 cells into academic and commercial labs. Another scientist might have told colleagues and companies that if they wanted more normal cells, they could make their own. All it involved was engineering and practice. This line of thinking, however, was completely foreign to the dogged, determined, ambitious man who was Leonard Hayflick, who was set on delivering a lasting human diploid cell strain to the world.

Besides which, he was now working under a new obligation. In February

1962, as Hayflick was watching his supply of WI-26 cells rapidly dwindle, the National Cancer Institute, part of the National Institutes of Health, the U.S. government's medical research agency, responding to the keen and obvious demand for Hayflick's new breed of cells, entered into a contract with the Wistar Institute. Under it the Wistar—meaning, in practice, Hayflick—committed "to produce, characterize, and store human diploid cell strains" and to distribute them to all qualified investigators.[16] From 1962 the Wistar began receiving from the National Cancer Institute annual contract payments of at least $120,000—nearly $1 million in 2016 dollars and about 10 percent of the Wistar Institute's income from grants and contracts.[17]

Koprowski's institute did need to make an initial outlay: In June 1962 the institute's board of managers approved the Wistar director's request for $8,000 to retrofit Hayflick's "new Diploid Cell Laboratory" with a power line, extra air-conditioning, and plumbing that the NIH contract didn't cover.[18] Given the size of the contract, they must have felt that the payoff would more than justify that investment.

The language of the contract may seem dull, but it will become important. It stated in part that when the contract was terminated, "the contractor agrees to transfer title and deliver to the Government, in the manner, at the time and to the extent, if any, directed by the Contracting Officer, all data, information and material which has been developed by the Contractor in connection with the work under this contract."[19] In other words, while Hayflick would be the developer of new diploid cells under the contract, the government would own them, and when the contract was up, he would need to transfer any cells he had developed to the National Cancer Institute or place them wherever the agency directed him to put them. (In the case of cells, privately held cell banks are often used as cell custodians.)

And so, working under this new contract obligation, Hayflick, after the failure of WI-27, asked Gard for another female fetus. And once again he waited.

Hayflick made one more decision in the wake of the WI-27 disappointment. With the next human diploid cell strain he was going to change the numbering, because some scientists seemed to be carelessly confusing the numbers of earlier diploid cell strains. In order that going forward from WI-27 there would be no more such confusion, he would raise the "2" to a "3" and the "7" to an "8." The next cell strain he created would be called WI-38.

On the warm, sunny morning of June 7, 1962, Eva Ernholm prepared for her job as an obstetrician/gynecologist at the Women's Clinic in a hospital near

Stockholm. One of few women in Swedish medicine in 1962, the thirty-seven-year-old Ernholm was an adventurer. As a medical student driven partly by an interest in Eastern religions, she had worked in a U.S. Eighth Army MASH unit in the spring of 1951, during heavy action in the Korean War, and returned to Korea with the Swedish Red Cross in 1953. She came back from her journeys bearing photos of herself bent over patients in a field hospital operating room and in fatigues surrounded by orphans, looking like she was having the time of her life.[20]

Ernholm was a blunt woman who knew her own mind and didn't hesitate to speak it. Whether people liked her or her views—and they inevitably felt one way or the other—was a matter of indifference to her. She was impulsive and decisive and passionate. And, like Hayflick, she was not someone who was turned aside easily. An accomplished pianist, she had insisted on using a construction elevator to lift her grand piano to the penthouse apartment where she had taken up residence one year earlier.

Ernholm had abandoned an early interest in neurology to study obstetrics and gynecology and was a newly trained specialist when she took her hospital job in 1961. At the Women's Clinic she shared a heavy workload with one other junior doctor, monitoring pregnant women, delivering babies, cutting out uterine growths, and tying women's tubes when they requested it. She also performed abortions.

Ernholm did not do this lightly. The following year, 1963, she would tell a newspaper reporter, "As a matter of principle, I am against all abortions that are not performed for medical reasons." But, she added, "evidently there are situations when the social circumstances are so valid they don't leave you any choice."[21]

On this lovely June day Ernholm was confronting such circumstances, in a woman in her thirties who throughout this book will be called Mrs. X. Her medical history was uneventful, apart from childhood bouts with whooping cough, measles, and scarlet fever. Really, her only problem was that her last menstrual period had been in late January. Mrs. X explained that she already had several young children, and that her run-down husband, a working man, was often out of town for his job. When he was home, he wasn't much use: he was an immature alcoholic who had done time in prison.

Ernholm took out her stethoscope and pressed it to Mrs. X's chest and back. Her patient's heart and lungs were clear, as they would need to be for the operation. Ernholm laid a hand on the woman's soft abdomen. It was painless, Mrs. X confirmed, when Ernholm pressed it. Now came the stirrups and the hard metal of the speculum. It couldn't be helped; one had to be sure that the patient was pregnant. Ernholm looked and then noted in Mrs. X's record that the cervix, the mouth of the womb, was bluish—a telling sign of pregnancy. By feel and by sight, Ernholm judged that she was sixteen or seventeen weeks pregnant. She recorded this and wrote: "Indication for abortion: General weakness." Then she added the details of Mrs. X's home situation.

Sweden, where abortion had been a capital crime one hundred years earlier, had legalized abortion in 1938. The law enacted that year stipulated that women could have abortions in three situations: in cases of rape or incest; if delivering the child would cause "sickness" or "weakness" that seriously endangered the mother's life or health; and for "eugenic" reasons, meaning that the mother or father was likely to transfer a serious hereditary disease to the child. (In this case the woman had to agree to be sterilized at the time of the abortion.) In 1946 the law was liberalized slightly to include "expected weakness" of the mother as one of the conditions that made an abortion permissible—the assessment that, given the mother's living situation and circumstances, her health could seriously deteriorate if she was to bear and raise the child.[22]

In practice, however, getting an abortion in Sweden in 1962 was far easier said than done. A woman who sought to end a pregnancy had two choices. She could apply to the Medicinalstyrelsen—the Royal Medical Board—which regulated abortions for the government from an imposing nineteenth-century building in Stockholm. (It memorialized Swedish King Oscar I, an honorary member of the Swedish Academy of Sciences and a prolific father who sired eight children—three of them by mistresses.)

Alternatively, a woman could try to convince two doctors that she needed to end her pregnancy. One of the doctors was often a psychiatrist. The other was the surgeon who would perform the procedure. Most Swedish doctors opposed abortion and refused to help women get them, leaving thousands of women's applications to grind through the slow-turning gears at the government's Royal Medical Board, often pushing abortions well into the second trimester.

Mrs. X was seeking to end her pregnancy at just about the worst time possible for a woman in Sweden who wanted an abortion. In 1951 the Swedish Medical Association had adopted ethical rule IV, which said that the physician "should consider his duty to protect and preserve human life from its implantation in the mother's womb."[23] The next year the chairman of the medical association, in the midst of a debate about whether to rescind the 1946 change that liberalized the law, said that abortion was in the same category as child murder.[24] Swedish abortion rates then fell markedly, reached a low point in 1960, and scarcely budged in the next two years.

The day after Ernholm examined her, Mrs. X was wheeled into an operating room, where Ernholm performed what was called a "minor Caesarean section." She cut through Mrs. X's abdominal wall, carefully dissected the bladder free from where it lay high on the uterus, and cut through the wall of the uterus. She removed the fetus and the placenta, being careful not to leave any tissue behind. "The cavern was cleaned. Suture of the uterus in stages," Ernholm wrote in the operative report. The fetus, she noted, "was 20 cm. long and female."

That fetus was wrapped in a sterile green cloth, handed to an aide, and taken to a car for transport to the virology department of the Karolinska Institute.

A few days later, on a gray, drizzly morning in mid-June, Hayflick sat down in one of the tiny "sterile" rooms in his lab. Following a routine that was now deeply familiar, he dipped a pair of tweezers in alcohol, flamed them in a Bunsen burner, waited for them to cool, and then lifted two small, purplish chunks of tissue from where they floated in a glass bottle of clear pink solution. He laid them on a petri dish. Using a pair of scalpels held at right angles, working them like an improvised scissors, he minced the lungs into innumerable pinhead-size pieces. He deposited them in a flask, where a trypsin solution would break down the connective tissue holding the cells together, releasing millions and millions of individual cells.

Later he poured that mixture into several small glass tubes, stoppered them, and loaded them into a centrifuge, a round machine that sat on a pedestal with wheels and could be moved here and there around the lab. He turned it on and the tubes began spinning so fast that they flew out at an angle horizontal to the machine. After twenty minutes or so, the cells, being heavier than the fluid, sank to the bottom of each tube as an off-white pellet.

He turned off the machine, recovered the tubes, and, back in the sterile room, repeatedly blasted the pellet in each tube with medium, using a glass pipette with a cotton stopper and the power of his lungs. Eventually the cells came loose from one another. He sucked them up in the pipette, transferring them bit by bit into a big glass bottle. Moving quickly now so the cells wouldn't stick, he poured the mix of cells and nutrient solution into several small glass bottles. He laid these carefully in the incubation room.

Some days later, after the cells had established themselves in the bottles, Hayflick handed a sample of them to Moorhead. This time the news that came back from his friend and colleague was good. Moorhead told him that the WI-38 cells' chromosomes appeared entirely normal.

Hayflick knew that if he could freeze a large enough quantity of WI-38 cells, he could provide vaccine makers with enough cells to make vaccines for decades to come. How was this possible, if the cells were mortal and would, sooner or later, die in their bottles? Hayflick had shown how the math worked in the landmark 1961 paper. It was all due to the extraordinary power of exponential growth.

Suppose, Hayflick wrote in that paper, you began with just one small glass Blake bottle. It was rectangular and roughly pint-sized, its larger, flat side measuring a mere 5.5 inches by not quite 3 inches. Such a bottle held roughly ten million cells when those cells had grown to confluence on its floor—really, on its side, because the bottles lay on their flat sides while they incubated. If at this point you split these newly planted cells into two bottles, and split the bottles again when the floors of those two bottles were covered, yielding four bottles; and if you then kept splitting the bottles in this way when the cells reached confluence, until the original cell population had doubled fifty times—roughly the Hayflick limit—you would produce, he calculated, 10^{22} cells, or 10 sextillion cells. Knowing as he did that 14.2 billion wet cells weigh about an ounce, Hayflick also calculated that the cells in that one original bottle would therefore produce twenty-two million tons of cells.[25]

Admittedly, this was a theoretical maximum. Real life wasn't nearly so neat. Sometimes Blake bottles got contaminated. Sometimes the ampules in which he froze the cells cracked or, worse, exploded during thawing because liquid nitrogen had leaked into microscopic holes in their closures while they were frozen. (Hayflick took to wearing goggles when thawing ampules.) Sometimes cells got lost during shipping. And sometimes, frankly, they got thrown down the drain when there were leftovers after shipping them out to scientists.

On the other hand, there was room for a few accidents and a little waste when working with a potential ten sextillion cells—a number that's so large it's difficult to grasp. One way to think about it is this: the freshly harvested WI-38 cells covering the floor of just one of Hayflick's pint-size Blake bottles, expanded until they have doubled roughly twenty times, would produce 87,000 times more vaccine than is made by a typical vaccine-making company, setting out today to make one year's worth of a typical childhood vaccine that it will ship to more than forty countries.[26]

The point Hayflick was making with his calculation in the 1961 paper was that a sufficient supply of cells, frozen and thawed when needed, bit by bit, would produce all the cells that the world would need for the foreseeable future. Using his method, he wrote, "one could have cells available at any given time and in almost limitless numbers."[27]

That summer, change was coming to America. As Hayflick cut the WI-38 lungs into minuscule pieces, the *New Yorker* published the first excerpt of Rachel Carson's classic *Silent Spring*, launching the modern environmental movement. As the cells first reached confluence late in June, the Supreme Court ruled that voluntary prayer in public schools violated the constitutional separation of church and state. When President Kennedy, speaking at Philadelphia's Independence Hall on a lovely, sunny Fourth of July, praised American democracy for encouraging dissent, three miles away the cells divided luxuriantly in ninety-six-degree heat. And as Hayflick set about creating a supply of frozen WI-38 cells that he intended to last for the foreseeable future, women's constrained access to abortion landed in the headlines, in the person of an actress in Phoenix named Sherri Finkbine. Finkbine, a mother of four and a host on the children's television show *Romper Room*, had taken the drug thalidomide to combat morning sickness early in her fifth pregnancy, unaware that it deformed fetuses. She could not get an abortion in Arizona, where state law allowed abortions only if the life of the mother was in danger.[28] She ended up flying to Sweden to terminate her pregnancy, the press following her every step of the way.

In deciding when to freeze the WI-38 cells, Hayflick had to strike a balance. He wanted to produce enough cells to fill plenty of ampules for future needs. He was keenly aware that two hundred ampules of WI-26 had not been enough. On the other hand, he didn't want the cells to get too old—to divide too many times—before he froze them. Vaccine makers wanted youthful cells that they could expand through many more divisions before they reached the end of their usefulness. They were also wary of older cells because each cell division increased the chance of chromosomal aberrations and thus, they feared, of the cells becoming cancerous.

And in fact one year after Hayflick launched WI-38, it would emerge that the chromosomes in WI-38 cells developed spontaneous abnormalities when the cells aged beyond forty divisions. These were not cancerous changes, according to the paper published in the *Proceedings of the National Academy of Sciences* by Moorhead and Eero Saksela, a young Finnish scientist then working at the Wistar Institute. "On no occasion," they wrote, did any of the cells with altered chromosomes develop abnormal shapes or start to divide abnormally quickly or open-endedly, all classic signs of cancer.[29]

But the abnormalities that Saksela and Moorhead documented would push vaccine makers and regulators alike to steer clear of WI-38 cells at high population doubling levels. By the late 1960s companies in the United Kingdom, one of the first countries to embrace human diploid cells for making vaccines, were beginning vaccine-making campaigns with cells no older than the thirtieth population doubling level.[30]

Hayflick decided to freeze the cells when they had been split into new bottles eight times. Eighth passage cells—so called because they had been "passed" into new bottles eight times—would be plenty youthful, *and* there would be plenty of them. Just one small Blake bottle of cells, split eight times, produced 256 bottles, each containing roughly ten million cells at confluence. And Hayflick typically placed the cells from a pair of lungs into four small Blake bottles at the outset.

Some bottles, Hayflick conjectured in 2013, were probably lost to contamination, perhaps early in the splitting process, eating into the total number of bottles available when the time came to freeze the cells. At any rate, there is no record of exactly how many bottles accumulated before Hayflick gave the order that the cells in them should be distributed into tiny ampules and frozen. What is certain is that the task at hand was huge.

Hayflick was not present in his lab on the day in late July 1962 that a number of Wistar technicians, some of them borrowed from other labs, assembled to do the job. Hayflick had traveled two weeks earlier to a World Health Organization meeting being held in Geneva to discuss the potential vaccine-making uses of his human diploid cell strains. Possibly he was still on the road.

The crew of lab techs faced long hours of monotonous work. They had first to loosen the cells from where they lay in single, sticky layers on the bottoms of the Blake bottles, using lung power and the pipette technology invented by Louis Pasteur nearly one hundred years earlier. They sucked up a mix of culture medium and cells, then blasted the fluid back into the bottle and repeated this over and over again, until the fluid turned milky white with floating cells.

Then they went to work with syringes, sucking up tiny portions of the fluid and injecting these into steam-sterilized gas ampules. It was delicate work. Each little wine bottle–shaped vial was about two inches tall and had an open neck roughly one sixteenth of an inch wide. Once loaded with cells, it needed sealing. The workers—some with more finesse than others—sealed the neck of each ampule by melting it with a pass through the flame of a Bunsen burner. Using tweezers, they pulled on the string of melting glass to work it into a blunt closure—all the while trying not to kill the cells in the ampule with too much heat. They were executing what would, with hindsight, become a critically important operation: getting the cells into the ampules without letting bacteria contaminate them.

Sterilizing procedure was nowhere remotely as good in the early 1960s as it is today, in large part because technologies now taken for granted didn't exist. For instance, today's ubiquitous laminar flow hoods, which prevent microbial contamination of the air over a scientist's work space, were just being developed. The ultraviolet lights then used at night had serious limits: the light they emitted killed organisms *on* surfaces but not *in* things—like cell cultures.

True, penicillin and other antibiotics were in wide use by 1962, and many labs used them liberally to protect cell cultures from bacterial contamination. But, Hayflick says, vaccine manufacturers were wary of antibiotics because of their potential for provoking allergic reactions in vaccinees. He decided to take the calculated risk of not using antibiotics. That decision would come back to bite him.

By the end of the day on July 31, 1962, the Wistar work crew had produced more than eight hundred ampules of WI-38. Each tiny vial held 1.5 million to 2 million cells. They were placed in a big, communal dry ice chest. A couple of months later Hayflick transferred them to more permanent digs in a liquid nitrogen tank in the Wistar basement. There they remained, tucked away at –320 degrees Fahrenheit.[31]

In October of 1962 a seasoned cell culturist named Robert Stevenson paid Hayflick a visit at the Wistar. The straight-shooting Stevenson was Hayflick's project officer at the National Cancer Institute, charged with overseeing the smooth and proper execution of the contract under which Hayflick was producing, storing, and distributing human diploid cell strains. Perhaps Stevenson learned during this visit that Hayflick had, ten days earlier, handed one hundred ampules of the new WI-38 cells to a visitor, Frank Perkins of Britain's Medical Research Council, the UK's top vaccine regulator. Perkins had taken them back to London on a transatlantic flight.[32]

On October 18, 1962, two days after he visited Hayflick's Wistar operation, Stevenson wrote a memo summarizing his visit and noting that he had made it clear to Hayflick that the cells were U.S. government property.

While the cells, he wrote, couldn't be subcontracted out to a non-U.S.-government agency for distribution, the Medical Research Council, or MRC, could function as a "distribution depot."[33] This the British agency promptly became, shipping the cells to scientists in Berlin, Madrid, Milan, Tehran, and Uppsala.[34]

At the end of the eventful year of 1962, Hayflick filed a progress report to Stevenson at the National Cancer Institute—a routine report owed to the government agency under the contract. It included a section entitled "Characterization of a New Human Diploid Cell Strain WI-38."[35]

As soon as the WI-38 ampules were first frozen, Hayflick went back to Sven Gard in Stockholm for the documentation that he knew that regulators would require. For the purposes of the Division of Biologics Standards, the unit at the National Institutes of Health that in those days licensed new vaccines, Hayflick's word would not be enough. He needed papers proving that Mrs. X, the mother of the WI-38 fetus, was as healthy as a horse and that no cancers, hereditary diseases, or infections were lurking in her or in the fetus's father.

Getting that information was a delicate task. Two months after the abortion, Mrs. X had no idea that her fetus had ended up anywhere but in a medical waste incinerator. In 1962 in Sweden, as in the United States, tissue from aborted fetuses was routinely used by scientists without the knowledge of the women who had the abortions. The stern Gard, aged fifty-six and childless, was not inclined to take up this job. He delegated it to someone he may have considered better suited to approach Mrs. X: a thirty-five-year-old physician named Margareta Böttiger. Böttiger was earning her PhD in his lab while running human studies monitoring the effects of the Swedish polio vaccine.

A dark-haired beauty with an oval face who hailed from a storied and powerful Swedish family, Böttiger was reserved, mannerly, and unthreatening. Like the gynecologist Ernholm, she confronted a medical profession in which just 14 percent of physicians were women. Unlike Ernholm, she had small children—her girls were three and six years old that August—and a physician husband who did no housework. She had survived by embracing the regular hours of a scientist in Gard's lab, rather than the pediatrics she had trained for immediately after medical school. Paying her invaluable nanny, Harriet, consumed almost her entire salary—but it never occurred to her to give up the work she loved.

The task at hand didn't quite fit that description. But Gard had asked her to do it, and she was not one to buck authority. Böttiger telephoned Mrs. X's primary-care doctor. Working with that doctor and the doctor's nurse, she pieced together as much as she could and got hold of the operative record from the hospital where the abortion was performed. The whole enterprise took some doing, and it wasn't until more than a year later—October 1963—that she wrote to Hayflick, attaching Mrs. X's medical record.

It cataloged her childhood bouts with measles and scarlet fever and whooping cough; her freedom from other infectious diseases since then; her several healthy children; and the absence of known hereditary diseases or tumors in her family.

In her accompanying cover letter Böttiger wrote that she believed Mrs. X to be perfectly fine. But, she added, the father appeared to be subpar mentally. Böttiger also warned Hayflick, who had apparently asked for blood samples from the couple, that when Mr. X got back from his out-of-town labor, it might be tough to get his blood drawn. She did not elaborate on why.

Asked about this in 2015, Hayflick could not recall why he would have requested a blood sample from Mr. X. He added that Mr. X's reported mental deficiency did not trouble him. Mental deficiency, he wrote, is not infectious and would not have been relevant to the cells' safety for vaccine making.[36]

CHAPTER SEVEN

Polio Vaccine "Passengers"

Bethesda, Maryland; Philadelphia; and Clinton Farms,
New Jersey, 1959–62

*Oh, they kept taking rooms away from me, and help. But I—my best people
stayed with me.*

—Bernice Eddy, former NIH vaccine safety scientist, 1986[1]

In the spring of 1960, at the National Institutes of Health in Bethesda,
Maryland, just outside Washington, DC, a feisty fifty-seven-year-old PhD
scientist with an open, square face and neatly coiffed dark brown hair was worrying about her alarming findings in a group of newborn hamsters.

Bernice Eddy had grown up in a family of physicians in Auburn, a town of
199 in rural West Virginia, and earned her PhD in microbiology from the University of Cincinnati. After a stint researching leprosy in Louisiana, she came
to the NIH in 1937.[2] By the late 1950s she had been working for more than
twenty years in the NIH division that was then the gatekeeper of the U.S. vaccine market. The Division of Biologics Standards (DBS) filled the role played
by a key branch of today's Food and Drug Administration, assessing new vaccines and issuing licenses when it deemed a product ready for market. Eddy
had done well in her work scrutinizing vaccine safety and effectiveness: in 1953
she won a "superior accomplishment" award and a pay raise from the NIH
director.

The next year Eddy was put in charge of the unit's polio vaccine safety
tests. It was a high-stakes job. On the heels of a 1952 polio epidemic that had
infected nearly 58,000 Americans and paralyzed more than 21,000 of them—
the country's worst polio epidemic ever—virologists were in the midst of an all-
out sprint to get a polio vaccine licensed. That first vaccine, newly invented by
Jonas Salk, was made from polio virus that multiplied in monkey kidney cells
and was then killed with formaldehyde. The Salk vaccine was entering human
trials in 1954 when Eddy isolated live polio virus from three lots of the vaccine
made by the California-based Cutter company. She injected monkeys with the
suspect vaccine and found that it paralyzed some of them. She reported the

findings to her bosses and sent them photos of the afflicted animals.[3] They were ignored, and in April 1955 the Salk vaccine was licensed.[4] Mass vaccinations began, and the Cutter vaccine was among those distributed. It paralyzed 192 people, many of them children.[5] Ten people died.[6] The government was forced to temporarily recall all polio vaccine, sowing public panic. Manufacturing changes were mandated, and the vaccine returned to the market. But public confidence took months to rebuild, a fact that was responsible for many of the more than 28,000 cases of polio in the United States that year.[7]

Despite her earlier efforts to warn her bosses about the flawed vaccine, and over her protestations, Eddy was pulled off polio vaccine work as part of the Cutter-episode fallout.[8] It was worse for the higher-ups: senior officials including the director of the NIH, William Sebrell, and the secretary of health, education and welfare, Oveta Culp Hobby, lost their jobs. So did William Workman, Eddy's boss.

Eddy, who would show amazing staying power over the years, persisted at the DBS and in 1959 landed in *Time* magazine with her friend and NIH colleague physician/scientist Sarah Stewart. The pair had discovered a mouse virus that was now named "SE polyoma"—"SE" for Stewart and Eddy, and "polyoma" meaning "many tumors." Stewart isolated the virus from tumors in three laboratory mice. Then the two scientists conducted experiments showing that fluid from the mouse tumors caused new malignant tumors not just when injected into other mice but also when injected into hamsters and rats.[9] The notion that viruses might cause cancer was resurfacing and gaining traction, and their discovery got the attention of other scientists in a big way.

But beginning shortly before the *Time* article was published that July, Eddy launched another experiment—one born of a nagging worry that her polyoma studies had prompted and that now wouldn't leave her alone.

Her concern was this: if virus-bearing fluid from a mouse tumor could so easily cross species lines to cause cancer in hamsters and rats, why couldn't a virus from, say, a monkey—a species much closer to *Homo sapiens*—cause cancer in humans? Her question wasn't academic. Salk's polio vaccine had been injected into more than 69 million Americans since its launch in 1955. The vaccine was grown in monkey kidney cells. And Eddy well knew, as did any vaccinologist who worked with them, that those kidney cells harbored plenty of viruses—"simian" viruses, in scientific parlance—from *simia*, the Latin word for ape.

These were viruses that lurked in apparently healthy monkeys, and especially in their kidneys. However normal the animals seemed, the viruses regularly killed their kidney cells in the lab, forcing scientists to jettison spoiled cultures. The Wistar's chief Koprowski, who had a keen interest in the matter because he was developing his own polio vaccine, noted that the viruses "are more often than not dormant in the intact [monkey], but go on a rampage when infected tissues are removed soon after the animal's death."[10]

A scientist at the drug company Eli Lilly, Robert Hull, had begun cataloging each new simian virus that was discovered, classifying it according to the cellular damage that it caused in monkey kidney cell cultures. In 1958 he reported that eighteen new simian viruses had been discovered in just the previous two years. "As long as primary monkey kidney cultures are used in the production and testing of virus vaccines," his paper concluded, "the problem of simian virus contamination will remain."*[11]

The assumption about these simian viruses, made by everyone from Salk to the NIH's top vaccine overseers, was that while they might be an annoyance in the lab or on the production line, they were not a danger to human beings, because they were killed by the same formaldehyde that killed the polio virus in the Salk vaccine. What was more, the reasoning went, even if simian viruses occasionally managed to survive the manufacturing process, they were clearly innocuous in humans: weren't there tens of millions of healthy Salk vaccinees walking the streets to prove the point?

Bernice Eddy couldn't so casually accept those assumptions. In June of 1959, without her NIH boss's knowledge, she launched a bold experiment. She took monkey kidney cultures prepared by the DBS—these were from rhesus monkeys, the species widely used in polio vaccine making, which will be important. She froze them, ground them up, put them through a very fine filter that strained out bacteria but not viruses, and injected a fraction of an ounce of the resulting fluid under the skin of newborn hamsters. Fully 70 percent of the 154 animals that she injected developed tumors, and every animal that did so died. What was worse, it wasn't just one or two of the ground-up cultures that were at fault: she had prepared twelve lots of ground-up monkey kidney cells, each derived from between eight and thirty-two monkeys. Nine of the twelve turned out to be cancer causing. The virus appeared to be common in the monkeys.[12]

---

*"Primary" means that the kidney cells were used once and then dispensed with, rather than multiplying through many generations in lab dishes, as Hayflick's human fetal cells did.

Eddy then went a step further, taking tumors from two of the sick hamsters, mincing them finely with scissors, and injecting tiny bits of tumor under the skin of forty more newborn hamsters. All but two of those hamsters got cancer and were dead in less than three months.[13]

It was by now the early summer of 1960, and Eddy was anxiously pondering her findings when she got word of a speech that had been given at a major polio vaccine conference in nearby Washington, DC. In the world of polio vaccinologists and public-health people, the talk, by Maurice Hilleman—a highly respected, tough-talking Montanan who headed vaccine research at Merck—landed with the explosive force of a hand grenade. It soon caused Eddy to gather her courage and approach her boss with her findings.

The leading polio vaccinologists who convened at the conference at Georgetown University that June were buzzing with optimism. The killed-virus Salk vaccine had already been on the market for five years, and the incidence of polio in the United States had fallen dramatically, from nearly 25 cases per 100,000 people in 1955 to fewer than 5 per 100,000 people in 1960.[14] But pockets of polio stubbornly persisted, particularly in poor communities. Still more worrying was the fact that the worst cases—the ones that paralyzed people—had not declined at as steep a rate as cases that sickened people but from which they recovered fully. Alarmingly, cases of paralytic polio more than doubled between 1957 and 1959 to more than 6,000. It seemed clear that the Salk vaccine was going to curtail polio but not conquer it. And so the assembled virologists were eagerly awaiting the licensing of a second solution—the first live polio vaccine—a vaccine that, they anticipated, would be cheaper, easier to administer, and more effective than Salk's. It would be swallowed rather than injected, and so would mimic the natural route of polio infection. It would generate robust levels of antibodies in the lining of the digestive tract, where the virus first encounters the human immune system, as well as in the blood—unlike Salk's killed-virus vaccine, which was injected into muscle.

The scientists at the conference knew that they were witnessing the home stretch of an intense three-way race to win a U.S. license for a live vaccine. The high-powered, sharp-tongued Albert Sabin was clearly in the lead. But Herald Cox, Koprowski's former boss at Lederle, had taken the vaccine that Koprowski developed while at the company and continued tweaking and testing it; he was fighting hard to stay in the race. And Koprowski himself had refused to give up on beating his archenemy, Sabin. Since leaving Lederle, the Wistar czar had adapted and renamed the vaccine he had brought with him from

Lederle and continued to test it, vaccinating more than 300,000 people in central Africa, mainly in the Belgian Congo.

But what the audience heard from Hilleman at that June 1960 conference put a damper on the buzz about the hoped-for licensure of a live vaccine. Hilleman shocked them with the news of an unexpected finding by Ben Sweet, a scientist he supervised at Merck's West Point, Pennsylvania, campus. Sweet had discovered a new, invisible simian virus contaminating Sabin's live vaccine and almost certainly Cox's and Koprowski's live vaccines too. All three vaccines were made using monkey kidney cells. And unlike Salk's killed vaccine, none of the live vaccines were treated with the formaldehyde that was presumed to kill simian viruses.

This new simian virus differed in an important way from the dozens of others that were now known to infect the monkey kidneys used by polio vaccine makers. The new virus didn't declare itself by damaging cultures of the cells. It didn't "go on a rampage," as Koprowski put it. Instead it sat quietly, leaving the monkey cells looking and acting perfectly normally in their bottles. It was impossible to detect, making it impossible for vaccine makers to jettison infected cultures. But it was there nonetheless.

How, then, did Sweet stumble on the new virus? A detailed rendering of his discovery comes from *The Virus and the Vaccine* by Debbie Bookchin and Jim Schumacher.[15] The authors recount that Sweet happened to be working with a different species of monkey from the species whose kidneys were normally used to make polio vaccine. When he exposed kidney cells from this different species, the African green (or grivet) monkey, to fluids from rhesus monkey kidney cells—the cells used to make polio vaccine—the African green kidney cells sickened and died. It emerged that the rhesus cells had passed to the African green cells a virus—a virus that lay silent and invisible in rhesus monkey cells but caused major damage in the kidney cells of African greens, bloating the cells and riddling them with holes.[16] (The virus was also discovered to live silently in the kidney cells of another monkey species also used in polio vaccine preparation: cynomolgus monkeys.)

If there was any good news in the Merck scientists' findings, it was this: Sweet had found that the silent simian virus was inactivated by the same formaldehyde that was used to kill the polio virus in the Salk vaccine. So while the virus may have been injected into the arms of some of the seventy million Americans who had received the Salk vaccine, it had also, it seemed safe to say, been dead on arrival.

The remaining days of the conference buzzed with talk of the new virus, which Hilleman dubbed the "vacuolating virus" because of the holes, or vacuoles, it made in the cells, leaving them looking like Swiss cheese. Formally the new, silent virus was named SV40. News of it soon reached Eddy at the nearby NIH.

Eddy didn't know what it was in the rhesus monkey kidney cells that had caused her hamsters' tumors. She hadn't had the tools on hand to identify whatever it was, so she had simply called it a "substance." But when she heard of Hilleman's vacuolating virus, she immediately suspected strongly that her "substance" and the new, silent virus, SV40, were one and the same thing.

On July 6, 1960, one month after Hilleman's much-noted talk at the Washington conference, Eddy sent a memo to her boss, Joseph Smadel, who was newly in charge of vaccine safety testing at the DBS. She titled the memo "The presence of a [cancer-causing] substance or virus in monkey kidney cell cultures."

Eddy wrote to Smadel that she had heard about Hilleman's SV40 virus— and that she herself had inoculated newborn hamsters with "specially prepared monkey kidney cells." Tumors occurred at the injection sites, she wrote. "Eventually, the animals die." She hoped, she added, to do some follow-on experiments as quickly as possible, to see if the "substance" causing her hamsters' tumors was in fact SV40.[17]

Smadel was a man's man, a foul-mouthed, no-nonsense, dictatorial virologist in his early fifties who had staffed an advanced World War II field laboratory in France after D-Day and later, working in Malaysia, discovered that an antibiotic, chloramphenicol, could effectively treat typhus and typhoid fever. Smadel had come to the NIH in 1956 on the heels of the disastrous Cutter incident. He had been a key advocate of the Salk vaccine, and he was acutely aware of the damage that another round of Cutter-like bad press could do to public adoption of *any* polio vaccine. So as Smadel read Eddy's note, he got angry. He summoned her and tore a strip off her in language he later acknowledged was "not even diplomatic." He dismissed the tumors in Eddy's hamsters as "lumps," called her data "inadequate," and shut down her "entirely unwarranted" suggestion that they might be related to SV40 or have implications for human cancer.[18]

Eddy wasn't fazed. She went back to her lab determined to identify the "substance" that had caused the cancers in her hamsters. It was completely conceivable that her "substance" had been injected, alive, into the millions of people in other countries and the roughly 10,000 in the United States who had participated in Cox's, Koprowski's, and Sabin's trials of live polio vaccine. And it was also possible, she was convinced, that it had been injected—and was still being injected, in the Salk vaccine—into the arms of tens of millions of U.S. schoolchildren.[19]

Hayflick recalls learning about the discovery of the SV40 contamination by Sweet at Merck even before Hilleman's bombshell speech at the Washington conference in June of 1960. Hayflick was a friend of Sweet's, and the circle of virologists in the Philadelphia area was a close one in which news was freely exchanged and traveled quickly. Similarly, Eddy's findings soon made their way to Hayflick through the scientific grapevine, as bad news always does.

Hayflick immediately grasped the headache that SV40 posed for polio vaccine makers and regulators. Admittedly, Hilleman and Sweet had argued that it wasn't an insurmountable obstacle. They had tested the blood of people who had swallowed live polio vaccines in trials and found no antibodies to the simian virus, strongly suggesting that SV40 did not proliferate massively in the human intestinal tract and from there invade the body. Still, they conceded, the possibility could not definitively be ruled out. Nor could the long-term possibility that, having invaded the body, such a virus could eventually cause cancer, "especially when administered to babies" with their less-developed immune systems. Going forward, the pair wrote in the paper that summarized their SV40 findings, "the simple solution" would be to ensure that the live polio virus seed stocks—the stocks that were amplified by companies to make production-scale quantities of vaccine—were not contaminated with SV40, and to throw away vaccine lots that were already tainted.[20]

The problem, Hayflick was convinced, was not so simply solved. For one thing, the testing of the live vaccine to ensure that it was free of SV40—testing that would clearly be needed going forward—would be time-consuming and expensive. And even if no ill ever came to those already vaccinated, Hilleman and Sweet themselves had admitted that "other undetected [simian] viruses might also await demonstration."[21] Surely there was a better answer than passively waiting to stumble on the next unwelcome "passenger" virus in monkey kidney cells.

In the summer of 1960 Hayflick was still fully two years away from launching WI-38 from the lungs of Mrs. X's fetus. As news of the discovery of SV40 landed, Hayflick was still working with the first twenty-five fetal cell lines, coming to realize that they aged and died in their dishes, with all that that portended. He was also finding them, to all appearances, free of lurking, unwanted viruses. At the same time, he was discovering that they could be infected with many other disease-causing viruses—including polio. And Moorhead, bent over his microscope, was reporting that their chromosomes were reassuringly normal. The obvious thing to do fairly shouted at Hayflick. He would make a polio vaccine using his normal human diploid cells. Then he would see if it worked.

Down the hall from Hayflick's lab and across the atrium, in the other wing of the *V*-shaped Wistar Institute, another brain was buzzing with the vaccine-making implications of the new human cells. Koprowski had been front and center at the June conference in Washington, DC, where the sudden specter of SV40 took top billing. There he had given a speech downplaying SV40's significance, and that of any other monkey virus that might be found contaminating live polio vaccine.

Koprowski pointed out that in the last several years of trials, millions of people around the world had already received various live poliovirus vaccines made using monkey kidney cells—his vaccine and Sabin's and Cox's—to no obvious detrimental effect. The discovery that SV40 had been lurking in those vaccines, he argued, "should hardly deter anybody from accepting the product." Silent viruses inhabited all kinds of animals cells, he argued, and doubtless would be found, for example, in the calf lymph that had been successfully used to make smallpox vaccine for two hundred years. Should we therefore stop vaccinating against smallpox?

He conceded that it would be desirable, going forward, to try to rid monkey cells of extraneous viruses, and he mentioned several possible approaches, including using—he didn't mention Hayflick or the Wistar by name—some newly available human cells.[22]

As he minimized the risks of SV40, Koprowski was still holding out the ambitious hope that the U.S. surgeon general—Leroy Burney, whom Koprowski had been sure to invite to the Wistar's gala opening symposium the previous year—was going to pronounce his live polio vaccine the favorite child of the U.S. government; the vaccine that would be chosen to move through licensing, leaving its two competitors in the dust.

Burney was under pressure to make a choice, soon, among the three vaccine candidates, not least because the United States was still exclusively using Salk's killed, injected vaccine while the country's bitter cold war rival, the USSR, had already fed eighty million people Sabin's purportedly superior, oral, live vaccine. Not only that, but there had been sporadic new outbreaks of polio in the United States in 1958 and 1959. These outbreaks had—fairly or unfairly, for most cases occurred in undervaccinated populations—undermined confidence in the Salk vaccine.

On August 24, 1960, Burney told a press conference that he had picked a winner: Sabin's vaccine had bested its two competitors in safety tests in monkeys, and he had chosen it to proceed through licensing. Any live vaccine licensed in the future would have to be as good as Sabin's, or better. Within hours, three major vaccine manufacturers announced that they would begin making the Sabin vaccine, and others appeared ready to follow suit. Koprowski's vaccine was done, to all appearances.

But for the ever-enterprising Koprowski, Hayflick's new human cells meant that the game was not necessarily over. In late October, Koprowski—joined by a twenty-eight-year-old Wistar physician/scientist, Stanley Plotkin, who had helped run trials of Koprowski's monkey cell–based polio vaccine—sent off a long letter of advice to a World Health Organization committee. The group was soliciting input on the standards that live polio vaccines should be held to.

Koprowski had minimized the SV40 monkey virus problem only four months earlier, when his own monkey kidney–based polio vaccine was still in the running for U.S. approval. Now, with Sabin's vaccine rolling quickly toward being licensed, he sounded more alarmed. The letter noted that monkey kidneys are riddled with simian viruses. Eliminating SV40 from the live vaccine, the writers observed, "may present insurmountable obstacles."[23] And yet hundreds of thousands of fresh monkey kidneys would continue to be needed to make live polio vaccine, increasing the chances of a potentially cancer-causing virus finding its way into some lots of vaccine and "making the case even weaker for the use of such a tissue."

By contrast, Koprowski and Plotkin wrote, cells were now available from normal human fetuses—cells that had normal numbers of chromosomes. They could be frozen and used for vaccine production when needed. And each line of such cells could be "scrupulously investigated" for hidden viruses. Koprowski and Plotkin laid out the pros and cons of monkey-kidney versus human cells in an accompanying table. Its columns had titles like "Possibility of freeing poliovirus from simian viruses" (with the human cells it was "possible"; with the monkey kidney cells it was "impossible") and "Procurement of tissue" (with the monkey kidney cells it was "difficult"; with the human cells it was "easy").[24]

Koprowski sent the letter off to Geneva. Then he waited for Hayflick to show just how a polio vaccine could be made—using the human fetal cells and not Sabin's but Koprowski's polio vaccine.

Hayflick first grew a small amount of Koprowski's polio vaccine virus in petri-dish cultures of his normal WI-1 cells.*

Then he inoculated fluid from these petri-dish cultures into bigger quantities of WI-1 cells in quart-sized bottles. Koprowski's polio vaccine virus destroyed the cells within two days, first causing the long, tapering fibroblasts to become round and then causing them to burst, or "lysing" them. Each lysed cell ejected up to ten thousand new virus particles into the nutrient fluid that bathed it.

Next Hayflick inoculated that fluid into still more WI-1 cells, this time in still bigger, gallon-plus bottles. These he left to incubate at body temperature. Five days later he filtered the resulting soup of culture medium, cellular debris, and live viruses to strain out bacteria. He froze it at –94 degrees Fahrenheit. He had just produced the first vaccine grown on his human diploid cells.[25] (It was not the first vaccine ever grown in human cells; Sven Gard's group in Sweden had done that when they made that fleeting polio vaccine using fetal skin and muscle cells in the mid-1950s—the vaccine that was impracticable because they couldn't make enough of it.)

As Hayflick prepared the vaccine, another stunning piece of news arrived about the silent simian virus, SV40. The previous summer, vaccine makers and regulators alike had taken heart from the finding by Sweet and Hilleman at Merck that SV40 was killed by the same formaldehyde used to kill the polio virus in the Salk vaccine.

But in March 1961 British researchers reported in the *Lancet*, a widely read medical journal, their discovery that, contrary to the Merck scientists' report, SV40 was resistant to formaldehyde: it was not inactivated by the chemical as quickly as poliovirus was. That meant that the simian virus could survive sometimes, alive, in Salk's vaccine. They argued for "an accumulating body of evidence that killed poliomyelitis vaccine in the past has contained [SV40], probably in the living state."[26]

---

*For simplicity this book will refer to "polio vaccine virus." However, Hayflick used a Koprowski polio vaccine virus called CHAT, which is a type I polio virus. Of the three polio virus types, type I causes the most disease. However, commercially marketed polio vaccines protect against all three types: I, II, and III. For an explanation of the distinctions and their discovery, see David M. Oshinsky, *Polio: An American Story* (New York: Oxford University Press, 2005), 117–21.

A few months later researchers from the Medical Research Council Laboratories in London reported in the *British Medical Journal* that they had found antibodies against the SV40 virus in eleven of twelve schoolboys who had received the full slate of three Salk injections.[27] The presence of antibodies did not mean that the virus was alive and replicating in the boys—the whole principle of injecting a killed vaccine is that the body forms antibodies even to a dead virus. The boys might well have received dead SV40 with their Salk injections. But the presence of SV40 antibodies in eleven of the twelve boys was an alarming indication of just how widely SV40 might have infiltrated the supply of Salk vaccine. And the more widespread it was, the more likely that some vaccinees had received injections in which the virus had survived alive.

The same month, March 1961, that the *Lancet* article appeared, a House of Representatives subcommittee held hearings on production of the live vaccine. Koprowski had a conflict and couldn't testify, but he wrote to the lawmakers, making the same argument that he had made to the World Health Organization a few months earlier. Given the ubiquity of the silent SV40 virus in the monkey cells used to make the live vaccine, it would be impossible to produce it in a cost-effective manner. It would be both cheaper and sounder scientifically to switch to human cell strains. There was, he added, "not a shred of evidence" that they caused cancer.[28]

This was the backdrop against which Hayflick pushed ahead with his human fetal-cell–produced polio vaccine. He finished making the vaccine in or about January 1961. Now, as the SV40 problem emerged as more widespread than anyone had first understood, Hayflick and Plotkin set about ensuring that the new vaccine would be safe to inject into the most fragile of human beings: newborn babies.

In the late 1950s there had been a bump in polio cases in babies aged six to twelve months old—a vulnerable period when protective antibodies inherited from a baby's mother have waned away but the baby's immune system is still immature and can't mount a full-fledged attack on foreign invaders. Koprowski and Plotkin argued in a 1959 paper that this meant that babies should be vaccinated as early in life as possible. The paper reported their success in a trial of Koprowski's live polio vaccine, which was then still a contender for licensure; it had boosted antibody levels in babies as young as one day old.[29]

Importantly, babies also provided a source of unvaccinated trial subjects in which to test Hayflick's new, live vaccine. Such subjects were otherwise tough to find: by 1961 an estimated 90 percent of the nation's children and adolescents had received the Salk vaccine, as had 60 percent of adults younger than forty years old.[30] Many among the other 40 percent likely had antipolio antibodies already, having been exposed to polio during decades of living. The presence of preexisting antibodies would make the results of a live vaccine test difficult to interpret.

Thanks to Koprowski's connections, the Wistar researchers had a ready-made source of newborns to vaccinate, at an unusual women's prison headed by an equally unusual woman.

Clinton State Farms was located in rural New Jersey, sixty miles northeast of Philadelphia. The campuslike institution, which housed women serving prison sentences of more than one year, was run by Warden Edna Mahan. She had been in charge of the prison for thirty-three years.

Mahan, then sixty, was a 1922 graduate of the University of California at Berkeley. She had a warm smile, an aquiline nose, and large, light, penetrating eyes. She was a 1960s-style liberal well before that decade arrived. An ardent advocate of rehabilitation, she banished handcuffs and allowed the several hundred inmates to earn trust by steps. They wore color-coded uniforms reflecting their degree of freedom, and the best-behaved were allowed to work in the surrounding community by day, at jobs as farmhands and domestics.

"The atmosphere at Clinton Farms is not that of a prison. No girl is locked in," former First Lady Eleanor Roosevelt wrote after attending an eighth-grade graduation ceremony for inmates in 1956.[31] Roosevelt remarked that the prisoners actually cheered when Mahan's name was mentioned and noted that up to four hundred girls were allowed to picnic on a hillside with a single attendant accompanying them.

That same freedom made an impression on Koprowski, who reported in interviews late in his life that the Clinton inmates, being red-blooded women, took advantage of their freedom by hailing eighteen-wheelers on nearby Route 78 and enjoying moonlit trysts in the truck cabs—a claim that makes for a good story about how babies came to be born at the prison but for which there is no evidence but Koprowski's word.[32] What isn't in dispute is that about sixty infants were born to inmates every year, either in the prison's own Stevens Hospital or at nearby Hunterdon County Medical Center. In 1960 there were fifty-four births at the prison.[33]

At the time that Hayflick produced his polio vaccine, Koprowski had already been vaccinating newborns at Clinton Farms with experimental monkey cell–produced polio vaccines for five years. The door to the prison opened for him because a colleague, a University of Pennsylvania doctor named Joseph Stokes, socialized with a member of the Clinton Farms Board, and because

Stokes's brother, Emelen, also a physician at Penn, sat on New Jersey's Board of Control of Institutions and Agencies. It didn't hurt that the strong-willed Mahan herself was a big believer in medical research. In the first experiment she approved, in 1946, inmates were infected with body lice daily to gauge the effect of their nutritional status on the pace at which lice bred on their bodies.[34]

Plotkin, who did much of the actual vaccinating for Koprowski in the late 1950s, recalls making the drive to the prison as often as twice a week in that era. The prison babies had the great advantage that they didn't disappear into the surrounding community within a week, as newborns did from city obstetrical wards. It was not unusual for babies to remain in the nursery at Clinton Farms for four to six months, allowing for follow-up testing to measure antibody levels in response to experimental vaccination.

Whether the inmates who were new mothers felt able to refuse to participate is an open question. "Dr. Agnes Flack, Medical Director, and Miss Edna Mahan, Superintendent, of Clinton State Farms, were extremely helpful in obtaining permission for vaccination of the infants," Koprowski and Plotkin wrote in the acknowledgments of one 1959 paper.[35]

There may have been a strong incentive for the prisoners to volunteer their infants for the studies: the chance to spend more time with their babies. In his 1999 book *The River*, writer Edward Hooper examined Clinton Farms birth records from the mid-1950s and determined that babies enrolled in Koprowski's polio vaccine trials stayed an average of four to six months at the prison, a figure also reported by Koprowski and his colleagues in the 1959 paper.[36] Hooper found that those not in trials were placed with a social welfare agency or with relatives after about four to six weeks.[37]

Before taking the new live polio vaccine made in fetal cells to Clinton Farms, Plotkin and Hayflick ran what they would describe in the resulting paper as "exhaustive" safety tests on it. First they needed to ensure that the vaccine wasn't contaminated with some virus or bacterium besides polio—a virus or bacterium that could theoretically be lurking in the WI-1 cells. So they injected the vaccine into dozens of mice, rabbits, guinea pigs, and hamsters, then watched them for symptoms of illnesses caused by microbes known to infect these species, like herpes simplex and the *Bacillus* that causes tuberculosis. None got ill. This test, they would later write, "presumably" ruled out the presence in the vaccine of such microbes.[38]

Next they neutralized the vaccine virus with antipolio antibodies and injected the resulting fluid into plates of monkey cells. If a hidden nonpolio virus was lurking in the vaccine, it would damage the kidney cells. Hayflick and

Plotkin were thinking of SV40, and also of a lethal herpes virus known as B virus that occasionally killed researchers and animal handlers who were bitten by monkeys. But the plates of monkey cells remained healthy. (To be sure that SV40's effects would be noticed, they used kidney cells from African green monkeys.)

Then they ran several tests to make sure that the genome of the vaccine virus hadn't mutated to a different, dangerous form when it was grown in WI-1 cells. Their tests indicated that it hadn't. This was decades before the advent of gene-sequencing techniques, so they had to infer the results from indirect laboratory tests. One test involved injecting the vaccine virus into the brains of five monkeys, then observing them for twenty-one days before euthanizing them and looking at slices of their brains and spinal cords under the microscope. The monkeys didn't get polio, and their brains appeared normal.[39]

After weeks of testing the vaccine, Plotkin and Hayflick felt that they had done all that they could do. Hayflick recalled in a 2014 interview being so confident of the new vaccine's safety that he at one point fed it to his own children, Joel, Deborah, and Susan, then aged four, three, and two years old respectively. He did not recall whether this occurred before or after the first trial of the vaccine at Clinton Farms.[40]

The first baby at Clinton Farms swallowed the human cell–based polio vaccine in the late spring or early summer of 1961. The vaccine was administered either by Plotkin or by Suzanne Richardson, a nurse assistant who often helped him.

The timing was apt. Bernice Eddy's paper about the lethal hamster tumors caused by a "substance" in rhesus monkey kidney cells had finally been published in May, after being held up for months by her boss, Smadel. Then, in late June, Koprowski used the high-profile annual meeting of the American Medical Association in New York City to put the SV40 issue on the radar of practicing physicians for the first time. He warned them that the "obsolete" method of making polio vaccine with monkey kidney cells risked more "virus surprises." He called Hayflick and Moorhead's human cells "the obvious choice" for making polio vaccine going forward.[41]

Not long after, the lay press picked up on SV40. Late in July, the Associated Press, in a story that appeared on page thirty-three of the *New York Times*, reported that both Merck and Parke-Davis had stopped making the Salk vaccine because it had been found to contain a monkey virus "believed harmless" by the NIH. The story did not mention cancer.[42] Nor did it report that Hilleman, at Merck, had insisted that the company drop its production of the Salk vaccine when the company's tests found live SV40 in its vaccine.[43]

The *National Enquirer* wasn't as sanguine. Despite its reputation for hyperbole, the tabloid that August ran an accurate, thorough story under the headline THE GREAT POLIO VACCINE COVER-UP. "The polio shots you have taken may KILL you," it began, beside a subheadline that screamed 70% OF HAMSTERS DEVELOPED CANCER IN LAB TESTS. The paper reported clearly on Eddy's findings and quoted Koprowski, again promoting Hayflick's cells and suggesting that companies were clinging to monkey cells only due to fear of change.[44]

Days before the *Enquirer* story appeared, the Division of Biologics Standards at the NIH began for the first time to require vaccine makers to sample their Salk polio vaccine lots to ensure that they were free of live SV40 before sending them out. The change was made eighteen months after the letter to the editor in the *Lancet* first reported that live SV40 could survive in the Salk vaccine. The DBS did not recall any of the Salk vaccine that was already on the market. Nor did it at this point require companies to make the expensive switch to using African green monkey kidneys for making polio vaccine. These monkeys don't naturally harbor SV40, so using kidneys from this species would have obviated the contamination problem.

By the end of the summer of 1961, six full-term Clinton Farms infants, aged between nine and fifty-seven days, had been vaccinated with Hayflick's human-cell-propagated vaccine.[45] That October Hayflick sent the resulting paper, with himself as first author—the others were Plotkin, Koprowski, and Koprowski's administrative deputy and lab manager, Tom Norton—to the *American Journal of Hygiene*, the premier academic journal of the day for public-health research.

Their findings were encouraging, the authors reported in the paper, which was published in March 1962.[46] The polio vaccine that was made using human fetal cells was free of extraneous, nonpolio viruses that critics might contend could have been living in the WI-1 cells in which it was produced. Five out of the six vaccinated babies excreted the virus in their feces for more than one week, a sign that the virus had established an active infection in their intestines. Such an infection was needed to provoke an antibody response. What was more, the vaccine appeared genetically stable—tests of virus from the feces of the five babies showed that it hadn't morphed into something more toxic during its trip through the infants' bowels.

An earlier draft of the paper, written before most of the babies had been vaccinated, conceded that all of the testing in the lab and in animals wasn't proof positive of the vaccine's safety. "It is possible," the draft read, "that the growth of a [weakened] poliovirus in human cells may increase its [ability to cause illness in humans] or change other of its characteristics. Only a large scale field trial, which we are presently organizing, can determine whether or not pathogenicity for man is increased."[47]

By the time the paper was published, however, the wording had been changed. "Our results with a [weakened] poliovirus vaccine grown in human cells indicates [sic] that the virus does not increase its pathogenicity for the human, or change any other of its characteristics," the paper stated. "We are presently organizing a large-scale field trial to further demonstrate the safety and efficacy of this material."[48]

The authors ended the published paper by singing the virtues of all of Hayflick's human diploid cell strains for making antiviral vaccines. These were superior, they wrote, in a dozen ways to the monkey kidney cells still being used to make both the Salk and the Sabin vaccines, and there was theoretically no end to the vaccines that could be made with them, against diseases as diverse as rabies, measles, chicken pox, and even the common cold.

And what became of SV40, and Eddy? Her dressing-down from her boss, Smadel, in the summer of 1960 didn't cow her. In fact, within weeks she used the occasion of a meeting of the New York Cancer Society to present her findings that a "substance" in rhesus monkey kidneys caused hamster-killing tumors. She hadn't cleared the content of her presentation with Smadel.

"I knew when I was doing it I'd be in trouble," Eddy recalled late in life. "And I didn't care much."[49]

Soon thereafter, Roderick Murray, the chief of the NIH's Division of Biologics Standards, where Eddy worked, told her in a memo that she was being freed of her "irksome" responsibilities trying to ensure vaccine safety.[50] Instead she would begin conducting her own independent research. In a separate memo Smadel let her know that she wouldn't speak at any more meetings without his reviewing and approving her remarks first.[51]

In the summer of 1961, shortly after her paper documenting the hamster tumors appeared, Eddy was downsized to two assistants and assigned to room 207 in building 29 at the NIH, which had until then been a supply room.[52] It measured about sixteen by fourteen feet. Ruth Kirschstein, who was a young DBS scientist during Eddy's demotion and who later rose to become the NIH's deputy director, would, late in her own life, describe the DBS's failure—including her own—to take Eddy's findings seriously as "not a terribly pretty story."[53]

Hilleman, whose Merck lab identified SV40 and who ended the company's production of the Salk vaccine, recalled years later that Eddy's NIH bosses "tore the hell out of her" for her findings because her studies failed to use rigorous controls. "But," he added, "she was right."[54]

Eddy's next paper, on which she labored for a year despite the punishing atmosphere, would demonstrate that SV40 and the monkey kidney "substance" that caused tumors in her hamsters were one and the same thing. It was published in May 1962 after Smadel again held it up for several months before allowing Eddy to submit it.[55]

The following month Koprowski rushed into print a paper by Wistar researchers and the surgeon Robert Ravdin showing for the first time that the silent monkey virus affected *human* cells—cells scraped from people's skin and the insides of their cheeks. The cells became abnormally shaped, divided faster, piled up on top of one another, and outlived their uninfected counterparts. The paper featured photos of the cells' disrupted, abnormal chromosomes at 2,500 magnifications.[56] A short time later the Nobelist John Enders and his colleagues Harvey Shein and Jeana Levinthal published similar findings in the *Proceedings of the National Academy of Sciences*. The trio had found that SV40 caused malignant changes in kidney cells from human fetuses, newborns, and three-month-olds.[57]

Nine months later, in March 1963, the NIH's Division of Biologics Standards began requiring that all polio vaccine be free of SV40 *before* its inactivation by formaldehyde, rather than allowing companies to sample vaccine once it was inactivated and ready for market. This in effect forced manufacturers to switch to using the kidneys of African green monkeys, which don't naturally harbor SV40, rather than continuing to use kidneys from SV40's natural hosts: rhesus and cynomolgus monkeys.

By this time 98 million Americans had received the Salk vaccine since its launch in 1955. Another 10,000 had potentially been exposed to SV40 when they participated in trials of live polio vaccines between 1959 and 1961. In addition 100,000 members of the military were potentially exposed to the silent monkey virus when they were injected with adenovirus vaccine between 1955 and 1961 to protect them from the respiratory infection, which travels easily in the close confines of military barracks.[58]

CHAPTER EIGHT

Trials

Philadelphia and Environs, Spring 1962
Bethesda, Maryland, 1955–63
Europe, 1962–63

My mind is not entirely free from the prejudice that our research teams want these infants because they are always available in quantity and under conditions which permit [a] wide variety of controls.
—John Cardinal O'Hara, archbishop of Philadelphia, June 26, 1959[1]

In March 1962, as Hayflick and his colleagues' new paper reported the launch of a polio vaccine made using human fetal cells, the Wistar Institute made a new hire.

Jimmy Poupard was the son of a paper box factory worker and a housewife. He was a devout Catholic and onetime altar boy in the rough center city Philadelphia neighborhood where he grew up. By the time he graduated from Roman Catholic High School for Boys in 1960, Poupard, a slight, dark-haired loner, knew he was finished with formal education. He loved maps and science and *National Geographic*, but he hated school and knew no one who had gone to college. His hard-drinking contemporaries were hanging out on street corners playing cards and were bound for either the army or jail. But Poupard's fussy young parish priest, Father Arthur Nugent, urged the graduating Poupard to take an intense one-year program in medical technology at the Franklin School of Science and the Arts in downtown Philadelphia. It was his idea of heaven: there were ten female students for every young man who attended the med tech course.[2]

As he learned to draw blood and process patient specimens, Poupard also became smitten with virology. He began reading everything about viruses and vaccines that he could get his hands on. The field was blossoming. He wanted to be part of it. He landed a job at the drugmaker Wyeth but grew restless when he realized it would take him years to move to the virology department. After a few months he wangled an interview at the Wistar Institute. When he was hired, he couldn't believe his luck; Hilary Koprowski inhabited the most exalted realms of virology.

Poupard went to work in Koprowski's polio lab. He inoculated rabbits, nursed HeLa cells used for testing stool samples for the polio virus, and occasionally descended the wrought-iron stairs to retrieve a test tube of polio virus from a communal walk-in freezer in the back of the building. The freezer was kept at −94 degrees Fahrenheit. Poupard would don a big blue parka, enter the freezer, open the door of the wire mesh cage into Koprowski's section, brush the frost off a Magic Marker label on a rack of test tubes, grab a test tube, and get out of the freezer as quickly as humanly possible. (Also resident in the cage were hundreds of stool samples from trials of polio vaccine that Koprowski had run in the Belgian Congo in the late 1950s, kept in small Dixie-like paper cups. They were eventually burned in the Wistar's courtyard incinerator.)

But mainly Poupard was on deck for the regular phone calls that came from the nursery for premature babies at Philadelphia General Hospital. When the lab phone rang and it was a nurse reporting that a new preemie was available at the hospital—the baby had to weigh at least two pounds, Poupard recalls, or it was deemed too vulnerable for the trial that was under way—he would retrieve a tiny, frozen test tube with a black rubber stopper, run it under cold water until the frozen vaccine it contained melted into a bright yellow liquid, and stick it in his pocket along with an empty test tube and a lancet wrapped in sterile packaging. (A lancet is a two-sided scalpel blade used for making small incisions.)

Then he would shed his white lab coat, descend the Wistar's broad front steps, turn right, and head down Thirty-sixth Street, passing the giant Hospital of the University of Pennsylvania on his left before the street dead-ended half a block farther on, in a T junction at a pedestrian path called Hamilton Walk. There he confronted a big, thick, black wall. It separated two worlds.

The wall marked the edge of the University of Pennsylvania campus. On the near side was the prestigious university hospital. On the other was Philadelphia General Hospital—PGH—the charity hospital for the city's poor, nicknamed "Old Blockley" after the family that once owned the land on which it stood. The City of Philadelphia had bought that land in 1832. There it built an insane asylum, a poorhouse, and the hospital that would become PGH.

In the 1880s the famous Canadian physician William Osler, then chair of clinical medicine at Penn, used to troop through the PGH wards followed by a gaggle of medical students, demonstrating patient examination at the bedside— a novel departure in those days from classroom-based medical education. Osler reigned over autopsies in a special "autopsy house" at the edge of the grounds backing on the big black wall.

In 1920 the insane asylum and the poorhouse were moved, leaving just the hospital that became PGH. It was enormous, comprising more than two dozen buildings—most of them big brick structures—and occupying twenty-six acres. Funded almost entirely by the city, PGH had a mandate to serve city workers: policemen, firefighters, and other city employees had their own ward. But the huge majority of Old Blockley's patients were poor people with no-where else to go. Most of them were black: African Americans made up three quarters of the hospital's walk-in cases in 1955.[3]

Black people felt welcome there because, relying as it did on city funding, PGH had been obligated to establish a clear nondiscrimination policy.[4] African American women especially liked the place and traveled there from all over the city, ignoring their neighborhood hospitals. Unlike at other hospitals, no ques-tions were asked about finances when patients arrived, and half of patients were treated for free. Every patient was respectfully addressed as "Mrs." or "Mr." And on the obstetrics wards single women giving birth rarely had to fend off prying, judgmental questions. The same went for the one in seven women on these wards who was seriously ill from a botched back-alley abortion.[5]

A new obstetrics building opened in 1955 to help cope with the 37 percent increase in births at PGH during the 1950s. It was followed the next year by two new nurseries—one for full-term infants and the other for premature ba-bies. The number of bassinets in the latter soon grew from thirty-eight to sixty-two. The obstetrics patients were overwhelmingly African American: 94 percent of the nearly five thousand births at PGH in 1960 were to black women.[6]

PGH was an ambitious place in the early 1960s: a ten-year report published late in 1961 boasted of new buildings, renovation of obsolete facilities, and the "beginning of the hospital's recognition as a major research center."[7] In 1955 the hospital had launched the PGH Research Fund to get its staff more actively engaged in research. With its heavy patient load and physicians from the city's medical schools under contract to provide patient care, it was "ideally suited" for research, the 1961 report went on, noting that, if several major grant appli-cations came through, PGH would vault into the "major leagues" of research institutions.[8]

Koprowski had already taken advantage of the pro-research agenda at PGH. Paul György, an eminent nutritionist who had discovered several of the B vita-mins and who was head of pediatrics at Old Blockley, opened the door there to the Wistar's chief, giving him and his researchers permission to use the babies in the hospital's premature infant nursery. Beginning in January 1959—when he was still bent on beating Sabin in the race to license a live polio vaccine— and continuing through June 1961, Koprowski, Plotkin, and other Wistar researchers, working in some cases with doctors from PGH and Philadelphia's Department of Child and Maternal Welfare, fed Koprowski's then–monkey cell–based polio vaccine to scores of premature babies weighing as little as 1.75 pounds.[9]

It is unlikely that the researchers sought parental permission for these experimental vaccinations, according to Plotkin.[10] In the papers that were published reporting on the trials, they do not mention doing so. They do thank György for making the nursery and its babies available and the PGH nursing staff and two physicians who were not authors of the papers for their assistance and cooperation.[11,12]

The scientists were seeking to understand, they wrote in one of the papers that resulted, why some newborns resisted intestinal infection with live polio vaccines—infection that was necessary for the vaccine to successfully generate protective antibodies. They were also examining how antibodies that the babies received from their mothers while in the womb influenced their responses to vaccination.

In their first study, begun in 1959, the scientists found that 92 percent of the babies became infected with the vaccine virus. But only 59 percent responded to the infection by producing protective antibodies, and their responses were "distinctly" weaker than those of babies that were even two or three months old.[13]

Nonetheless, the takeaway from that trial, Plotkin recalled in a 2015 interview, was that having 60 percent of premature newborns even somewhat protected from polio was considerably better than having none of them protected.[14] While viral infections were rare in preemies, the authors conceded in one of the papers, unusually severe infections in these babies "assuredly do occur."[15]

Three years later their calculus hadn't changed. With naturally occurring poliovirus still circulating—988 Americans were paralyzed by polio in 1961— they felt that vaccinating newborns, including preemies, was a priority.[16] Admittedly, maternal antibodies that lingered in the babies' systems for the first six months of life could make vaccination less successful by eliminating the vaccine virus before it could prompt an immune response that would generate the baby's own antibodies. But the trade-off was this, Plotkin said in 2015: if the babies weren't vaccinated while they were a captive population in the nursery, the chance to vaccinate them could be gone for good.

So it was that, armed with the new Koprowski polio vaccine made with Hayflick's human diploid cells, the Wistar researchers began vaccinating dozens of full-term and premature newborns. The preemies they found were at PGH, where pediatrics chairman György was still in charge.

Poupard was hired to vaccinate the preemies and to take their blood. Their stays in the hospital would last for weeks—far longer than full-term babies stayed—and would, if their mothers cooperated, be followed by checkups in the outpatient clinic for several months, allowing for follow-up blood draws to measure antibody responses to the vaccine.

When Poupard first arrived in PGH's premature infant nursery, he immediately appreciated what he was up against. Guarding the nursery like a mother bear was the head nurse, a skinny woman in her fifties whose starched white cap perched upon gray hair often done up in a bun. She never cracked a smile. She took one look at the nineteen-year-old Poupard and without uttering a word made it clear that she was offering him access to her babies—she always called them "my babies"—only under duress.

"She would stand there while you washed your hands to her specifications and while you gowned," Poupard recalled in a 2014 interview.[17] "Then she would bring the baby." Poupard would pour the tiny tube of yellow vaccine down the baby's throat. Then, still under the head nurse's killing stare, he would uncork the empty test tube and unwrap the lancet.

With the baby lying on a table or being held by someone else, Poupard would cut an incision of about one quarter inch in the soft flesh of the baby's heel. He would quickly set down the lancet, grab the test tube, and begin repeatedly scraping the edge of its mouth along the length of the incision, collecting every drop of blood he could squeeze out. He needed at least one milliliter of blood in order for the baby's antibody levels to be tested. That's about one tenth of a teaspoon, but with a three-pound baby, this is easier said than done. At first the baby would inevitably be crying. This boosted its circulation, so that the cut would bleed. But problems arose if the baby, as often happened, then went to sleep. At that point blood flow from the wound would slow and then stop.

"The last thing you wanted to happen is the baby goes to sleep, 'cause then they won't bleed," Poupard recalled. So he would wait until the head nurse turned her head and then flick the baby behind its knees "to cause as much pain as I could to the poor little kid to wake them up." Some babies would keep sleeping anyway, and he would get back to the lab without enough blood. But he succeeded often enough.

Leonard Hayflick as a young man (right) with his friend Norman Cohen.

The Wistar Institute of Anatomy and Biology soon after Hilary Koprowski took over in 1957. The building on the University of Pennsylvania campus in Philadelphia dates to 1894.

I.S. Ravdin, surgeon-in-chief at the Hospital of the University of Pennsylvania, 1959. Ravdin made it possible for Hayflick to obtain aborted fetuses from the hospital.

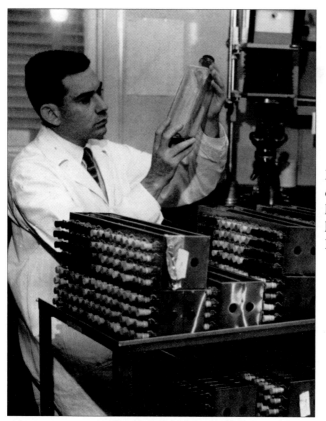

Leonard Hayflick examines normal human fetal cells in the lab at The Wistar Institute, circa 1960.

BELOW LEFT: A microscopic view of young WI-38 cells—fibroblasts from the lungs of a Swedish fetus aborted in 1962. The oblong shapes are cell nuclei. The long, tapered cell bodies are much lighter. The very dark clumps (center near the bottom) are the chromosomes of a cell preparing to divide.

BELOW RIGHT: Old, or "senescent," WI-38 cells, which have stopped dividing and have lost their slender, compass-needle shapes. Their nuclei are stained with a dye that selectively stains senescent cells.

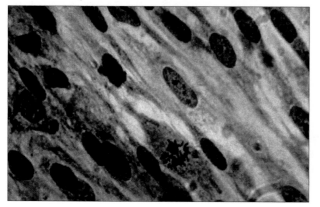

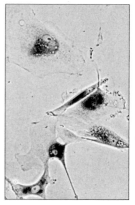

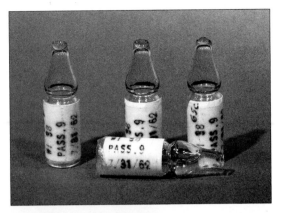

Young WI-38 cells were stored for freezing in these ampules on July 31, 1962.

NIH polio safety scientist Bernice Eddy (left) pictured with her colleague Sarah Stewart, in the 1950s. Eddy was demoted after she discovered— and spoke openly about—a cancer-causing virus in the monkey kidney cells used to make polio vaccine.

While the issue was ignored by the mainstream press, the *National Enquirer* reported prominently on the silent, cancer-causing monkey virus that scientists later estimated contaminated tens of millions of doses of the Salk polio vaccine. This cover story appeared on August 20, 1961.

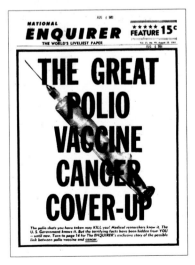

The nineteen-year-old Jim Poupard gave premature babies at Philadelphia General Hospital experimental polio vaccines made with human fetal lung cells while working for Hilary Koprowski at The Wistar Institute in 1962. He is pictured during his medical technician training in 1961.

The new Philadelphia General Hospital Premature Infant Nursery in 1956. African American women trusted the hospital and 94 percent of babies born there in 1960 were black. The hospital's senior pediatrician permitted The Wistar Institute scientists to use the babies for polio vaccine trials.

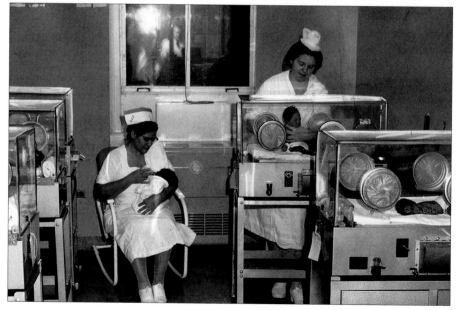

Philadelphia General Hospital, circa 1964. The hospital served mainly African Americans. The black wall that separated PGH from the University of Pennsylvania is visible in the foreground.

Roderick Murray led the National Institutes of Health division that approved vaccines for the U.S. market from 1955 until 1972. For a decade after Hayflick derived the WI-38 cells in 1962, Murray refused to approve vaccines made using them.

The Pfizer polio vaccine made in Hayflick's WI-38 cells won FDA approval in 1972, ten years after the cells became available. Supply shortages and a campaign by Lederle Laboratories to sow distrust in the vaccine led Pfizer to withdraw it from the U.S. market in 1976.

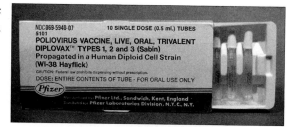

Sir Norman McAlister Gregg, the Australian ophthalmologist who discovered that rubella damages fetuses. Gregg listened deeply to patients, leading to his classic 1941 finding.

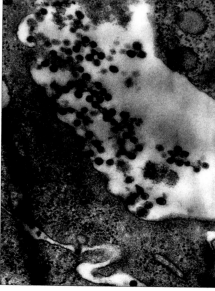

Rubella virus particles in a space between two cells, as seen through an electron microscope. On the left, particles are budding from the surface of a cell. When a cell is invaded, the virus co-opts the cell's machinery to make many more viruses.

Stanley Plotkin uses a pipette to transfer rubella virus at The Wistar Institute, circa 1965. Today most pipettes are operated by thumb-controlled pistons, much like syringes.

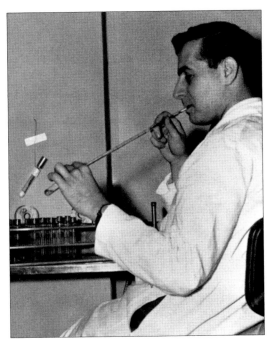

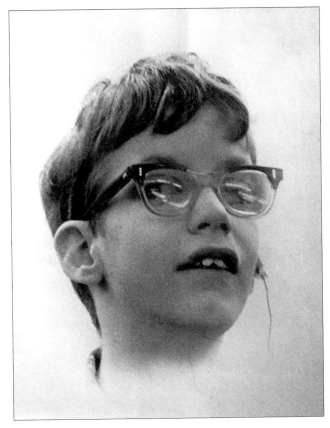

Stephen Wenzler as a child, circa 1972. Stephen's mother, Mary, had rubella early in her pregnancy. Stephen was born blinded by cataracts and profoundly deaf. His heart was also damaged by the virus.

St. Vincent's Home for Children in 1971. Stanley Plotkin tested his rubella vaccine on children at the home from 1964 to 1967, with the approval of John Joseph Krol, the Roman Catholic archbishop of Philadelphia. The Home was owned by the archdiocese and staffed by the Missionary Sisters of the Precious Blood.

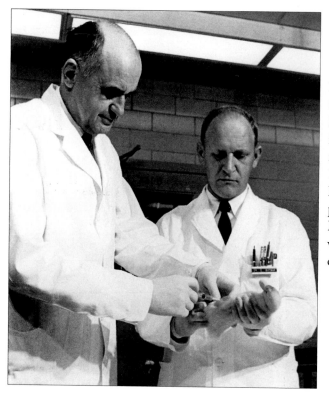

Maurice Hilleman (left) with colleague Eugene Buynak, injecting a duck with rubella in the late 1960s. Unlike Plotkin's human-cell-propagated vaccine, Merck's first rubella vaccine was made in duck embryo cells.

A March of Dimes poster in 1970 invoked medical authority and fear to urge people to get vaccinated against rubella with the new vaccines.

This government-sponsored billboard was part of a campaign to immunize millions with the newly approved rubella vaccines, before an epidemic that was expected as soon as 1970.

Dorothy Horstmann, a Yale University pediatrician, challenged the effectiveness of the rubella vaccines that were licensed in 1969 and 1970. She finally persuaded Merck's Maurice Hilleman to switch to Plotkin's superior vaccine.

Stanley Plotkin in his office at the Children's Hospital of Philadelphia, circa 1979. His rubella vaccine, made in WI-38 cells by Merck, was approved by the FDA in 1978.

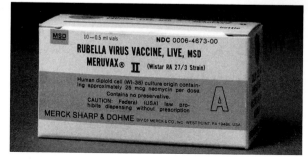

Merck's new rubella vaccine as it entered the U.S. market in January, 1979. The box notes the use of Stanley Plotkin's RA 27/3 rubella virus strain and its production in the WI-38 cells derived by Leonard Hayflick.

Leonard Hayflick at Stanford circa 1975, the year that the National Institutes of Health investigated his stewardship of the WI-38 cells.

James Schriver headed the Division of Management Survey and Review, the NIH's internal auditing office. Early in 1976, he issued a damning report on Hayflick's handling and selling of the WI-38 cells.

Anna MacConnell at eighteen months old in 2002. Scarring of her windpipe after open-heart surgery meant that she had to breathe through a tube until she was three years old. Anna was deaf, blind, and had a four-part heart defect called tetralogy of Fallot. Her mother had rubella while pregnant.

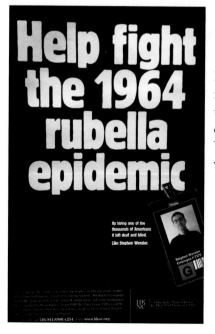

In 2004, the Helen Keller National Center for Deaf-Blind Youths and Adults marked the fortieth anniversary of the 1964 rubella epidemic with a poster featuring Stephen Wenzler. It urged employers to hire those whom the epidemic left deaf and blind.

Bullet-shaped rabies virus particles are shown here magnified about 70,000 times with an electron microscope. Beginning in 1962, Hilary Koprowski and Tadeusz Wiktor used Hayflick's WI-38 cells to develop a much-improved rabies vaccine.

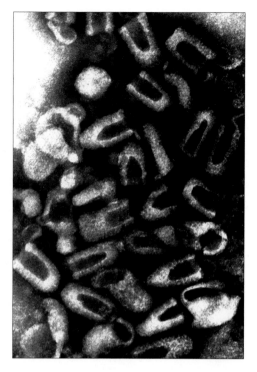

Hilary Koprowski being "vaccinated" by Stanley Plotkin (left) during the first human trial of The Wistar Institute's WI-38-propagated rabies vaccine in 1971. The actual vaccination happened moments earlier. This one was for the camera. The vaccine's co-inventor, Tadeusz Wiktor, is "restraining" Koprowski.

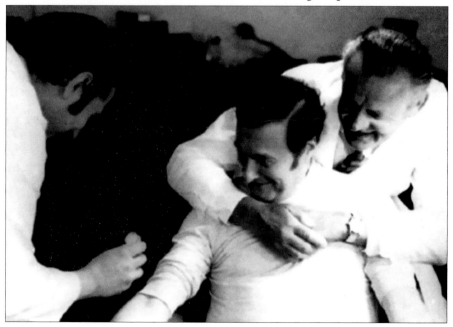

Stanley Plotkin at age eighty, in 2012. "I am fond of saying that rubella vaccine has prevented thousands more abortions than have ever been prevented by Catholic religionists," he says. Today, Plotkin consults for vaccine companies and nonprofits and writes frequent articles urging countries to establish an international fund for vaccine development.

Leonard Hayflick and his wife, Ruth, at home in the Sea Ranch, California, in 2013. Hayflick kept a liquid nitrogen refrigerator of WI-38 cells in his garage on bluffs above the Pacific until 2006, when he donated the cells to the Coriell Institute for Medical Research. "It was time," he told *Nature*, "that my children—now adults—should leave home."

Poupard asked himself many times, as he walked on his errands between the Wistar and PGH, passing through a gap in the big black wall that separated it from the university, why Koprowski would be putting an experimental polio vaccine in premature babies. But he was a kid with a high school education, and the orders were from the godlike man himself. If Koprowski wanted it done, there must be a good reason. Poupard proceeded without ambivalence.

His work wasn't confined to PGH. The Wistar researchers were preparing data for a first-of-its-kind meeting to be held in Geneva that July, where experts convened by the World Health Organization were going to examine the potential of Hayflick's human diploid cells for vaccine production and make recommendations to WHO's director general. Immunization data would be important for the experts to have in hand, and the more of it the better.

So that spring of 1962 Poupard also drove his pea green 1956 Impala with a "med tech" parking sticker on the windshield to Clinton Farms, to collect pre- and postvaccination blood from full-term babies. He was also sent to a third destination, three miles down Woodland Avenue, a streetcar-clogged commercial artery that ran from the university to southwest Philadelphia. There, in Hayflick's old childhood neighborhood, he visited a brand-new, redbrick building: St. Vincent's Hospital for Women and Children.

The hospital was run and financed by the Archdiocese of Philadelphia. It had begun as a Catholic home for "unfortunate infants" in 1858 and been expanded to include a maternity home in 1885. But that huge, rambling Victorian building was torn down in 1959. People called the new replacement St. Vincent's Home for Unwed Mothers, although this was not its official name. One four-story wing had dorm-style rooms for the pregnant girls and women who kept it at or near capacity. The other wing was a three-story, twenty-two-bed maternity hospital where they gave birth.[18]

In an era when pregnant girls and women often disappeared suddenly from their communities only to return months later with tales of European sojourns or visits to sick aunts, the Home for Unwed Mothers hosted around sixty girls and women on any given day.[19] They came from every kind of situation. There was a rape victim who was perhaps twelve or thirteen years old; a stunning flight attendant who arrived by taxi in a pink Chanel suit with a pink pillbox hat; and the underage daughter of a hugely prominent Catholic politician, who quietly arranged with the archbishop to have his daughter admitted under an assumed name.

Under the watchful eyes of the black-habited Medical Mission Sisters who staffed the home and hospital, the young women went to school, did their homework, took crafts classes, grumbled about the endlessly recycled maternity clothes, went to Mass, and took their prenatal vitamins. They looked forward to their weekly escape from the home on Sundays, when they were allowed to walk up Woodland Avenue to buy soap and toothpaste and shampoo. They talked one another out of running away—rumor had it that they would wind up in jail if they did.

Many knew that they would never hold, or even see, their babies: girls and women putting their babies up for adoption would learn only the sex of their child, so that they could bestow a name on the baby.[20]

St. Vincent's Hospital was, like Clinton Farms, an attractive spot for the Wistar researchers because of the follow-up that it allowed. In addition to the newborn nursery, there was a sixty-cot nursery for older babies, some of whom lingered in the hospital for months while the Catholic Children's Bureau looked for foster homes or adoptive parents. In 1962, the hospital was beyond overflowing, with an average of eighty-four babies on any given night.[21] If infants reached one year of age, they were transferred to St. Vincent's Home for Children, which stood back to back with the Home for Unwed Mothers, on Greenway Avenue, the next street over. (The Home for Children too will play a role in this story.)

Sister Mary Jacob, a thirty-five-year-old Medical Mission Sister with a ski-jump nose and a mischievous smile, was in charge of St. Vincent's Hospital several years earlier, when Koprowski's Wistar scientists first sought to vaccinate newborns there with their boss's monkey cell–propagated live polio vaccine. It was 1959 and she was worried, she told them, about the safety of the live vaccine. She had recently read comments from the surgeon general saying that it wasn't ready to be licensed.[22] So she was declining their request.

She also had a contract to keep, she wrote to her boss, Archbishop John Cardinal O'Hara, Philadelphia's top prelate, after the Wistar scientists appealed her decision to him. "Every child admitted to our care has the written consent of his parent for 'vaccination or any necessary operation.'" That covered only "normal" procedures, not experimental ones, she argued.[23]

The seventy-one-year-old archbishop, a former president of the University of Notre Dame who was known to answer his own doorbell ("How else can I meet the poor?" he once asked),[24] agreed with Sister Mary Jacob and said so in his reply to her.

"If I am acting *in loco parentis* for the children at Saint Vincent's I have qualms of conscience about the use of the children for the testing of remedies and procedures," O'Hara wrote to the young nun. "My mind is not entirely free from the prejudice that our research teams want these infants because they are always available in quantity and under conditions which permit [a] wide variety of controls."[25]

Early in 1960 Koprowski tried again, writing directly to O'Hara, urging him to permit a trial of his monkey cell–propagated vaccine. "Our interest in vaccinating the infants ... at St. Vincent's Hospital would be primarily to protect them," he wrote, noting that some 25 million people in various trials had by then been vaccinated with experimental live polio vaccines. In babies at Philadelphia General Hospital, he added, "it has been proved that such vaccination provides longlasting—probably lifelong—immunity." He would be happy to come and discuss the matter in person with the archbishop.[26]

Koprowski was at his manipulative best in this letter; he could not accurately have described "longlasting" or "lifelong" immunity in babies that had been vaccinated only one year earlier.

O'Hara turned to Sister Mary Jacob for her opinion.

"This type of study should not be conducted without the specific written permission of each parent," she persisted. "Since this is not possible under the present circumstances, we do not think it is advisable to recommend this study."[27]

"My dear sister," O'Hara replied to the thirty-five-year-old nun late in March 1960. "I am deeply grateful for your letter. . . . Since you do not recommend the project of Dr. Koprowski, I am writing to tell him it is not authorized."[28]

O'Hara died six months later in August 1960. He was replaced by John Joseph Krol. One year later, in late summer 1961, the U.S. government licensed Sabin's live polio vaccine. Perhaps because this legitimized live vaccines, or perhaps because Krol was a staunch conservative and a defender of hierarchical authority, by the spring of 1962 Wistar researchers had gained access to St. Vincent's Hospital; the fetal-cell–produced experimental polio vaccine was being given to the infants there; and Poupard was drawing their blood to measure their antibody responses. (At St. Vincent's Hospital and Clinton Farms it was nurses who administered the vaccines, pouring the yellow fluid into the babies' mouths. At PGH Poupard did this because, he recalls, the researchers didn't trust the head nurse to vaccinate the babies.)[29]

By July 5, 1962, polio vaccines made using Hayflick's human cells had been fed by the Wistar team to 132 premature and full-term infants ranging from just born to several months of age. Plotkin, Norton, and another Wistar virologist and veterinarian, Richard Carp, reported on the vaccinations in a working paper they submitted on that date to the experts meeting in Geneva at the World Health Organization.

They noted that two thirds of the babies had excreted the vaccine virus in their feces—a sign of the active intestinal infection that was necessary for the baby to generate protective antibodies. As for untoward effects, they wrote, "No reactions were observed in any of the 132 infants." They reported that the vaccines appeared to be genetically stable, meaning that lab tests showed that the virus excreted in the feces of the vaccinated babies hadn't reverted to a more virulent form of poliovirus.[30]

As for Poupard's painfully gathered blood samples, the data they yielded on the babies' antibody levels—a rise in antibodies is the gold standard of a vaccine's effectiveness—never did see the light of day. The Wistar scientists didn't report on antibody levels from the vaccinated infants either to the World Health Organization or in any paper in the medical literature. It's not clear why. Nonetheless, at that meeting in Geneva in the summer of 1962, Hayflick's human diploid cells met a warm reception.

He attended that Geneva meeting shortly after his thirty-fourth birthday. For a man not yet old enough to run for president, he had accomplished an extraordinary amount in the course of the previous three years. He had created the first self-replicating human cells shown by exhaustive microscopic examination to be normal diploid cells. He had seen what biologists for the past fifty years had failed to notice: that such normal lab-dish cells were mortal and inevitably aged and died in culture. He had posited that these normal, mortal cells had an analog: the normal, aging cells of living humans. He had shown that these normal cells launched from human fetuses appeared to be free of hidden viruses and that they could be infected with dozens of disease-causing viruses. He had put them forth as a promising new tool that could serve biologists in the lab and that could allow vaccine makers to get away from using monkey cells. And he had done so at precisely the time that a silent simian virus, SV40, was discovered to be contaminating the polio vaccine that had been injected into tens of millions of Americans.

Hayflick had even made an apparently safe and promising polio vaccine using the human cells. And as the WHO committee met in Geneva, he was in the midst of launching what he hoped would be a lasting human fetal cell strain: WI-38. With luck and persistence it might just become the standard cell strain for vaccine making.

As he returned from Geneva that summer, Hayflick had every reason to be pleased with his accomplishments and optimistic that his new vaccine technology would help the industry to turn a corner. What he hadn't banked on was one inveterate obstacle. He worked at the NIH and his name was Roderick Murray.

As the director of the National Institutes of Health's Division of Biologics Standards, Murray was the top U.S. vaccine regulator at a tumultuous time. An introverted, Harvard-educated physician with a square jaw and thin lips, he had been born in New Zealand and raised in Scotland and South Africa and spoke with a noticeable if hard-to-place accent. Murray was as colorless and buttoned down as the edifice he inhabited: building 29, a five-story, red-brick minimalist structure that sat on a slight rise on the NIH campus in Bethesda, Maryland. The government had erected the new building in a hurry in 1960 to house the growing DBS. After the Cutter debacle in 1955, when live polio virus paralyzed scores of children who received the Salk vaccine and then spread in their families and communities, it became clear that government vaccine regulators were outgunned. So the DBS expanded dramatically. Between 1955 and 1963 its staff grew from 54 to 249. Murray presided over the division from a corner office on the first floor of the new building; he had been promoted to head the DBS within months of the Cutter incident.

As the assistant director of vaccine regulation during the Cutter episode, Murray was a reluctant recruit to the top job. But he was obligated to take the position when it was offered, because he served as a member of the Commissioned Corps of the U.S. Public Health Service, a quasi-military operation; to turn it down would have been tantamount to disobeying orders. James Shannon, the new NIH director, may have quickly realized that Murray was ill suited to the position, for Shannon soon placed the tough Joseph Smadel in the DBS to keep an eye on Murray. Essentially, Smadel told Murray what to do, and Murray did it—until Smadel's untimely death in 1963 from kidney cancer.[31]

Long before the Cutter affair, Murray had been a close witness to another vaccine disaster involving hepatitis B, then an especially frightening disease. The chronic, often-deadly liver infection was known to be caused by a virus passed by bodily fluids, but the virus had not been isolated in the lab, and there were no blood tests for hepatitis B. This meant there was no way of screening the blood supply to rid it of the virus, since people could carry it for long periods without showing symptoms of the disease. (When it did show up, it could turn the skin and the whites of the eyes yellow, a condition known as jaundice. The yellowing occurs because the diseased liver can't package for excretion a yellow pigment called bilirubin.)

During World War II Murray, then a major in the U.S. Army, served for several years with a medical lab in the South Pacific. He was charged with establishing standards for production of plasma—the fluid, noncellular component of blood, which was in great demand for injured soldiers. As such, he would have been keenly aware of the tragic episode in 1942 when some 50,000 army personnel contracted hepatitis B. The soldiers had received a vaccine against yellow fever that was tainted with blood plasma from carriers of hepatitis B; the plasma was used to stabilize the vaccine. Between 100 and 150 of the men died.[32] (The tainted vaccine was produced at the Rockefeller Institute in New York City with blood serum donated by New Yorkers and by students, staff, and faculty at Johns Hopkins University.)[33]

After the war Murray joined the NIH office that was the predecessor of the Division of Biologics Standards and made himself into an expert on hepatitis B. In 1951 he and colleagues from that office, called the Laboratory of Biologics Control, demonstrated that blood plasma transfusions from seemingly healthy carriers of the virus—people with no symptoms of hepatitis such as yellowing of the skin or eyes—did indeed cause the then-untreatable disease. They showed this by injecting blood from donors suspected of having hepatitis B into sixty young, healthy "volunteer" prisoners in U.S. penitentiaries at Mc-Neil Island, Washington, and Lewisburg, Pennsylvania. Twenty-one of the prisoners got hepatitis B. The liver tests of another six participants suggested that they, too, had contracted the disease, although they did not become jaundiced. The paper reporting this was published in 1954, the year before Murray was promoted to head the DBS.[34]

It is shocking today to read of physicians deliberately infecting people with a damaging and sometimes fatal virus. In the case of the Koprowski and Hayflick polio-vaccine experiments, the researchers could at least argue that the polio virus had been weakened and that the vaccinations could conceivably benefit the infants. There was nothing in Murray's prisoner experiment that could have benefited the healthy men, aged twenty-one to thirty-five years, who participated, and who were put at serious risk of harm. The experiment flew in the face of the Hippocratic oath that Murray and the four physicians who coauthored the paper with him had sworn to uphold. (It is a common misconception that the oath pledges the new physician to "first, do no harm." But it did commit them to "utterly reject harm and mischief.")

Yet it is also true that the prisoner trial emanated from the Laboratory of Biologics Control at the National Institutes of Health; that it was funded by the Department of the Army and facilitated by the Department of Justice, specifically, by the Bureau of Prisons; and that it was published in the *Journal of the American Medical Association*, a leading medical journal. The medical research establishment, in other words, openly approved and sponsored the experiment.

The Murray experiment was just one of scores that make shocking reading today and that were carried out by U.S. medical researchers during World War II and the decades that followed it. Most of the time their experiments were conducted on marginalized groups: poor people on hospital charity wards, people of color, prisoners, people dying of cancer, and orphans and the intellectually disabled in institutions. Most of the time the subjects of these experiments had no idea they were being experimented upon or being put at risk of harm and no capacity or opportunity to give free and informed consent to participate.

The historian David J. Rothman, in his excellent book *Strangers at the Bedside*, explains how the urgent wartime atmosphere, and all that was at risk during World War II, quickly and easily led medical researchers, funded by the government, to treat individual human beings as expendable in the pursuit of supposedly higher, humanity-benefiting ends—namely, keeping soldiers on the front lines healthy and ready for action at all costs. The public too took the World War II–era experiments in stride; everyone, even prisoners and the institutionalized mentally ill, could and should do their part for the war effort.[35] Press accounts lauded the heroics of prisoners who "volunteered" for this higher purpose.[36]

So during World War II medical researchers on the home front tested experimental influenza vaccine by intentionally infecting young offenders; tried to give prisoners gonorrhea to study the effectiveness of preventive treatments; and injected teenagers at the Ohio Soldiers and Sailors Orphanage with heavy doses of *Shigella* bacteria, which causes dysentery, a violent diarrhea, in an attempt to immunize them. (The teens did develop antibodies but also became so extremely ill that the vaccine idea was jettisoned.)[37] The researchers running these experiments were not outliers, dispatched to carry out a necessary evil that their mainstream colleagues shunned. They were accomplished scientists, like Werner Henle, an eminent virologist at the Children's Hospital of Philadelphia. Or they were on their way to fame, like Jonas Salk.[38]

After the war the great advances of the 1940s—the new availability of penicillin the poster child among them—fueled a full-on charge by the U.S. government, enthusiastically backed by Congress, to beat down disease, just as Hitler had been beaten down. And the new, heroic soldiers in that war were the medical researchers. Their primary funding agency was the government's National Institutes of Health, which in 1953 erected an enormous five-hundred-bed Clinical Center on its campus—a hospital reserved for research on people who were at once patients and experimental subjects.

The NIH made no distinction between the role of the researcher motivated by the quest for discovery and that of the physician whose purpose is to heal the patient and defend his or her welfare. The agency held that one person could play both roles at once, without any conflict of interest. So the NIH promulgated no rules about informed consent, and individual researchers at the Clinical Center—like those non-NIH scientists around the country whose projects were funded by the NIH—were under no obligation to inform their patient-subjects about the risks of the procedures being carried out on them, or even that they were subjects of experiments.[39]

Given this backdrop, it is not surprising that few if any eyebrows were raised when Roderick Murray was elevated to head the government's vaccine-regulating branch, the DBS, one year after he and colleagues published, in the prominent *Journal of the American Medical Association*, the paper announcing that he had infected healthy young men with a life-threatening virus, establishing that hepatitis B was indeed a blood-borne disease.

"The men volunteered for this purpose," Murray and his coauthors wrote in the pages of *JAMA*, adding in a footnote: "The service rendered by the volunteers is gratefully acknowledged."[40]*

Ruth Kirschstein, a young physician and scientist who later became as close to Murray as anyone at the DBS, recalled late in her own life that "Dr. Murray . . . had had an unfortunate experience. A number of people had died of hepatitis following his studies." She added immediately: "He was a very fine researcher."[41]

---

*Today, under rules first put in place by the United States in 1978, extra protections pertain to prisoners in human experiments funded by the government. For instance, the ethics boards that approve human trials must include a prisoner or a prisoner representative if prisoners are to be subjects in a study. The prisoners' participation in a trial, or their decision not to participate, may not be allowed to affect parole decisions. And the types of study allowed must pertain to prisoners particularly—for example, they may be studies of diseases that occur disproportionately in prisons or studies of the causes and effects of incarceration.

Nor is it surprising that Murray, if he hadn't already been so, had become an extremely cautious man by 1959 and 1960, when Hayflick's first twenty-five human diploid cell strains were launched, followed by WI-26 in 1961 and WI-38 in 1962. By then Murray had witnessed at close quarters the 1942 yellow fever vaccine crisis. In the early 1950s he himself had caused hepatitis B in at least a score of previously healthy young men. He had weathered the 1955 Cutter polio vaccine incident. He had seen his predecessor at the DBS, William Workman, forced out because of that debacle. Now he was in the throes of handling the fallout from the discovery of the silent monkey virus, SV40, in the Salk vaccine. He had every reason to be risk averse.

And he was just that, according to Paul Parkman, a virologist who worked in the DBS beginning in 1963. Murray was famous within the DBS for the stack of papers, interspersed with purple sheets of carbon paper, that sat to his left on his desk. It was the pile of action items awaiting decisions—decisions that in the view of many at the DBS Murray was agonizingly slow to make and yet not at all inclined to delegate.

When obstacles arose, Murray's approach was "Don't feed the problem and it'll go away," according to John Finlayson, a blood products expert who worked in the DBS for decades, beginning in 1959.[42]

Murray was also tough to approach. "He was an elitist, and very proud," recalled Kirschstein.[43]

As an expert, Murray knew something about hepatitis that may have been worrisome when Hayflick's human fetal cells first came to his attention with the publication of the landmark paper in December 1961, describing the first twenty-five human diploid cell lines that Hayflick had derived and their inevitable aging and deaths in culture. What Murray knew was this: babies were sometimes born with hepatitis B. It's now known that newborns contract the disease from their mothers during birth and not while in the womb. But that wasn't known at the time. In Murray's mind it must have been at least conceivable that any given strain of Hayflick's human diploid cells might have come from a fetus infected with hepatitis B. There was no laboratory test for hepatitis B, so it was a fear that Hayflick did not have the tools to disprove.

Beyond the risks of hidden hepatitis infection, Murray was likely troubled too by the idea that the human cells might transmit cancer to vaccinees through a hidden, cancer-causing virus of the sort that the agency was spending huge amounts of money in that era to discover.

Whether religious or moral views also colored Murray's thinking about the use of aborted fetal cells in vaccine making isn't known. But it was common knowledge in the DBS that Murray was profoundly anticlerical; this emerged whenever the deeply Catholic James Shannon, the NIH's director, insisted that prayers be offered at public ceremonies.

Whatever his views, Murray was in a position of public accountability, and he was faced with a challenge the moment that Hayflick and Moorhead's landmark paper appeared in print. For the paper proselytized for the use of the new cells in viral vaccine making, "in particular poliovirus vaccines." In the midst of the SV40 crisis as he was, Murray needed to respond to Hayflick. If he was not going to give the nod to companies to use the new human fetal cells, he needed a very good reason why.

As many people in power who are faced with thorny decisions continue to do today, Murray outsourced the problem. In January 1962, the month after the Hayflick-Moorhead paper was published, he convened a panel of eight NIH experts to study the cells and to weigh in on their promise, or perils, for use in making virus vaccines—as opposed to using monkey kidney cells. The group was both illustrious and powerful—it included people like Wilton Earle, a giant of cell culture, and Smadel, the Salk vaccine booster who had done all he could to stifle Bernice Eddy's findings that something in rhesus monkey kidney cells caused cancer in hamsters.

One year later, in January 1963, the group published its answer to Murray in the form of a dense, six-page paper in *Science*, the high-profile journal that is the unofficial house organ of American science.[44]

While the report conceded the recent burst of interest in the human diploid cells and said that "further investigation" of their potential use in vaccine making was justified, it cautioned that they were "not precisely characterized, genetically or otherwise, and may be subject to random fluctuations in properties." And it made clear that such "random fluctuations" might not be good ones.

"Continuously cultured cells eventually develop characteristics suggestive of malignant change, which theoretically might be attributable to some as yet undefined viral activity," the committee declared. "Even though no virus has been shown to be involved in these particular changes, there can be no absolute guarantee that a given strain of continuously cultured cells will never yield a previously unknown virus . . . that is infective and [disease-causing] for some cells . . . under some conditions."[45]

One could not, in other words, prove a negative. And Hayflick's cells were going to be held to that standard, regardless of the real, demonstrated, and costly deficiencies of freshly harvested monkey kidney cells.

Hayflick fought back in the strongest way he knew how. He enlisted his former mentor from Galveston, the cell-culture guru Charles Pomerat, and the respected chromosome expert T. C. Hsu as coauthors with him and Moorhead on a *Science* paper that appeared a few months later, rebutting the committee's findings point by point.[46]

But the damage was done. Murray now had official approval for resisting Hayflick's cells, in the form of the combined wisdom of eight NIH experts, published in the top science journal in the country.

Across the Atlantic the reception of the human diploid cells couldn't have been more different. Even while the just-hatched WI-38 cells were incubating in the room next to Hayflick's lab in the summer of 1962, the World Health Organization expert meeting on human diploid cells convened.

Eight months earlier the WHO had launched a study of Hayflick's cells by scientists in six countries.[47] (They studied WI-26, since the first twenty-five WI strains had been lost in the freezer accident and WI-38 hadn't yet been derived.) Then, during three perfect mid-July days on Lake Geneva, the scientists hunkered down and scrutinized their results. The meeting was chaired by Sven Gard, the chair of virology at the Karolinska Institute, and attended by Hayflick and, among others, Frank Perkins, a big, outgoing, silver-haired British scientist who carried index cards with jokes, some of them raunchy, written on them. He would pull one out and read it during a meeting or over dinner, as the occasion demanded. Perkins was impressed with the human diploid cells and was fast becoming an important Hayflick ally. His opinion mattered, for he was the United Kingdom's top vaccine regulator.

The group of experts convened by the WHO concluded in a written report that human diploid cells were promising for helping researchers to diagnose viral diseases—an important use for an organization whose job was to track infectious diseases worldwide.[48] Using the cells, disease detectives could take, for instance, a sputum sample from a patient with a respiratory infection and, based on the damage it did to the cells in a dish, identify the viral culprit. By this time scientists had cataloged nearly one hundred human viruses that infected the cells. The cells' huge "virus spectrum" set them apart from other lab-based cells, which were not so hospitable to a broad array of viruses.[49]

In the context of vaccine making, the WHO group wrote, the cells' apparent freedom from contaminating microbes and cancer-causing properties "are of particular importance." Yes, they might possibly harbor unrecognized infections or cancer-causing capabilities that current tools simply couldn't detect.

Nonetheless, the scientists concluded, compared with the alternatives, human diploid cells "represent at the present time the nearest approach to an acceptable system" for manufacturing vaccines.[50]

They urged the WHO's director general to develop manufacturing standards for vaccines made using the cells, "as it may be anticipated that [they] will increasingly be used in vaccine production."[51]

In Sweden, Switzerland, and Croatia they were already in use. Late in 1961 Hayflick had produced more of Koprowski's live polio vaccine, this time using his recently launched WI-26 cells as microscopic vaccine factories. The resulting vials he handed to Koprowski, who promptly sent them on to two European colleagues: Margareta Böttiger, the polio-trial point woman in Sven Gard's Stockholm lab; and a short, thin, intense pediatrician named Fritz Buser, in Bern.

A few months later Hayflick would make still more Koprowski polio vaccine, this time using his brand-new WI-38 cells. This he sent to another Koprowski collaborator: a tall, extremely laconic Croatian named Drago Ikić (pronounced EEK-ich), who tended to utter his rare words with a slight smile.

Ikić, a physician/scientist who ran the prominent Institute of Immunology in Zagreb, had just left a job as the country's top vaccine regulator and retained close ties to that office. He would become a major proponent of human diploid cell vaccines. In 1967 and 1968, due to his influence, Yugoslavia would become the first country to license WI-38–propagated vaccines, against polio and measles.

The three European physicians—Ikić, Buser, and Böttiger—set to work running vaccine trials in infants and children with Hayflick's new human cell–produced polio vaccines. They had a big stage when they were ready to announce their results: in September 1963 ninety-six delegates from eighteen countries met in the Croatian seaside resort town of Opatija for a conference on the study and uses of human diploid cells. By then the trio had vaccinated nearly six thousand infants, preschoolers, and schoolchildren.

They were enthusiastic about what they had found. The vaccine generated antibodies just as well as monkey cell–based vaccine, Böttiger reported after feeding the vaccine, in juice, to 125 elementary school children in Uppsala.[52]

"We did not observe any untoward reactions in vaccinated individuals, neither virus associated ones, nor any side effects," proclaimed Buser, who had fed the vaccine to eight hundred Swiss infants and children.[53]

"One of the most important objects of our observations was infectious hepatitis," Ikić said after vaccinating five thousand Croatian preschoolers and schoolchildren in the spring of 1963. (He first tested the vaccine on 179 staffers at the Institute of Immunology.) "It can be concluded on the basis of the information collected that human diploid cell strains (Wi-38)[sic] are free of infectious hepatitis virus."[54]

Perhaps with these results buoying his inveterate optimism, Koprowski huddled in his office with Hayflick and Plotkin in December 1963 to compose a letter to his fellow Polish expatriate C. Mackowiak—"Macko"—the director of the French vaccine-licensing agency. There the Institut Mérieux, a vaccine maker based in Lyon, was inquiring about making a Koprowski polio vaccine using WI-38 cells.

"Dear Macko," wrote Koprowski, "WI-38 has met all the requirements for a human diploid cell strain to be used for vaccine production." American companies weren't using it, he added, only because the vaccine-regulating Division of Biologics Standards "has chosen to ignore" the deficiencies of monkey kidney cells. "We are firmly of the opinion that if a vaccine manufacturer [using WI-38 to make polio vaccine] were to apply to [the DBS] for a license, it would be impossible for Dr. Murray to refuse."[55]

Back in Bethesda the taciturn Murray was giving no indication that this was the case. And U.S. vaccine makers weren't about to bet that he would change his mind.

PART TWO
Rubella

CHAPTER NINE

An Emerging Enemy

Australia, 1941
London, England, 1962–63

Although one was struck with the unusual appearance of the cataracts in the first few cases, it was only when other similar cases continued to appear that serious thought was given to their causation.
— Norman McAlister Gregg, Australian ophthalmologist, 1941[1]

Early in 1941 a tall, athletic eye surgeon with a thriving practice in Sydney, Australia, noticed an alarming uptick in the number of blind babies being sent to his office.

Norman McAlister Gregg had a mind as sharp as his impressive abilities in cricket, golf, and tennis. He had finished with first-class honors in his medical class at the University of Sydney in 1915 before departing for World War I, where in France he served as a captain in the Royal Army Medical Corps. He was decorated for "conspicuous gallantry" after searching out and tending to the wounded while under heavy enemy fire.

After the war Gregg completed a residency in ophthalmology in the United Kingdom. He then returned to Sydney, where he launched a successful private practice. He was a man with little tolerance for slackers or fools, but he was kind and compassionate to patients and had a habit of listening deeply to their stories. He was also infectiously enthusiastic, with a curious, penetrating mind.[2]

By early 1941 Gregg, balding and bespectacled at age forty-nine, had become the senior eye surgeon at the Royal Alexandra Hospital for Children in Sydney. In the first half of that year, the blind babies began turning up, one after another. By June he had seen thirteen of them—an unusually high number in a city of around one million residents.

All the babies had cataracts: milky white opacities in what should have been the transparent lenses of their eyes. (The lens is an elliptically shaped, bloodless, nerveless structure that measures about one centimeter in diameter in adults. It sits behind the pupil, shape-shifting to help the eye focus on near or distant objects.) The white opacities, usually in both eyes, had been present from birth, their parents said. They made what should have been black pupils appear as if they were white.

When Gregg put drops in the babies' eyes to expand their pupils, the pupils responded weakly and sluggishly to light. The older babies—those more than three months old—also displayed coarse, jerky, purposeless eye movements. "It was a searching movement of the eyeballs and indicated the absence of any development of [focus]," Gregg would write.

But it was the extent of the cataracts themselves, which were unlike any congenital cataracts that he had seen, that caught Gregg's attention as he examined baby after baby. The opacity was densely white in the dead center of the lens, but toward the periphery its density lessened, changing to a cloudier, smoky appearance: a "whitish haze." Finally, there was an unaffected zone at the very edge of the lens.

Gregg knew that during embryonic development the center of the lens grows first and that its peripheral layers are laid down later in pregnancy, like the outer layers of an onion. Whatever had caused the cataracts must have done so during early embryonic life.

The babies' eyes weren't their only problem. They tended to be small and poorly nourished. They had difficulty breast-feeding, a problem often seen in newborns with heart defects. Gregg asked a pediatrician colleague, Margaret Harper, to examine eight of the babies. She heard a harsh murmur along the breastbone in every one of them. It would ultimately emerge that twelve of the thirteen had been born with heart anomalies. Gregg was disturbed, and suspicious that what he was seeing wasn't a mere coincidence. The received wisdom held that all birth defects were inherited, transmitted from parent to child in the genes. To suggest that environmental factors might play a role was considered patently unscientific. But this sudden "epidemic" of cataracts in newborns, the similarity of the babies' unusual cataracts, and the co-occurrence of the babies' eye and heart problems led Gregg to suspect a common, possibly environmental cause.

One day two mothers of babies with cataracts were sitting in Gregg's waiting room talking about their babies. One mentioned to the other that she had had German measles while she was pregnant. She was worried that it had affected her baby. The other mother said that she too had had German measles while she was expecting. During her baby's appointment, each woman mentioned this fact to Gregg and asked him if the disease might be to blame.[3]

Gregg had been searching for a clue, and this one all but shouted at him. He took careful histories from the two mothers and began contacting the mothers of the other eleven babies to ask if they had suffered from German measles while they were pregnant. He also contacted close colleagues—fellow Sydney-based ophthalmologists—and asked them how many cases of congenital cataracts they had seen recently. Could they ask the mothers whether they had had German measles while pregnant? He put the same questions to ophthalmologists around the rest of eastern Australia from Melbourne to Brisbane.

The medical name for German measles is rubella. It comes from the Latin and means "little red." Rubella is generally a mild disease that is transmitted by droplets coughed, sneezed, or otherwise expelled from the mouth or nose of an infected person. It is characterized by fever, swollen glands, and a rash. It owes its popular nickname to the fact that it was first described by a German, Friedrich Hoffmann, in 1740.[4] (The first English description was penned by British physician William Maton in 1815.)[5]

A Scotsman, Henry Veale, serving with the Royal Artillery in India, coined the term "rubella" in 1866, after carefully documenting the course of an outbreak in a Bombay boarding school. His study also distinguished the disease from classical measles, a distinctly different malady, which was present in the school at the same time.[6]

Before Veale's paper was published, and for most of a century afterward, rubella was viewed by doctors as an annoyance—a sort of "bastard measles"—although thanks to the work of Veale and others, it was formally recognized as a distinct entity by an international medical congress in 1881. It was a trivial disease, they thought, an irritant that could confuse the diagnosis of other, more dangerous rash-inducing diseases, particularly classical measles and scarlet fever, which killed children with regularity.

In fact, rubella is so mild that up to two-thirds of infected people aren't even aware they have it.[7] Those who do have symptoms may experience a low-grade fever and swollen glands where the jaw meets the neck and at the hairline on the back of the neck—symptoms that set in twelve to twenty-three days after they are first infected with the virus. In some people, especially young women, rubella can cause aching joints, or even an arthritis that makes joints red, hot, stiff, and swollen and that can continue or recur for months. In about one in five thousand cases, and typically in adults rather than children, rubella causes encephalitis, an inflammation of the brain that is fatal in one of five cases.[8]

Rubella's hallmark is a pink or red rash that starts about two weeks after exposure. It often begins on the face and travels down to the trunk and limbs. It's composed of flat pink or red patches and can have small, raised bumps. Occasionally it is itchy. It lasts about three days.

Rubella is not as wildly contagious as classical measles. But it definitely spreads. While people are at their most contagious when the rash is new, they can pass on the virus for a week before and a week after the rash appears. If they don't have a rash or other symptoms, they can spread the virus anyway.

When Gregg began seeing the blind babies, Australia had been at war since September 1939, with large numbers of young men living in closely packed military camps in preparation for shipping out to Europe and Africa. Infectious diseases circulated easily in the crowded barracks.

In 1940 Australia experienced a widespread rubella epidemic. Unusually, it was a severe rubella that knocked down fully grown adults. Many had throbbing wrists and ankles and raw, sore throats. Others were simply laid low. In the new infectious disease unit at Sydney's Prince Henry Hospital, the average rubella patient stayed eight days.[9]

Medical experts have hypothesized that conditions were ripe for the Australian outbreak not only because of the crowded military camps but also because many of the recruits in wartime Australia came to large cities from rural areas where they had likely never been exposed to the disease. They therefore hadn't developed antibodies against the virus, making them a prime breeding ground for an epidemic. As the war went on, the continuing influx of recruits to the cities provided a constantly replenished source of nonimmune soldiers.[10] They then went home on leave, taking the disease back to their families, wives, and girlfriends. The situation may have been aggravated by the employment of young women in munitions factories, offices, and the armed services.

After his two patients first asked if their babies might have been damaged by rubella, Gregg checked the records of rubella admissions to the infectious-disease unit at the huge Prince Henry Hospital. What he found confirmed his suspicion: the peak period of rubella admissions had occurred between mid-June and August of 1940, seven to nine months before the bulk of the unlucky babies were born in March, April, and May of 1941.

When Gregg interviewed the mothers of his eleven other infant patients with cataracts, only one of them said that she had *not* had German measles while she was pregnant. She also told him that she was kept so busy looking after her ten children that she couldn't remember any details of her pregnancy, except that she was ill at about the sixth week.[11]

The answers that came in from his ophthalmologist colleagues in Sydney and the rest of eastern Australia were as close to definitive as Gregg could have hoped. His colleagues had diagnosed sixty-five babies with cataracts. This brought the total, including the babies that Gregg had seen, to seventy-eight. Of these, the mothers of sixty-eight reported having had German measles while pregnant. Among the remaining ten, five said they didn't know, or hadn't had rubella. In a couple of cases the ophthalmologists didn't get around to asking the question. In another the mother reported "kidney trouble" while pregnant.

Of the sixty-eight women who were sure they'd had rubella, the vast majority had been ill during the first or second month of pregnancy. For most of these that meant July or August of 1940.

For Gregg the case was clinched. In October of 1941 he stood before the Ophthalmological Society of Australia and reported on his cases, stating plainly that rubella during pregnancy had caused not only the cataracts but also the heart defects. By that time fifteen of the babies were dead. Their autopsies had revealed a number of heart defects, most commonly the failed closure of a fetal blood vessel connecting two major arteries near the heart, a condition called "patent ductus arteriosus" (PDA).

Gregg published his findings that same year in *Transactions of the Ophthalmological Society of Australia*, in a now-classic paper entitled "Congenital Cataract Following German Measles in the Mother."[12]

While his discovery was taken seriously and quickly followed up and confirmed in Australia, elsewhere Gregg's findings were slow to be picked up, in part because people were distracted by the war. He also took his share of disdain for bucking the received wisdom of the day by suggesting an infectious cause for a set of congenital defects. One editorial in the British journal the *Lancet* in 1944 noted that the study was retrospective and had relied on women's word-of-mouth accounts of having had rubella. Gregg, it intoned, "cannot yet be said to have proved his case." The editorial writer went on to assail the lack of statistical rigor in a key 1943 follow-up study by other Australians, which linked maternal rubella during pregnancy to cataracts, deafness, heart disease, and microcephaly—an abnormally small head, which is frequently accompanied by intellectual disability.[13] The *Lancet* writer concluded that, if rubella was a real problem in pregnancy, it would likely have been noticed long ago: "The lay public have always held that congenital malformations have an extrinsic explanation—from being frightened by a dog to falling down stairs—and it will be strange if the influence of a mild illness in the first months of pregnancy, accompanied by a rash, has escaped attention."[14]

In 1946 an editorial in the *Journal of the American Medical Association* fully accepted Gregg's findings and their serious implications but conjectured that the particular rubella virus that caused the severe epidemic in Australia in 1940 might have had unique abilities to affect the fetus and might be responsible, through travelers, for cases that had since been reported in the United States and England.[15] American women were paying attention, and many decided to take no chances. One study followed 104 women in New York City who between 1949 and 1955 were diagnosed with rubella during the first three months of their pregnancies. Forty-five chose to have abortions because of their infections.[16]

It would take study upon study in the 1950s to win full acceptance of Gregg's findings by the medical profession. The follow-up studies documented a broad range of damage that the virus did to fetuses in early pregnancy—*any* rubella virus, not just the Australian virus of 1940. They confirmed that rubella's ruinous results included deafness, cataracts, heart defects, microcephaly, and associated intellectual disability. Later, autism would be recognized as another sign of the brain damage in congenital rubella survivors.[17] Any combination of rubella-induced problems would come to be dubbed congenital rubella syndrome—CRS for short. Rubella, an exclusively human virus no more than three millionths of an inch in diameter, was, it was at last quite plain for all to see, a menace to life in the womb.

And there was no defense against it. The rubella virus had not been isolated in the lab. Until it was captured, there could be no vaccine.

In late September of 1962 a brief article appeared in the *British Medical Journal* under the title "Rubella, 1962." The report, by three doctors in general practice in Beckenham, a London suburb, described "a widespread epidemic of rubella" between March and July that year.

The three general practitioners had seen 355 patients with rubella in that short space of time—nearly 6 percent of the patients in their practice of about 6,500. But, they wrote, this was likely an underestimate. They suspected that another 200 people had been infected but had not been seen in the office—people who had telephoned them but not come in; others who had mentioned having the disease after the fact; and still others who they suspected had been infected but had not noticed or reported the disease.

Children aged five to ten years old were by far the most often affected, the trio reported. Fully 25 percent of the practice's patients of this age had been diagnosed with rubella. But, the doctors added, "in this group it is very probable that over 50 percent were infected. Local schools have confirmed that more than half of classes were absent during this period, presumably because of rubella."[18]

What the Beckenham doctors saw was repeated in untold numbers of doctors' offices in the United Kingdom during the spring of 1962 and again in the spring of 1963. (In the United Kingdom, as in the United States, rubella infected people throughout the year, but infections peaked in the early spring.)*[19]

Rubella seemed to be everywhere: The mayor of Rugby and his three children were down with German measles and confined to their home, the *Times* of London reported in mid-March 1962.[20] "Between 20 and 35 Eton College boys have German measles and about 20 are in the school sanatorium. The others are recovering at their homes," the newspaper added in a report about the famous prep school later that month.[21]

The star batter R. E. Marshall of the Hampshire cricket team "had contracted German measles and had taken them home with him—so his colleagues hope," the London newspaper the *Guardian* reported in June 1962.[22]

And women who, early in their pregnancies, knew or suspected they had the disease wrestled with a terrible choice. They could carry the fetus to term, accepting the high risk that it would be damaged by the virus. Or they could seek an abortion.

One woman, identified only as "A Mother," wrote to the British newspaper the *Guardian* in August 1963. After she contracted rubella very early in her pregnancy, her doctor told her there was only a one-in-three chance that the baby would be born undamaged.

"When he gave me the chance to enter the hospital immediately and have the pregnancy terminated, I felt I had no choice," she wrote, adding that she nonetheless felt "a deep elemental repugnance for what I was doing." She added: "Now when people say, or I hear myself saying, how lucky I was, I feel simultaneously a twist of revulsion. It is not lucky to have disposed of a life. What was lucky, however, was the chance that I was sent to a humane doctor."[23]

Many pregnant women weren't as certain as the letter writer that they had had rubella. For them, decisions about whether to continue a pregnancy were

---

*There are no records of how many cases occurred during the British epidemic, but one expert, Elizabeth Miller, notes that in nonepidemic years during this era, two hundred to three hundred children were born with congenital rubella syndrome in England and Wales. Miller estimates that there could have been ten times as many born on the heels of an epidemic. Interested readers should see Elizabeth Miller, "Rubella in the United Kingdom," *Epidemiology and Infection* 107 (1991): 34.

perhaps still more agonizing. What would have served them—and what wasn't available—was a definitive lab test to identify rubella infection.

The first vital step toward developing such a test came in October 1962, when two groups of American researchers published papers in the journal *Proceedings of the Society for Experimental Biology and Medicine*. More than twenty years after Norman Gregg recognized the link between rubella and fetal damage, the virus had finally been isolated in the lab.

One pair of physicians, Thomas Weller and Franklin Neva at the Harvard School of Public Health, had cultured the virus by inoculating plates of human amnion cells with the urine of Weller's ten-year-old son, who was ill with rubella. But their method of identifying the virus in a lab dish took so long—from two and a half to four months—that it rendered it useless for worried pregnant women.[24]

The more practical achievement came from Paul Parkman, the mild-mannered son of a post office clerk and a homemaker from the tiny town of Weedsport, New York. Parkman was a young physician and virologist at the Walter Reed Army Institute of Research in Washington, DC. With his colleague Malcolm Artenstein and his boss, Edward Buescher—and with throat washings from a score of young military recruits who were hospitalized with rubella at Fort Dix, New Jersey, in February and March of 1961—Parkman had devised a way to get around a vexing property of the rubella virus.

Unlike viruses such as polio, which tears through cells in culture, exploding them and leaving chaos and debris in its wake, rubella was seemingly indolent in a culture dish, leaving no clear signs that it had infected cells. Parkman's group came up with an indirect test to prove that the virus was present in culture. They first inoculated throat washings from the Fort Dix recruits onto African green monkey kidney cells, keeping uninoculated monkey cell cultures as controls. Seven to fourteen days later they added another virus, a gut virus called ECHO-11, to all of the cultures. ECHO-11 destroyed the uninfected control cells in two or three days. But rubella virus blocked ECHO-11's effects, leaving the kidney cells intact.[25] It was a cumbersome way to identify a virus, but it was certainly quicker than Weller and Neva's method. And in the face of an epidemic, it was a welcome development.

In London that autumn of 1962, a pediatric resident at the Hospital for Sick Children on Great Ormond Street—nicknamed "GOSH," for "Great Ormond Street Hospital"—read the new papers on the isolation of the rubella virus with keen interest.

Stanley Plotkin, who had helped test Koprowski's polio vaccine while at the Wistar Institute in Philadelphia, had paused his research career to complete the training in patient care that would qualify him as a pediatrician. That summer, just after his thirtieth birthday, he had finished a first year as a pediatric resident at the Children's Hospital of Philadelphia. This second year of residency, which he was spending at GOSH, would see him through to full qualification as a pediatrician. After that, he had reassured Koprowski, he intended to return to the Wistar.

Plotkin was born in 1932 in the Bronx, the son of a telegrapher and a bookkeeper—first-generation U.S. immigrants whose own parents had fled the hostile climate for Jews in eastern Poland. A slight, bookish, precociously intelligent child, he was nearly felled by pneumococcal pneumonia in 1936, before antibiotics were available. He was plagued by asthma and at age nine was sent alone to the National Home for Asthmatic Children in Denver, where he contracted influenza, was hospitalized, became comatose, and again nearly died. He emerged months later weighing forty-three pounds, the left side of his face stilled by a facial nerve paralysis called Bell's palsy.[26] (The paralysis was transient in Plotkin's case; it isn't always.)

Plotkin was a quiet, studious boy who frequently skipped grades in school. He graduated from the Bronx High School of Science at sixteen, after working furiously to keep up with peers who were two years older. He says that to this day he has never inhabited a more competitive academic environment.[27]

As a teenager Plotkin read voraciously, regularly raiding the public library a few blocks from his family's two-bedroom apartment on East 178th Street. At fifteen he stumbled on two books that changed his life. The first was *Arrowsmith*, a 1925 novel by Sinclair Lewis, which chronicles the career journey of a young physician who tries his hand as a small-town doctor but eventually becomes an immunologist and vaccine researcher under a larger-than-life mentor named Max Gottlieb, who is based at a thinly disguised Rockefeller Institute. The second book, *Microbe Hunters*, published in 1926, was a best-selling nonfiction rendering of discoveries by great biologists like Louis Pasteur. It was written by Paul de Kruif, a Rockefeller Institute microbiologist-turned-writer who had cowritten *Arrowsmith*, although his name wasn't on it.

Both books dripped with a romantic view of science. *Microbe Hunters* declares on its first page that it is the story of "bold and persistent and curious explorers and fighters of death" and reminds readers that scientists' achievements "are on the front pages of the newspapers." *Arrowsmith* is flush with the thrill of discovery, the agony of being bested by a competitor, and the eventual rewards of long, painstaking hours in the lab. (The central character, Martin Arrowsmith, is also confronted with the corrupting temptations of wealth, fame, and the flesh.)

Inspired, Plotkin attended New York University on a fully funded state scholarship, then sat a three-day exam trying to win one of thirty-five sought-after scholarships that would pay his way through any in-state medical school. Without the scholarship, financing an MD would be impossible for his family.

Plotkin recalls that he applied to half a dozen medical schools while he was awaiting the exam results. In an era when Jews were not welcome as medical students, he heard back from none of them. Then he heard from the state. Plotkin had placed fifteenth among the test takers and won full funding at any New York medical school.

"We definitely have to accept you, since you've won this scholarship," he recalls being told by an administrator at the State University of New York's Downstate College of Medicine in Brooklyn, the only medical school that accepted him.

By the time he landed in medical school in 1952, Plotkin the undergraduate had fallen in love with Shakespeare, studied philosophy, and dissected a cat, a shark, and a fetal pig. None of which quite prepared him for Kings County Hospital—the huge, busy hospital in the Flatbush section of Brooklyn that was the teaching hospital for the Downstate College of Medicine. It was a world apart where, as Plotkin recalls, "after dark the medical student was king."

He rotated through the specialties and soon knew that he didn't want to be a surgeon. He would avoid holding retractors in the operating room by swapping duties with another classmate so he could instead care for patients on the ward. Yet the prospect of specializing in internal medicine didn't thrill him either. It seemed like a road to treating unhealthy adults just to keep them in a holding pattern.

Pediatrics was different. There he might influence the whole of a life. And there vaccine research was desperately needed. That realization came home to him as he cared for children who were brain damaged and deafened by meningitis—an inflammation of the membranes that enclose the brain—caused by the bacterium *Haemophilus influenzae*. By the end of his third year in medical school, Plotkin knew that he would become a pediatrician. He was also determined to be a research scientist.

For his rotating internship—a mandatory yearlong boot camp for newly minted doctors—he chose to go to Cleveland Metropolitan General Hospital, because there the director of pediatrics was Frederick Robbins, who two years earlier had won a Nobel Prize. Plotkin found that he was too busy taking care of patients to do any research, but he enjoyed the proximity to the man who, with Enders and Weller, had discovered that poliovirus could be grown in nonnervous tissue, opening a whole new world to virologists.

As he finished his internship in Cleveland, Plotkin's next step was clear. Because he was between eighteen and twenty-six years old, the Selective Service Act of 1948 required twenty-one months of military service from him. The only way to avoid this was to go to work for what was, in the letter of the law, a branch of the U.S. military: the Commissioned Corps of the U.S. Public Health Service. So he signed up with the Epidemic Intelligence Service, the "disease detective" branch of the CDC, itself part of the Public Health Service.*

After introductory training in Atlanta, Plotkin surprised his CDC supervisor, Alexander Langmuir, by requesting assignment to an anthrax investigations unit in Philadelphia. But Plotkin, as usual, had done his homework. He had been reading groundbreaking papers by a polio vaccinologist named Koprowski, who was just taking over the Wistar Institute. And the CDC's anthrax project was based at the Wistar. His move there, he says, was the single most important decision of his professional life because it landed him in the lab of his own "Max Gottlieb."

Plotkin remembers vividly the moment, shortly after he arrived at the Wistar in August 1957, when he first presented himself in Koprowski's office, hoping, in addition to his CDC work on anthrax, to talk his way into a polio research position in Koprowski's lab. Displayed prominently on Koprowski's desk was a cartoon depicting a particularly brutal-looking Neanderthal man. Its caption read: "We welcome your suggestions." It set Plotkin to laughing, which in turn made him fear that the illustrious Wistar director would think he was an idiot.

Koprowski did not think so and made room for the twenty-five-year-old Plotkin in his nascent polio lab. In August 1957 this lab consisted, until the Wistar's renovations could be completed, of a big semicircular second-floor

---

*The name of this government agency has since changed several times; today it is the Centers for Disease Control and Prevention. But the acronym, CDC, has remained unchanged.

room without air-conditioning. There Koprowski's first hire, the young lab technician Barbara Cohen, sat measuring the amount of poliovirus in clear tubes of chimpanzee stool that Koprowski's team had sent back from the Belgian Congo. (Feeding experimental polio vaccine to chimpanzees was Koprowski's prelude to vaccinating hundreds of thousands of people in central Africa.)

The less-than-perfect lab space did not deter Plotkin, who was thrilled to be taken on by a man he judged to be not only highly intelligent but also highly cultured, a man with a breadth of vision and a zest for life that Plotkin simply wanted to be around.

And as the Wistar's face-lift progressed under its new chief, so did Plotkin, eventually moving into a third-floor lab, where he worked on anthrax when not laboring in Koprowski's second-floor polio operation. There he learned how to grow and count polioviruses, how to isolate different strains of the virus, and how to weaken them for use as vaccines. His name began appearing on Koprowski's polio papers.

In the spring of 1959 Koprowski offered Plotkin the opportunity to go to the Belgian Congo himself, to work with Koprowski collaborators—expatriate Belgians—who were vaccinating tens of thousands of children in Léopoldville with Koprowski polio vaccine. Plotkin jumped at the opportunity. Plotkin's CDC bosses, loath to appear to be taking sides in the vaccine race, made Plotkin take a leave of absence from his CDC duties during the two-month trip.

"The culture shock engendered by a visit to an undeveloped country was unforgettable," Plotkin wrote later. "More importantly, it taught me that vaccine development did not end in the laboratory and that field studies were not only essential but difficult and even dangerous."[28]

When he wrote "dangerous," he meant it. At one point during the vaccination campaign, Plotkin and his Belgian colleagues were vaccinating infants in the city of Kikwit when they were surrounded by an angry crowd who believed that the researchers were desexing their children because they were drawing blood, to determine the babies' antibody status, from the femoral vein, located in the groin. In an effort to calm the crowd, one of the expatriate Belgians drew blood from his own child's femoral vein, in full view of the crowd. This failed to quell the anger. The scientists were forced to call the local army base, which sent a unit of soldiers in trucks to escort the researchers out of the area.

Koprowski made the most of his protégé's African adventure, inviting print, radio, and TV reporters to a buffet luncheon and press conference just after Plotkin returned to Philadelphia in June 1959. It would mark the "first announcement of effectiveness of Wistar Institute oral polio vaccine during a recent epidemic in the Belgian Congo," the press release announced. And it would feature Plotkin, "who has just returned from making a survey of the mass inoculation."[29]

Plotkin, who as a fifteen-year-old had dreamed of being a microbe hunter, had, at the tender age of twenty-seven, become a certified member of that club.

Plotkin began his one-year residency at the Great Ormond Street Hospital in central London in July of 1962. He was working at a mecca for what he and other doctors called "clinical material," meaning people with diseases, in this case children referred from all over the southern part of England. Because GOSH was one of only two children's hospitals in a city four times as big as Philadelphia, kids with run-of-the-mill earaches and sore throats were few and far between.

Instead children with serious childhood ailments, from congenital heart disease to cystic fibrosis, turned up at the hospital, giving an ambitious young pediatrician all the disease exposure that he could dream of. To boot, it was not he but the housemen—brand-new doctors, called "interns" in the United States—who were responsible for looking after the patients staying on the wards. That left Plotkin with duty at the outpatient clinic—and time for research.

Plotkin had chosen GOSH with a view to working with one member of Koprowski's far-flung network of colleagues, Alastair Dudgeon, a brisk, impeccably dressed virologist with an upper-class accent and an omnipresent bowler hat, who had twice been decorated for bravery during World War II, when he commanded a company of the British army's Seventh Battalion Rifle Brigade in North Africa. Dudgeon was interested in congenital infections—infections acquired in the womb and carried in newborn babies into the world. As Plotkin began working with Dudgeon in July, across the Atlantic Hayflick was in the process of launching WI-38. In September the report from the doctors in the London suburb of Beckenham was published, documenting a glut of patients with rubella. In October the papers from the American virologists were published, announcing that they had captured rubella in lab dishes. It was as if an unseen hand had now put in place all the elements necessary for the drama that would follow.

Plotkin read the Parkman and Weller papers reporting on the lab isolation of rubella virus with great attention. The implications were impossible to miss. If rubella could be captured in a lab bottle, it could perhaps be weakened in a lab bottle to produce a vaccine. Koprowski's gut-wrenching loss in the race to license a live polio vaccine had been painful for Plotkin too. But the isolation of rubella opened a whole new opportunity to create a lifesaving preventive. And as the months passed in London, Plotkin saw firsthand what the absence of a vaccine meant in human pain and suffering.

Beginning in late 1962—nine months after cases of rubella began to surge in March of that year—babies with congenital rubella could be found on the wards of GOSH on practically any day. There was a two-month-old with cataracts and the heart anomaly called patent ductus arteriosus (PDA). A common heart defect in congenital rubella, it can cause a baby's heart to fail without a surgical repair. There was a five-month-old with microcephaly. There was an eleven-month-old who was deaf and blinded by cataracts, and whose heart was hobbled by a hole in the wall separating the ventricles, the two chambers that pump blood out to the lungs and the rest of the body.

There were also slightly older children, victims of the rubella that had circulated at lower levels prior to the epidemic, like the deaf and blind four-year-old who also suffered with a condition called tetralogy of Fallot. This four-part heart defect starves the blood of oxygen, sometimes turning sufferers blue.[30]

When he wasn't seeing patients, Plotkin worked in Dudgeon's lab, using Paul Parkman's new technique to grow the virus from throat swabs of patients with active rubella. He often got positive results from swabs taken during the first week of the rash and especially during the first few days. The problem was, while this test could confirm a rubella infection, a negative test could not rule out infection.

The new technique for isolating the virus also allowed Plotkin to measure the change in rubella antibody levels in blood taken from patients with active disease and blood drawn from the same patients two or three weeks later, when, he found, their antibody levels had risen significantly. There was one kind of patient who desperately wanted laboratory confirmation of whether she had had the disease: the worried pregnant mother.

"The fact that it is now possible to diagnose rubella infection by virus isolation from throat swabs and by [blood tests] is of practical importance in relation to rubella in pregnancy," Plotkin, his British boss Dudgeon, and another colleague, A. Melvin Ramsay, wrote with what can only be called understatement in the resulting paper, published in the *British Medical Journal*.[31]

Plotkin, Dudgeon, Ramsay, and another colleague, N. R. Butler, also studied what was happening in the immune systems of babies, toddlers, and children with congenital rubella syndrome—patients on the wards at GOSH or sent to them from elsewhere by other doctors. Some scientists had wondered if babies infected in the womb failed to "see" the virus as foreign and make antibodies against it—a phenomenon called immunological tolerance.

But twenty-two out of twenty-five children with congenital defects who were older than six months of age—and therefore didn't have maternal antibodies lingering to confuse the test—had rubella antibodies in their blood, meaning that they had responded normally to the presence of the virus, "seeing" it as foreign and making antibodies against it.[32]

There was so much that still wasn't known. Was the blood test not sensitive enough to pick up low levels of antibody that might nonetheless have been present in the other three children? Had the youngsters with antibodies generated these only after birth, when the damage was already done? If they had in fact produced their own antibodies while in the womb, when exactly during pregnancy had their immune responses kicked in? Clearly they couldn't have been effective in those first vulnerable weeks of embryonic life.

Plotkin and his colleagues were mapping uncharted terrain. Today it's known that rubella is a spherical virus made of a single strand of the genetic material RNA wrapped in a protein coat that is in turn surrounded by a fatty envelope. This envelope is studded with protein spikes of two types, labeled E1 and E2. When the body's immune defenses react to rubella, it's mainly these protein spikes, especially E1, that they are reacting to.

Rubella is, as viruses go, quite small. At fifty to eighty-five nanometers in diameter, it's about half the size of the HIV virus.[33] And it's some one thousand times smaller in diameter than the human cells it invades. Of course, size isn't what matters. What matters is what a virus does in the body.

Rubella begins by colonizing the nose and throat, where it lives and multiplies in surface cells and local lymph nodes for several days before invading the blood. From there, and before a nonimmune pregnant woman has mounted her immune response, it travels to multiple tissues, including the placenta.

Hunkered down in the placenta, the virus evades the maternal antibodies that soon obliterate it from the mother's blood. It can persist and replicate in the placenta for months.[34]

Probably due to damage that the virus causes to the blood vessels of the placenta, the virus is frequently able to infect the embryo, likely traveling in toxic clumps of cells that slough off from the inside of placental blood vessels and enter the embryo's circulation. And the embryo, during at least the first twelve weeks of pregnancy, doesn't have the tools to mount its own immune response. Instead it must rely on the mother's own antibodies for protection.[35]

But movement of maternal antibodies across the placenta isn't very efficient early in a rubella-infected pregnancy. Even by the middle of pregnancy, fetal levels of the mother's rubella antibody are only 5 percent to 10 percent of what they are in the mother's blood. So the growing embryo, in the crucial, earliest stages of its life, is left mostly defenseless against the virus, which circulates widely in its blood and can take up residence in virtually any organ.[36] Molecular tools and imaging methods available today have identified rubella virus in the livers, kidneys, lungs, hearts, spleens, lymph nodes, brains, and eyes of fetuses aborted due to maternal rubella.

Rubella is different from other agents that cause birth defects, in that it doesn't usually affect the carving out and shaping of organs and other structures. In rubella-affected infants, you won't find the shortened, deformed limbs that marked thalidomide babies. You won't find cleft palates or club feet or the exposed spinal cord that marks the failure of an embryonic structure called the neural tube to close. Instead the virus homes in on newly formed structures: the long, thin fibers of the lens of the eye; the delicate inner ear, the seat of hearing; the lining of the heart; the small blood vessels that feed what should be a growing brain with oxygen and nutrients.[37]

Virologists have found that rubella doesn't immediately kill the cells that it invades. Rather, it slows them down. They don't replicate as quickly as uninfected cells; in fact, the virus prompts them to make a protein that inhibits mitosis.[38] Eventually they die, sooner than they should, prompted by the virus. So it makes sense that organs of rubella-infected fetuses and infants have been found to have fewer cells than normal and that affected babies weigh on average 65 percent of normal. Why do any cells survive? Because the virus doesn't by any means infect all the cells in an organ. As few as 1 in 100,000 cells may be invaded. These infected cells occur in patches that are scattered in affected organs.[39]

There are very few embryos that escape rubella once it strikes in early pregnancy. There's a 90 percent risk of fetal damage with a rubella infection during the first two months of pregnancy and a 50 percent risk of such damage during the third month.[40] And during particular windows the growing embryo is exquisitely vulnerable: one prospective study found that ten out of ten embryos became infected when pregnant mothers had a rubella rash between three and six weeks after their last menstrual period.*[41]

---

*One reason that critics were initially skeptical of the Australian Norman Gregg's 1941 findings was because his study was "retrospective," meaning that he asked the mothers about their history of rubella infection only after their babies had been born and diagnosed with

Once a pregnancy is into its fourth month, the odds of rubella infection doing damage to the fetus diminish significantly. The reason: the increasingly active fetal immune response combines with antibodies from the mother to keep the virus in check.

After a baby with congenital rubella is born, the virus can remain living in certain tissues and continue to do damage. Long-term problems can include the persistence of virus in the cerebrospinal fluid that bathes the brain and spinal cord, leading to bouts of brain inflammation called encephalitis.[42] And whether because of direct viral damage to the pancreas or because the virus triggers an autoimmune reaction, causing the body's own antibodies to attack the pancreas's insulin-producing cells, babies with congenital rubella grow up to get type 1 diabetes at many times the rate of the general population.[43] Congenital rubella sufferers also endure eye problems that go well beyond cataracts and include glaucoma—elevated pressure in the eyeball that damages whatever limited vision a person may have—and chronic inflammation of the iris and its appendages.[44] It's not known how often the virus itself continues to live in the confined chamber of the eye, doing damage. It certainly does in some cases: in 2006 living rubella virus was captured from the eyes of a twenty-eight-year-old man who was born blind and deaf from congenital rubella after a British rubella epidemic in 1978.[45]

Papers documenting the long-term problems of rubella-affected children wouldn't begin to appear until the late 1960s, when Australian virologists published a twenty-five-year follow-up on the damaged infants who were born in 1941.[46] But Plotkin had plenty of other rubella-related problems occupying him as he finished his residency in London in June of 1963.

It was clear to him that rubella was going to plague generations of newborns if a vaccine was not developed. Rubella epidemics recurred cyclically, predictably, every three to five years in the United Kingdom and every six to nine years in the United States.[47] The next one was only several years down the road. Those were years, he was convinced, in which a vaccine could and should be made.

That summer Plotkin and his wife, the former Helen Ehrlich, whom he had married as he graduated from medical school, set off on a three-month tour of Europe—along with their newest family member, one-year-old Michael. They

---

congenital rubella. By contrast, the study that found ten of ten embryos affected, published in the *Lancet* in 1988, was prospective: scientists followed the pregnancies from the time of the diagnosis of rubella in the mother in early pregnancy and tracked the babies' outcomes.

drove a blue Ford two-door and otherwise lived, as Plotkin would write a few months later, "like gypsies."[48] They traversed France, Switzerland, and Italy and ended up on the Croatian Riviera. Finally, in September, they boarded a Great Holland Line steamer for home.

Several months behind them, another traveler would cross the Atlantic, arriving in time for spring on the U.S. East Coast. It was the rubella virus.

CHAPTER TEN

Plague of the Pregnant

Philadelphia, 1963–64

... rubella does not seem to invoke the fascination of thalidomide despite the fact that in a single epidemic in the United States it caused more birth defects in one year than thalidomide did during its entire time on the world market.

—William S. Webster, University of Sydney Medical School, Australia, 1998[1]

The Wistar Institute was still a bustling biological crossroads, when Plotkin reoccupied his third-floor lab there in October 1963. Koprowski had expanded his core scientific staff to thirty-nine. Eighteen graduate students were working in the labs, and Koprowski was playing his usual charming host to a near-constant stream of international visitors from Helsinki and Zurich, Paris and Milan, Tehran and as far away as Sendai, Japan. Some stayed perched in Wistar labs for weeks or months. Plotkin and Helen and baby Michael put down roots too, moving into a townhouse not far west of the university at 11 University Mews.

Koprowski was busy nurturing the freewheeling science that set the Wistar apart. Plotkin's mentor may have been an autocrat, but he was an autocrat who didn't micromanage his hires, who left them to do their creative best, protecting them from administrative hassles and money worries while he presided over the whole impressive dance alternating benignity and charm with storms of temper and deviousness.

What was more, Koprowski understood—a fact that still dazzled Plotkin—that life was more than science, that art and history and poetry and music and the enjoyment of beautiful women and excellent food and fine wine were as important as breathing. That Christmas parties in the atrium with him playing Chopin on a grand piano imported for the occasion were de rigueur.

"According to the notice which I received on my desk yesterday, a Christmas party is scheduled shortly," Plotkin wrote in a memo to Koprowski that December. "As I have been asked in the past to take care of some of the casualties, I would like to personally suggest that a tank of oxygen be in readiness for use in those who have embibed [sic] not wisely, but too well."[2]

Hayflick, by contrast, seemed to occupy a world apart, all seriousness and focus in his second-floor lab. A recollection from this era from Hayflick's late colleague Vincent Cristofalo is telling. It was 1962, and Cristofalo had just completed his PhD in physiology and biochemistry. He was being recruited by the Wistar and was being shown around. The young Cristofalo knew of Hayflick from the increasingly famous 1961 paper declaring that normal cells aged in the lab, so he was a little awed when his tour guide stopped in Hayflick's lab.

Cristofalo wrote forty years later: "Here I was looking upon this man, sitting at a desk in the center of his laboratory, with people bustling to and fro on all sides of him, going to the incubators or to the sterile rooms. Seemingly unperturbed by this frantic activity, Leonard was dictating letters to a tape recorder. My host, David Kritchevsky, interrupted him to introduce me. Len looked, for all the world, annoyed at the interruption. His demeanor signaled that he wished I would go away and not return. Nevertheless, he was minimally cordial; he gave me a reprint of his 1961 paper with Paul Moorhead and returned to his dictation."[3]

For Plotkin, freshly returned from London, the availability of Hayflick's WI-38 cells meant the opportunity to study the rubella virus in the lab—and, with luck, the chance to create a rubella vaccine. He was already a convert to the use of the fetal cells for vaccine making; it was he who had worked alongside Hayflick developing and testing the polio vaccine that Hayflick first made with them in 1961. It was he who, with Koprowski, had written to the World Health Organization, urging the use of the human fetal cells instead of monkey kidney cells for making live polio vaccines. Now he would see if the fetal fibroblasts could be co-opted to generate a rubella vaccine.

Plotkin knew that he wanted to make a live, weakened rubella vaccine—as opposed to a killed vaccine like Salk's polio vaccine. There were several reasons why.

First, he was more familiar with making live vaccines because of his work with Koprowski's live polio vaccine. Second, scientists were learning that it was extraordinarily difficult to kill the rubella virus and have it nonetheless maintain the ability to induce an effective antibody response. Third, live vaccines tended to generate longer-lasting immunity. And if rubella was going to be a childhood vaccine, the immunity it generated would need to last for decades: from girlhood through a woman's childbearing years. The decision worked out well for Plotkin. Merck experimented early with a killed vaccine and failed strikingly.[4]

It was a fortuitous time for a young U.S. medical researcher. Government coffers were benefiting from the booming postwar economy, and there was a new enthusiasm for medical research in Congress, spurred by World War II–era advances.[5] So the National Institutes of Health, the country's prime medical research–funding agency, was flush with cash and able to make more and more research grants to scientists all over the country as its budget grew from $36 million in 1955 to $436 million in 1965. The happy beneficiaries nicknamed the agency the National Institutes of Wealth, and a running joke among biomedical scientists went, "While you're up, get me a grant."

Koprowski had received a generous multiyear grant in the late 1950s from the NIH's National Institute of Allergy and Infectious Diseases to support work on his polio vaccine. When he renewed the grant for five years in the early 1960s, he redirected the money to cover Plotkin's rubella vaccine research. On his return from London, Plotkin began receiving $130,000 annually from the NIH for studying rubella and trying to develop a vaccine—more than $1 million in 2016 dollars.[6] Plotkin also sought out a foundation with a keen interest in projects related to disability: the Joseph P. Kennedy, Jr. Foundation, established in 1946 to memorialize the late President John F. Kennedy's older brother, who was killed over Suffolk, England, during a secret bombing mission in 1944. Between 1964 and 1967 the foundation would steer $180,000 to Plotkin for his rubella work.[7]

First Plotkin needed rubella virus—and he needed to see if it would grow in WI-38 cells. More than that, he wanted to know if several rubella strains—viruses collected from different, geographically dispersed people and thus possibly differing subtly from one another—would all find WI-38 cells hospitable to their growth. To collect them, he turned to colleagues.

Maurice Hilleman, the vaccine czar at Merck, sent Plotkin the West Point strain, named after Merck's big campus just outside Philadelphia. Paul Parkman, the good-natured young virologist who had first captured the virus in the lab, sent Plotkin M-33, the virus he had isolated from the throat of one of the young military recruits at Fort Dix, New Jersey. And Plotkin got still another strain of the virus, called Marshall, from Dudgeon, his former mentor at the Great Ormond Street Hospital, who sent it, ensconced in dry ice, on Pan Am flight 107, which flew direct from London to Philadelphia on Tuesdays and Thursdays.

"The viruses . . . arrived Thursday. Many, many thanks," Plotkin wrote to Dudgeon in early November 1963. He added: "There is a fair amount of rubella [vaccine] activity going on in the States, though this statement is mainly based on rumors rather than . . . publications. Merck is said to be experimenting with the vaccine, but they are very secretive about it."[8]

Merck, under the ever-ambitious Hilleman, was indeed working on a rubella vaccine. Hilleman had been happy to share his company's West Point strain with Plotkin because Hilleman had his team using a different strain, called Benoit, to develop a vaccine. It had been isolated in 1962 from the throat of an eight-year-old Philadelphia-area boy with that last name. (The French pronunciation of Benoit is "ben-WAH"; the American scientists called it "ben-OYT.") The Merck scientists grew the Benoit virus through multiple generations on African green monkey kidney cells and then in duck-embryo cells, aiming—as with any live virus vaccine—to weaken it to the point where it provoked an immune response without causing illness.

Plotkin had another competitor, in the Division of Biologics Standards, the vaccine-regulating arm of the U.S. government, where the taciturn Roderick Murray had now been the director for nearly a decade. Paul Parkman, at thirty-one years old, was exactly the same age as Plotkin. He had just moved from the Walter Reed Army Institute for Research to a job in the DBS. From there he was working full tilt on his own live vaccine under the mentorship of his boss, Harry "Hank" Meyer. Meyer, only a couple of years older, was a tall, ambitious, take-charge virologist with close-cropped, prematurely gray hair who had been at the DBS since 1959. He had recently been working in West Africa, helping to administer an experimental vaccine to millions of children in the face of an epidemic of classical measles.[9]

Parkman and Meyer were using African green monkey kidney cells to try to weaken the virus that Parkman had isolated from the Fort Dix soldier. They would end up naming the resulting vaccine HPV-77, for "high-passage virus," and because they had inoculated it sequentially into seventy-seven cultures of African green monkey cells, injecting first one culture of kidney cells, then, once the virus had multiplied in those, taking fluid from that first culture and inoculating a second culture, and so on until they reached what was commonly called the seventy-seventh "passage"—since fluid had been "passed" from one culture on to another seventy-seven times.[10]

Importantly, Parkman's new job had placed him at the power center of U.S. vaccine regulation: the DBS approved vaccines for the U.S. market. And

Parkman's boss and partner, Meyer, had just been named chief of the DBS lab charged with viral vaccine work.

Plotkin soon realized that he was fighting an uphill battle against Parkman, with his inside track at the Division of Biologics Standards, and Hilleman, with his years of vaccine-making expertise and all the resources of a huge drug company behind him. But Plotkin was stubborn, and young, and confident. And he thought he could make a superior vaccine with Hayflick's better, cleaner cells. What he needed to do, and quickly, was to establish whether those cells would become infected by rubella; if not, they would be useless for making a vaccine. He also wanted to study how exactly rubella wreaked its damage in infected cells.

So, working with André and Joëlle Boué, a Parisian husband-and-wife team who then were guest scientists at the Wistar, Plotkin injected bottles of Hayflick's WI-38 cells with the three strains of rubella virus that he had collected from the other scientists. He also grew uninfected control cultures of WI-38. As he had hoped, all three rubella strains readily infected the WI-38 cells.

However, once infected, the cells didn't act like the uninfected cells in the control bottles. In the control bottles the WI-38 cells, as expected, began to vigorously multiply. They reached confluence and needed to be split into new bottles every three or four days. They kept dividing for months. In the bottles that were full of infected cells, the cells stopped dividing after being split into new bottles just one, two, or three times. They ground to a complete halt within a few weeks.

When he looked at the uninfected control cultures under the microscope, Plotkin saw a field densely packed with the elongated, compass needle–shaped WI-38 cells. When he examined similar cultures of infected cells—and virtually all the cells were infected after a few days—they had the same tapered shape and appearance as healthy WI-38 cells. The virus wasn't exploding them, filling them with holes, or otherwise leaving distinct footprints.

But what was striking was the amount of white space. The infected cells were few and far between. The virus was somehow shutting down their ability to replicate. Plotkin wondered immediately if this damper on division in a lab dish translated to the fetus—and to the damage he had seen in babies. Two years later he and a Finnish collaborator, Antti Vaheri, would use WI-38–infected cells to identify, and publish in *Science*, their discovery of a protein responsible for inhibiting cell division in the rubella-infected cells. They dubbed it RVIMI, for "rubella virus-induced mitotic inhibitor."[11]

The good news for Plotkin was that, in the process of dividing just two or three times, the WI-38 cells had produced reams of new rubella virus particles: one million or more times the number of viruses he had initially inoculated into the bottles. For vaccine-making purposes, the fact that the WI-38 cells stopped multiplying after a few divisions didn't matter. In the vaccine-making context the cells' job was to produce virus, not to reproduce themselves.

In August 1964 Plotkin and the Boués submitted a paper to the *American Journal of Epidemiology*, reporting their finding that rubella somehow suppressed the division of WI-38 cells—but that those cells nonetheless produced huge quantities of the virus during their short lives in lab bottles. The WI-38 cell, the authors concluded, "might be used as a [cellular factory] for rubella vaccine." They ended the paper by citing the by-now-familiar list of the advantages of Hayflick's human diploid cells for vaccine making: their cleanliness; their normal chromosomes; their reassuring, noncancerous behavior; the fact that they could be frozen and later pulled out of the freezer and expanded into huge quantities.

Even before submitting the paper, Plotkin had reported his good news to Koprowski. The institute's chief couldn't make it to the June 1964 meeting of the Wistar's board of managers. But Art Stern, the Wistar's research administrator, brought the board into the loop at that meeting. "We have been successful in getting rubella virus to grow in diploid cells and are now attempting to [weaken] it," he told them. "A vaccine could be the possible end product of this work."[12]

On March 27, 1964, the CDC devoted the first half of its *Morbidity and Mortality Weekly Report*—the influential publication alerting doctors to circulating infectious diseases—to one subject.

"A nationwide epidemic of rubella (German measles) appears to be in progress," the article began. "A rise in reported cases was first noted late last fall in the northeast, with peak incidence being observed only during the past few weeks. The outbreak appears to have spread rapidly to the south and west."

The eight-page report laid out the local damage.

"Cases are about five times greater than for the comparable period in 1963, one of Massachusetts' high years for this disease."

In New York City, with nearly 8,700 cases already in 1964, "rubella cases are reported at about 17 times the number for the comparable period a year ago."

In Kentucky cases exceeded by 60 percent the comparable period in 1963; in Illinois the figure was 86 percent.

In Allegheny County, home to the city of Pittsburgh, the number of cases through March 13 already exceeded the number in the whole of 1963. The same was true in the state of Colorado and in Maryland.

The city of New Orleans had noted a "sharp increase" in rubella cases, starting, apparently, in the week of February 17.

Only the non-Colorado mountain states and the Pacific states appeared to have been spared. This wasn't altogether surprising. The *MMWR* noted that in the past the Pacific states had experienced rubella epidemics one year later than the rest of the country. (In 1965 the epidemic would indeed visit the Pacific and mountain states.)[13]

The epidemic would become the worst one since authorities in some U.S. states had begun keeping records nearly forty years earlier.[14]

Plotkin, primed for an epidemic onslaught by his experience in London, began tracking the babies born with congenital rubella at the University of Pennsylvania Hospital, Philadelphia General Hospital, and other area hospitals. He made an index card for each affected baby. "R-1," he labeled the first. With time, he would label card R-60, then R-105, and then R-132.

The cards had notations, including:

Baby M. Cyanotic.
Baby F. Congenital heart.
Baby T. Cataracts.
Baby-HUP-Autopsy. Died of congenital heart disease.[15]

Ultimately Plotkin compiled a list of rubella patients—the babies and their mothers—that ran to 1,700 entries.[16] He and colleagues also tallied all of the babies born with congenital rubella at Philadelphia General Hospital. From this they estimated that between April 1964 and March 1965 at least 1 percent of black newborns in Philadelphia were affected by congenital rubella—more than triple the number in the preceding twelve months.*[17]

Soon after Plotkin's return from London, it had become known among Philadelphia-area physicians that the young virologist at the Wistar was adept

---

*Plotkin and his colleagues reported on the number of black newborns because virtually all babies born at PGH were black. They actually measured the number of babies born with heart and eye defects—a proxy for congenital rubella syndrome. Because deafness, its most common manifestation, is hard to diagnose in newborns and wasn't captured in their survey, they estimated that in actuality closer to 1.4 percent of newborns at the hospital were affected.

at running the diagnostic blood test that would tell a pregnant woman if she had had rubella. His lab was soon deluged with requests for help.[18] He began meeting with pregnant women and couples in an office adjoining his lab.

Most often prospective parents were sent to Plotkin via their doctors. Sometimes they contacted him directly. The following letter, in cursive handwriting on a plain half sheet of white paper, arrived from a woman in nearby Pennsauken Township, New Jersey.

> April 2
> University of Penn.
> Wister [sic] Lab

Dear Sir:

Could you let me know is it dangerous or fatal. I have a daughter pregnant 3 months and had the measles. Is the danger over of injury to baby. It is no [sic] 4 month. Please let me know?

> Thank You.[19]

The testing that Plotkin could offer was far from perfect and could rarely provide the kind of definitive answers that women were seeking. He could confirm the disease by isolating rubella from a nose or throat swab. But swabs were reliably positive only if taken during the first day or two that a woman had had a rubella rash, and women seldom found him that quickly. Even if they did, isolating the virus from a throat swab with the cumbersome interference test using the ECHO-11 virus took ten days. Growing the virus for longer, to be certain his result was right, consumed another two or three weeks.

Blood tests, which he used more frequently, were no speedier. Plotkin first tested a blood sample taken during the acute phase of a pregnant woman's illness—or as soon as possible thereafter, and ideally within two weeks of the rash. If she had waited longer to see a doctor, the odds of getting a definitive diagnosis diminished markedly.

But that first blood test alone wasn't enough. Next Plotkin needed to wait two or three weeks, take blood again, and measure the woman's antibody level again. A rise in antibodies from the first test to the second confirmed recent rubella.

Start to finish, Plotkin's blood testing could easily stretch to four weeks. The wait was excruciating. Some women, and some doctors too, couldn't bear it.

"Many physicians do not feel such a delay is tolerable," Plotkin wrote in a paper published in the *Journal of the American Medical Association* that October. He pointed readers to patient number 13 in table 3. The twenty-eight-year-old mother of two was eight weeks pregnant when she developed a rash not typical of rubella: it was only on her abdomen, and her lymph nodes weren't swollen. Her blood was taken quickly, on the second day of the rash, and then again three weeks later. When Plotkin got back to her doctor with the clearly negative results, it was too late. The woman had already had an abortion.[20]

At the other end of the spectrum, some patients with unambiguous positive test results continued with their pregnancies. Patient number 19, aged twenty, had developed a rash, a fever, and swollen lymph nodes when she was nine weeks pregnant with her first child. Plotkin's blood tests showed that her antibody levels had soared eightfold over two weeks. She opted not to have an abortion because of "religious objections," the paper said in a footnote.[21]

Plotkin himself was a reserved man who, when he sat down with parents, was loath to push them either way in their decision making. But in his own heart he had no objection when they opted to abort a potentially damaged fetus. And it rankled him when antiabortion critics lambasted abortion as a means of avoiding possible congenital rubella, as did the authors of two letters to the editor of the *British Medical Journal* in the autumn of 1964.[22]

One critic, a pathologist at a regional hospital in Ipswich, England, wrote that logical consistency demanded that, if one were to abort fetuses likely to be blind, deaf, or intellectually disabled, the same should be done with those likely to suffer asthma, diabetes, high blood pressure, myopia, or personality disorders. This, he concluded, would allow the country to "finish up with a one-party State and no prisons."[23]

Plotkin was furious. The critics' "sense of ethics is apparently so fine that it enjoins others to suffer," he shot back in a letter that the journal did not publish. "What justification is there for increasing the burden of a family by a child with heart disease or deafness? . . . We have no right, because of personal moral or religious imperatives, to demand acceptance of an unnecessarily high risk of congenital abnormality."[24]

This isn't to say that Plotkin thought any parent's decision easy or obvious. "What I remember is the anguish, of course, of the parents, trying to make a decision about whether to continue with the pregnancy," he said in a 2012 interview. "Some were already decided when they came to see me. . . . Whatever the risk was, they didn't want to take it. Others, of course—particularly those with religious beliefs—were conflicted about what to do."[25]

In many cases women who wanted to abort weren't given the option. Saul Krugman, who headed the Department of Pediatrics at New York University, reported to an American Academy of Pediatrics meeting in May 1965 that the therapeutic abortion committee at NYU had "many situations where a woman had had rubella which she has ignored for a week or more until she realizes she has missed a period." The NYU committee, he noted, "does not approve abortions in such cases," despite the likelihood that the woman would give birth to a damaged infant.[26]

Robert Hall, an obstetrician at Columbia University in New York City, wrote that at first only "the wives of insistent physicians" received therapeutic abortions for rubella, "under psychiatric guise."[27] As the epidemic escalated, more women succeeded in obtaining hospital abortions. These were almost always well-positioned women. In 1965 Hall estimated that for every nine privately paying patients—that is, wealthier women—who obtained rubella-related abortions, one poorer, publicly supported patient was able to terminate her pregnancy.[28]

For every woman who managed to obtain an abortion, there were scores of pregnant women worrying, in the absence of any episode of fever or any telltale rash, that their own fetus might nonetheless be infected. They scanned their memories, hoping that an ill-recalled childhood episode of German measles might mean that they were immune.[29] Public health officials told newly pregnant women—often mothers with several children already at home—to stay away from children, who were the most likely carriers of the disease.[30] How exactly a mother of young children might manage to do this was unclear.

"In epidemic years, the specter of tragedy hung over virtually all pregnant women," writes the rubella historian Leslie J. Reagan. "No one could rest easily knowing that the epidemic had spared her; perhaps she had failed to recognize the disease."[31]

Plotkin's regular conferences with anguished parents continued for most of 1964. The follow-up letters to their doctors were, occasionally, a joy to write.

"Many thanks for your letter regarding Mrs. [F.]," he wrote to an Allentown, Pennsylvania, doctor in February 1965. "It's a great pleasure to hear that her baby was entirely normal and I am glad that I had some part in giving her the confidence to continue the pregnancy."[32]

But far more often the testing left a painful uncertainty. One pregnant patient's antibody levels rose between September 11 and 25, but only somewhat. Plotkin wrote to her doctor in mid-October that the levels "are not diagnostic

and could reflect either infection in the recent past or the distant past. I suspect the problem is that we did not see the patient until 18 days after the onset
of the rash.[33]

In the end, the rubella epidemic that swept the United States in 1964 and
1965 infected an estimated 12.5 million people, or 1 in 15 Americans. More
than 159,000 of these infections included joint pain or arthritis, typically in
women. Roughly 2,100 people developed encephalitis, a brain inflammation
with a 20 percent mortality rate.[34]

Some 6,250 pregnancies ended in miscarriages or stillbirths. An estimated
5,000 women chose to get abortions. Still another 2,100 babies died soon after birth.

Roughly 20,000 babies were born, and survived, with congenital rubella
syndrome. Of these, more than 8,000 were deaf; nearly 4,000 were both deaf
and blind; and 1,800 were intellectually disabled. About 6,600 babies had
other manifestations of congenital rubella, most typically heart defects. Often
babies were born with several of these disabilities.[35]

These numbers are, at the very best, approximations. They come from a
1969 CDC report whose authors stressed that it was not until 1966 that physicians were required to report rubella cases to authorities. Before then data were
gathered by many, but not all, states and were voluntarily passed on to the
CDC. The authors noted their data's incompleteness, called it "preliminary,"
warned that it was intended primarily for disease-control experts, and wrote
that it should be "interpreted with caution."[36] Nonetheless, these have become
the "official" numbers and are reported on the CDC Web site and elsewhere to
this day. One top rubella expert who was deeply involved with patients in the
1960s and for decades afterward says that the numbers are "SWAG"—a scientific wild-ass guess.

As he undertook his first steps toward making a rubella vaccine late in 1963,
Plotkin decided that his vaccine virus was not going to be captured, as Parkman had done at Fort Dix, from the nearest convenient patient who had the
virus colonizing his or her throat. That approach, he reasoned, risked contamination with other "passenger" viruses also resident in the patient's throat. Instead, like Hayflick did when he first sought to grow normal, clean cells in
culture, Plotkin turned to the likeliest source at hand for isolating rubella virus
and *only* rubella virus: infected fetuses.

On January 23, 1964, Plotkin wrote to Franklin Payne, the head of obstetrics and gynecology at the Hospital of the University of Pennsylvania.

"Dear Dr. Payne, as you know, we have been collecting aborted fetuses from your service for some time now and using them for tissue culture. We have now entered a new phase of the project in which we are trying to isolate viruses from the fetuses which may be associated with abortion. . . . We hope that you will be able to help us in this project."[37]

Plotkin did indeed receive Payne's help, and that of the obstetrics and gynecology departments at other area hospitals. The most help, though, came from the rubella epidemic that was then bearing down on the eastern United States.

As the epidemic unfolded, Plotkin began to receive a steady stream of aborted fetuses. Some were delivered in plastic bags from area hospitals. Some he fetched himself, carrying them across the street from the university hospital in glass lab bottles. He would dissect and mince their organs, plant minuscule bits of tissue in culture bottles, put them in the incubator, grow them, and then test them for the presence of rubella virus. It worked to his advantage that rubella infected virtually every fetal organ.

When a fetus came from a patient whom Plotkin hadn't personally encountered, it's not clear whether the doctors involved asked the woman's consent for its use. When a fetus came from a woman whom Plotkin had seen and who had decided to abort based on his laboratory findings from her throat swab or blood, he did ask her permission for its use. In at least one case that led to knowledge that must, for the woman involved, have been wrenching.

"Dear Dr. Eisenberg," Plotkin wrote to a Cherry Hill, New Jersey, physician two days before Thanksgiving in 1964. "In answer to your letter regarding our studies on the conceptus of Mrs. [F.], please be informed that as of the moment, we have not yet recovered rubella virus from the foetus. . . . In Mrs. [F.]'s case the virus had probably not penetrated to the foetus."[38]

By the time he wrote that letter, Plotkin had received and tested thirty-one fetuses for rubella. He found that seventeen were infected with the virus.[39]

Of the thirty-one fetuses, it is fetus 27 that made its way into scientific history. It came from a twenty-five-year-old woman who was exposed to rubella eight weeks after she missed her period. Sixteen days later her lymph nodes were swollen and she developed a blotchy rash. A swab taken from her nose on the second day of the rash tested positive for rubella. Two weeks later the mother of fetus 27 had an abortion.

The fetus was dissected immediately, Plotkin would write in the paper that put the scientific world on notice that he was a contender in the vaccine race. "[Tiny pieces from] several organs were cultured and successful cell growth was achieved from lung, skin and kidney. All . . . were found to be carrying rubella virus."[40]

The kidney tissue from fetus 27 produced virus that grew particularly well—there was a lot of it present in the fluid bathing the cells. He named this virus RA 27/3. The "RA" signified "rubella abortus." The "27" denoted that its source was the twenty-seventh fetus he received during the 1964 epidemic. And he chose the "3" because the kidney was the third organ that he harvested from fetus 27. It was this virus that Plotkin chose to develop into a vaccine.

Like Hayflick before him, Plotkin grew the kidney cells from fetus 27 in glass bottles, bathing them in medium, placing them in the incubator at 95 degrees Fahrenheit, and splitting the cells into two new bottles when they covered the floor of the first one. When he had split the kidney cells into new bottles four times, he took the virus-filled fluid that bathed them and inoculated it onto fresh cultures of WI-38 cells that Hayflick provided.*

He repeated this inoculation of fresh WI-38 cultures three more times, every ten days or so. From the final set of WI-38 cells he harvested all of the yellow, virus-laden fluid and divided it into aliquots, distributing one-teaspoon amounts into screw-top test tubes and freezing these at –76 degrees Fahrenheit. It wouldn't be many months until he removed the vaccine, thawed it, and prepared to put RA 27/3 in human beings for the first time.

---

*Despite rubella's stunting effect on cell division, the kidney cells from fetus 27 did grow long enough in their lab bottles to be split four times. Plotkin surmised that they were capable of doing so because relatively few of the kidney cells were infected with rubella when he first planted the cells in bottles.

Rabies

Philadelphia, 1962–64
Morocco, 1964
Wabasha County, Minnesota, 1964

The currently available rabies vaccines are indisputably the crudest biological products injected under the human skin.

—Hilary Koprowski, speaking to the First International Conference on Vaccines Against Viral and Rickettsial Diseases of Man, Washington, DC, November, 1966[1]

Rubella was one among many. The chase for new virus vaccines in the 1960s was as hot as today's quest to unravel the profound mysteries of the human genome. Virology and vaccinology attracted the best and most ambitious scientists. And if they succeeded, vaccine inventors could expect to see the fruits of their labors saving lives not merely in their lifetimes but, in the best cases, within several years.

Its chief, Hilary Koprowski, was the most ambitious of the many ambitious virologists at the Grand Central Terminal of virology that was the Wistar Institute in the 1960s. And he was a man in a hurry when it came to one vaccine, for a disease that had preoccupied and fascinated him since he observed a rabid vampire bat—its brain was later dissected to look for the microscopic hallmarks of rabies—when he was working for the Rockefeller Foundation in Rio de Janeiro in the early 1940s. His visit with the eminent rabies scientist J. L. Pawan in Trinidad in 1944 only stoked that interest.[2] Koprowski was lured by rabies even in the midst of his race to beat Sabin to an oral polio vaccine in the 1950s. He published 21 papers on the terrifying disease in that decade alone—and 187 in the course of his lifetime, nearly four times as many as he wrote on polio.

Rabies is the most lethal infectious disease to afflict humans. Once a person develops symptoms, death virtually always follows.[3] Globally it is a serious burden. In 2013 the World Health Organization estimated that the virus kills one person every nine to twenty minutes, conceding simultaneously that this is likely an underestimate, due to underreporting and misdiagnosis.[4] A great number of the deaths are of children, most of them poor, and almost all of them are in rural areas of the developing world. In India, for instance, rabies kills somewhere between thirteen thousand and thirty thousand people every year.[5]

Rabies is a bullet-shaped virus carried in animal saliva. It is almost always contracted through a penetrating wound from a rabid animal.[6] (Very occasionally transmission by inhalation has been documented—for instance, in explorers of bat-infested caves.[7] And rarely people have contracted rabies by receiving corneal transplants from people dying of unknown causes that turned out to be rabies.)[8]

Unlike rubella, rabies virus does not invade the blood. Rather, it has a strong predilection for nerve cells. After a person is bitten, the virus migrates to the nerve cells at the site of the bite and travels up them, toward the brain. When it arrives there days, weeks, months, or even years later, it causes a uniquely horrifying set of symptoms. They begin mildly, with symptoms that can include a fever, a sore throat, muscle aches, and a loss of appetite. Then, as the virus disseminates rapidly within the brain, victims go downhill. Commonly they become hypersensitive to bright lights, loud noises, and touch. They become agitated, confused, and combative. They may experience hallucinations and convulsions. While lucid spells are typical, these grow briefer over time.

From the brain the virus also travels down nerve cells, away from the brain to the kidneys, lungs, liver, skin, and heart—and to the glands that produce tears, sweat, and saliva, which it puts into overdrive. At the same time, the virus can make swallowing hard and painful, which, in combination with the extra salivation, causes foaming at the mouth. In about half of affected people, the sight, sound, thought, or swallowing of water causes painful, jerky spasms of the diaphragm, throat, and respiratory muscles. Thus the terror of water that is seen in rabies and rabies alone has given the disease its nickname, "hydrophobia."

In some people the limbs become paralyzed and then the paralysis climbs until it reaches the respiratory muscles and suffocates the victim. The others lapse into a coma and die. The whole process, from the onset of symptoms, takes from five days to three weeks.

At the midpoint of the twentieth century, rabies started to appear with markedly increased frequency in wild animals in some U.S. states, alarming public health authorities and terrifying people in affected communities. In Indiana laboratory-verified cases of animal rabies jumped from 156 in 1952 to

374 in 1954. In New York State over the same two-year period, the number of confirmed cases of animal rabies grew from 340 to 511. In Minnesota, which had reported 5 cases of animal rabies in 1948, there were 264 cases of animal rabies in 1952. That was the year of a huge polio epidemic, but in the Land of 10,000 Lakes people were more afraid of rabies. Cases were verified in cattle, cats, gophers, groundhogs, muskrats, foxes, horses, and raccoons, in addition to the better-known carriers: skunks and dogs. And they weren't confined to rural areas. In Minneapolis a rabid skunk was captured at 55th and Colfax and a rabid cat at 42nd and Pillsbury. There were reports of children being chased by rabid skunks.

"We seem to be sitting on a powder keg here," remarked Dean Fleming, the state's director of disease prevention and control.[9]

In 1952 twenty-four people in the United States died of rabies.[10] Remarkably, none of them died in Minnesota.[11]

The rabies vaccine that was then available in the United States differed little from the crude formulation that the French microbiologist Louis Pasteur had famously invented in 1885.[12]

It was made from the pulverized, dried brains or spinal cords of rabies-infected rabbits, in which the virus had been killed with a chemical or with ultraviolet light. The vaccine was administered after a bite in a series of daily injections of increasing potency, in the abdomen, for fourteen to twenty-one days.

The injections were painful and risky. While the rabies virus in them was supposed to have been killed, it wasn't always. In Fortaleza, Brazil, in 1960, sixty-six people were injected with rabbit brain–produced vaccine in which the virus had supposedly been inactivated with the chemical phenol but in fact was still alive. Eighteen of them—all thirty years old or younger—developed rabies and died.[13]

Other people who received the brain tissue–generated rabies vaccine were weakened, paralyzed, rendered comatose, or killed by a debilitating allergic reaction to the nerve tissue in the vaccine.[14] Called allergic encephalomyelitis, it affected one in every five or six hundred vaccinees and resulted when the body's immune system, activated by exposure to the animal brain protein in the vaccine, attacked that same protein, called myelin basic protein, in the vaccinee's own brain.[15]

People bitten by animals that were only questionably rabid were therefore confronted with an agonizing choice: submit to the vaccine and risk a possibly fatal allergic reaction, or refuse and gamble with their life that the animal had not been rabid.

In the mid-1950s scientists working at the Eli Lilly company in Indianapolis took what seemed to be a step forward, developing a rabies vaccine in which the virus was grown in fertilized duck eggs—duck embryos—before being killed with a clear, colorless liquid chemical called ß-propiolactone.[16] "The use of [duck embryos] in place of rabbit brain for propagating the virus eliminates the myelin," the authors wrote, referring to the substance that causes the severe allergic reaction. In making the vaccine, the virus-ridden embryos were harvested a few days before they hatched—and their heads were cut off. The result was a diminished risk of reactions and a vaccine offering "distinct advantages."[17]

Like its predecessor, the duck-embryo vaccine had to be injected into a bite victim beginning as soon as possible, for fourteen to twenty-one days running, depending on the severity of the bite. The initial course of injections was followed by booster shots at intervals. And while it only rarely provoked allergic reactions after it was brought to market by Eli Lilly in 1958, it would emerge that the duck-embryo vaccine too had a serious problem: it was not as effective at generating antirabies antibodies as vaccines made in rabbit brains.[18]

In the meantime the wild animal problem wasn't going away. In the United States in 1961, there were nearly 2,700 confirmed cases of rabies in animals, not counting pet cats and dogs—nearly triple the number that there had been twenty years earlier.[19] The associated human deaths—121 between 1950 and 1961—didn't come anywhere near the fatalities due to, for instance, measles or meningitis.[20] But they were invariably horrifying in their particulars, and, as is the case in the developing world today, the virus disproportionately killed children: half of the U.S. victims between 1946 and 1965 were fifteen years old or younger.[21]

From practically the moment that Hayflick derived the WI-38 cells in June 1962, his boss, Hilary Koprowski, planned to use them to try to invent a safe, effective rabies vaccine—a vaccine that would make for a badly needed improvement on the vaccines then available. A recent, important advance told him that using the WI-38 cells should be eminently possible: in 1958 a virologist named R. E. Kissling working in a CDC lab in Montgomery, Alabama, had shown that rabies virus didn't have to be propagated in ground-up animal brains and spinal cords, or even in whole duck embryos that were then guillotined: it could be grown in cell culture, and even in nonnerve cells.[22] By

November 1962 Koprowski had seen to it that rabies virus was being grown in WI-38 cells at the Wistar Institute.[23]

The heavy lifting of developing the vaccine—the long, workaday hours in the lab—he delegated to an amiable émigré and Polish cavalry veteran named Tadeusz "Tad" Wiktor. Wiktor had earned a PhD in veterinary medicine at the renowned National Veterinary School in Alfort, France. In 1955 he and Koprowski met at a course on rabies in Muguga, Kenya. At the time, Wiktor was working for the government Veterinary Service in what was then the Belgian Congo. There, it's said, he himself had been bitten by a wild dog and submitted to the long course of brain tissue–manufactured rabies vaccine injections.[24] Koprowski began a campaign to lure Wiktor to the Wistar, where he landed in 1961, as he turned forty years old.

Koprowski was normally a hands-off manager. But when it came to rabies, his keen interest led him to closely supervise Wiktor and another Wistar scientist, a stout, often-smiling doctor of veterinary medicine with slicked-back hair named Mario Fernandes, who had recently moved to the Wistar from the National Institute of Veterinary Investigation in Lisbon, Portugal.

At first the men struggled to get the virus to grow in the WI-38 cells. When they bathed fresh, uninfected cultures of the cells with the fluid medium from a previously infected culture, it was tough to infect the new cells. In 1963 they got past this problem by using trypsin, the "jackhammer" enzyme, to break up the cell sheet in an infected bottle before moving the *cells*, not just the fluid bathing them, into fresh bottles of uninfected WI-38 cells. It emerged that the virus was passed by cell-to-cell transmission and not via the fluid medium.[25]

Their success was such that by September 1963, scarcely one year after Hayflick launched the WI-38 cells, Koprowski was in a position to deliver an upbeat report to a group that could readily appreciate what a new rabies vaccine would mean: scores of virologists from eighteen countries who had converged at the Croatian resort town of Opatija for a symposium on the new human diploid cells and their potential for viral vaccine making.

Koprowski reported to the assembled scientists that he and his colleagues had succeeded in adapting rabies virus to grow vigorously in WI-38 cells. After the WI-38–grown virus was killed with the colorless liquid chemical ß-propiolactone and the resulting vaccine was tested in mice, it protected the animals when they were later injected with live rabies virus. In fact, its protective effect far exceeded the minimum level that regulators required in this mouse test.[26]

Mice are not men, but the signs were all good. "The production of effective and safe antirabies vaccine for man now seems feasible," Koprowski declared.

On May 8, 1964, an eight-year-old boy named Billy (a pseudonym), the son of a member of the U.S. Air Force stationed in Rabat, Morocco, was bitten on the face by his own dog, a stray German shepherd that the family had adopted off the street one year earlier. They had not immunized the dog against rabies and had let it roam at will. The boy's father shot the dog and took his son to the hospital at the U.S. Naval Air Station at Kenitra, thirty-three miles from Rabat. They were at the hospital within three hours of the attack. Doctors there counted eight muscle-deep puncture wounds in Billy's left cheek and chin. They cleaned the wounds and injected a dose of U.S.-manufactured duck-embryo rabies vaccine into the sixty-pound boy's abdomen.[27]

The next day, nineteen hours after the attack, Billy was back at the hospital as instructed. He was injected with another dose of duck-embryo vaccine. Doctors also injected Billy's wounds, and his buttocks, with antirabies antibodies from horses. This is called "passive immunization" and gives victims ready-made immunity to protect them until their own immune responses, prodded by vaccination, kick into full gear. Passive antibody injections are a crucial part of postbite rabies treatment to this day. (In rich countries human antirabies antibodies have replaced those from horses, which frequently provoke allergic reactions.)

Two days after the attack the dog's brain was examined at the Institut Pasteur in Casablanca. Looking at the canine brain cells under the microscope, pathologists found them scattered with deeply purple-staining, circular densities. They were called Negri bodies. They were full of virus, and they are the microscopic hallmark of rabies.

For twelve more consecutive days, Billy returned to the hospital and was injected with duck-embryo vaccine. Ten and twenty days after these injections were completed, he received booster shots. Then, in mid-June, he resumed the life of an eight-year-old boy.

RABIES THREAT WORSENS ran a headline in the *Kingsport Times-News*, published in a town in the mountains of northeastern Tennessee, on May 9, 1964, one day after Billy was bitten in Morocco. "Rabies among the grey fox population of Hawkins County is getting worse, with nine confirmed cases being found among the critters during the past month," the article began.

"The situation is getting so bad here that a health department official is advising persons walking in hilly or wooded areas to carry clubs to defend themselves with."[28]

One man reported that a gray fox had come out of hiding underneath his front porch and tried to attack him. The animal escaped after a fight with the man's three dogs. They were unvaccinated and had to be put down.[29]

In Greene County, directly south of Kingsport, some 1,100 unleashed dogs would be shot and more than one hundred people bitten by rabid foxes in the coming months. Farmers carried shotguns. When the days shortened later that year, schools began to open late, so that children would not have to wait for buses before daybreak.[30]

In 1964 cases of animal rabies in the United States jumped by 20 percent to nearly 4,800. The increase was driven by wild animals. While rabies in domestic dogs and cats was falling steadily as more people vaccinated their pets, that decline was more than offset by rabid skunks and foxes in particular, along with other wild animals. Wild creatures accounted for 4,155, or 86 percent, of the animal cases that year.[31] Some scientists speculated that a decline in the hunting of animals for pelts had caused their populations to boom and provided a large reservoir for rabies.[32]

In Mississippi, where no animal rabies had been reported in the prior two years, fifty-seven residents were bitten by rabid bats—which were also identified in forty-three other states. In southern Georgia, where no cases had been reported for seventeen of the previous eighteen years, 107 rabid raccoons were identified. Northeastern states including Maine, New Hampshire, and Connecticut saw their first significant increases in animal rabies in two decades.[33] "Frankly, we're faced with a problem for which we see no solution," James H. Steele, the chief of the veterinary section at the CDC, told the *Wall Street Journal*.[34]

In the same article Koprowski was quoted saying that he was developing an improved vaccine made in human cells and was tentatively preparing to test it in seventy "volunteer" inmates at the state prison in Vacaville, California. It appears that this trial never came to fruition, for there is no report of it in the medical literature.

On August 3, 1964, three months after his dog attacked him in Rabat, Billy became irritable. He complained of a sore throat and ran a low-grade fever. A doctor looked at his throat, saw that it was red, took a swab to see what bacteria might grow from it, and prescribed penicillin.

Over the next three days Billy became more testy. He withdrew. He lost his appetite. He reacted to changes of temperature with cries or muscle spasms. He became afraid of water and swallowed the penicillin pills without it. He complained that it hurt when he swallowed.

On the third day of these symptoms, his parents took him back to the hospital at the Naval Air Station, where he was admitted. He was alert, but he didn't want to be touched. When a doctor tried to examine him, he withdrew forcefully.

Over the next two days Billy deteriorated. He refused to drink anything. He began salivating excessively. He wouldn't swallow the extra saliva, so he drooled. Forty hours after he arrived at the hospital, he got on his hands and knees on his hospital bed. He told the nurses that this relieved the irritating feeling of the sheets on his skin. He began trying to bite his attendants and kept trying, even after being sedated. He bled into his digestive tract. He battled severe muscle spasms and struggled to breathe. Finally he had several uncontrollable seizures and died, forty-seven hours after arriving at the hospital and ninety days after his dog attacked him.

An autopsy was performed. The experts at the Institut Pasteur in Casablanca found no Negri bodies in Billy's brain. (They are missing in at least 20 percent of rabies cases, and their absence doesn't rule out the disease.) So they took his fresh brain tissue and injected the brain of a rabbit with it. Sixteen days later they sacrificed the rabbit and looked at its brain cells. There they found Negri bodies.

The doctors who later described Billy's case in the *Annals of Internal Medicine* did not declare it an open-and-shut case of vaccine failure. Perhaps the vaccine had been potent when it was shipped from the United States to the Naval Air Station in Morocco but was damaged in transit. Perhaps nineteen hours had been too long to wait to inject the horse antibodies at the site of the wounds, and the virus was already making the relatively short journey from Billy's facial nerves to his brain by then; some experts believed that, once in the nerves, the virus became inaccessible to attacking antibodies. Perhaps the horse antibodies had destroyed enough of the rabies virus in the duck embryo vaccine that Billy did not generate a powerful immune response of his own. Such interference with the vaccine by the horse antibodies had been shown to occur in humans, which is why at least fourteen injections of the vaccine and two booster doses were required. Perhaps Billy should have had more than fourteen injections, or his booster doses should have been spaced out further in time after the initial injections, keeping his immune system gunning. They would never know.[35]

While Billy was dying in Morocco, a ten-year-old boy named Gary Sprick was camping in a tent on his family's farm in Wabasha County, Minnesota, with his sister and four-year-old nephew. On August 5, as he slept, a skunk came into the tent and bit him on his right wrist and on the pointer and pinkie fingers of his left hand. The skunk would not let go until Gary's sister shone a flashlight at it. It escaped. The resulting bites, doctors would later note, "were not clean puncture wounds but appeared to be chewed."[36]

Gary was injected with duck-embryo vaccine on the day that he was bitten and for thirteen more days running. It is not recorded whether he also received an injection of equine antirabies antibodies. On August 25, twenty days after the attack, Gary noticed that his right forearm was numb. He developed a fever and his muscles ached. He complained that his neck was stiff. Three days after his symptoms began, he was admitted to a hospital in Rochester with paralysis that was climbing up his body from his limbs. He was hallucinating, uncoordinated, and running a fever of 104 degrees Fahrenheit. He died on September 1.

One year later the CDC published an assessment of Gary Sprick's death. It noted that the incubation period of his disease, the time from the bites to the first symptoms, had been, in the context of rabies, very brief—twenty days. "As experience has shown that rabies vaccine is usually effective in preventing rabies only when . . . the incubation period exceeds thirty days," they concluded, "this is not considered to be a case of vaccine failure."[37] The circularity of that reasoning apparently didn't faze the authors.

The agency's opinion had changed by 1966, when it published a similar report involving another ten-year-old boy; another backyard tent, this one in South Dakota; another rabid skunk; another timely series of twenty-one duck-embryo vaccine injections; and another agonizing, protracted death.

The South Dakota boy's demise, the CDC authors wrote, showed the current therapies' stark limitations. "In spite of nearly ideal management including thorough cleansing of the wounds, [injection of the wound with antirabies antibodies from horses], and a full course of vaccine, the patient developed rabies in less than 30 days from the time of the bite."[38]

In 1964, the year that rabies in wild animals surged to the highest level in at least a generation, roughly thirty thousand people in the United States were vaccinated after encounters with potentially rabid animals; the CDC would later estimate that only one hundred to two hundred of the animals were actually rabid.[39] Among the bite victims Gary Sprick was the only person to die. In 1965 just one American died after being bitten by a rabid dog: a sixty-year-old West Virginia woodsman who had also received a prompt fourteen-dose course of vaccine.[40] The boy in South Dakota, who also promptly received the duck-embryo vaccine, was the only person to die of rabies in the United States in 1966. Clearly the duck-embryo vaccine worked in many cases. But for several heartbroken families, it didn't work often enough.

In September 1964, the month that Gary Sprick died, Koprowski, Wiktor, and Fernandes published for wider circulation, in the *Journal of Immunology*, the exciting findings that Koprowski had presented one year earlier at the symposium in Opatija, Croatia, showing that they had developed a rabies vaccine using WI-38 cells, and that it successfully immunized mice.[41] By that time they had tested the new vaccine in a species much closer to humans: fifty rhesus monkeys at the National Center for Primate Biology at the University of California at Davis.[42] The results were encouraging. All of the animals promptly developed antirabies antibodies, and the levels of those antibodies peaked at fourteen days after vaccination. After this they declined just slightly, then remained consistently high compared to control animals for the three months that Koprowski's team followed them before publishing the paper.[43] Of course, the vaccine hadn't yet passed the only test that mattered: protecting the animals when they were injected with real-world, disease-causing rabies virus.

That September Koprowski also filed with the U.S. Patent and Trademark Office an application for a patent on a "method of producing rabies vaccine."[44] It named him, Wiktor, and Fernandes as the inventors of a new vaccine that was both more effective and safer, by virtue of being grown on the human fetal cells that Hayflick had provided.

Because a grant from the NIH's National Institute of Allergy and Infectious Diseases had supported the trio's work to develop the improved rabies vaccine, the Wistar Institute would normally have been prevented from applying for a patent: in this era the U.S. government claimed ownership of all inventions discovered using its funding. However, the government did grant waivers on a case-by-case basis, if U.S. officials could be convinced that allowing an institution like the Wistar to patent an invention and then license it to a company was the only way to speed it to market for the public good. In the case of the potentially much-improved rabies vaccine, the United States granted a waiver, allowing the Wistar to become the owner if its application was successful.[45]

A patent would give the Wistar the exclusive right to make and sell the new vaccine for seventeen years going forward. Of course, the Wistar wasn't a vaccine company. But a patent would also give the institute the right to negotiate licenses with companies to make and sell the new vaccine—in exchange, they would have to pay royalties to the institute. And as director Koprowski would have the power to steer a portion of those royalties to the personal pockets of the inventors: himself, Wiktor, and Fernandes.

Hayflick was not named on the rabies vaccine patent application. In fact, he was completely unaware of it.

CHAPTER TWELVE

Orphans and Ordinary People

Philadelphia, 1964–65
Toms River, New Jersey, 1964–68

The principle of medical and surgical mortality, therefore, consists in never performing on man an experiment which might be harmful to him to any extent, even though the result might be highly advantageous to science, i.e., to the health of others.

—Claude Bernard, French physiologist, 1865[1]

The St. Vincent's Home for Children, at 6900 Greenway Avenue in southwest Philadelphia, was made of red brick, stood three stories tall, and took up most of a city block. Its two symmetrical wings were marked by long rows of rectangular windows. Broad steps led from the sidewalk to two sets of swinging front doors. The pediments high above these were crowned with stone crosses, for the Roman Catholic Archdiocese of Philadelphia owned and operated the home, where the cornerstone had been laid in 1937.

While many called the St. Vincent's Home an orphanage, not all the parents of the sixty-five children who lived there in November of 1964 were dead.[2] Some were sick, or destitute, or in jail, or heading that way. Or they were unmarried girls and young women who had chosen, or been forced, to give up their babies.[3] Many of those young mothers gave birth at the home and hospital for unwed mothers across the lane, where more than four hundred babies would be born in 1965.[4] Those babies born at St. Vincent's Hospital for Women and Children who weren't adopted or placed with foster families by the time they turned one year old—often black or mixed-race infants who were difficult to place for adoption—were transferred the fifty yards to the Home for Children.[5]

Five nuns staffed St. Vincent's Home, helped by a corps of hired child-care workers who dressed and fed the toddlers, changed diapers, played with the children, and bathed them each evening, assembly-line fashion. There were also two cooks, two adopted stray dogs named Jamie and Steve, and a grumpy maintenance man named Mr. Messina.[6]

The nuns belonged to the Missionary Sisters of the Precious Blood and wore friendly if impractical white habits up whose wide sleeves more than one stream of pee found its way during diaper changing. They lived in simple single rooms on the third floor and worshipped in a small chapel on the ground floor.

The rest of the lower two floors housed the children. It was a Spartan place, with hard terrazzo floors and stall-less bathrooms with tiny toilets. The nuns tried to make up in love what the building lacked in physical warmth, making sure the kids got out every day on the small playground surrounded by chain-link fence, reading to them at story time, and walking them to nearby Cobbs Creek Park. Sister Agape, the administrator, was a fierce advocate for the children. She arranged a week at a beach house in the summer and used the fact of occasional visits by state inspectors to squeeze every penny she could out of the archdiocese for things like new bedspreads that might brighten the place. She was known for not countenancing staffers who were only there to collect a paycheck.

Still, the nuns worried about the children. Sister Damiane, a petite, Austrian-born nun in her late twenties, mused that they were *children*, after all, and there were so many of them. Every one of them needed one or two adults to belong to—a level of attention she couldn't possibly provide.

There were crushing moments: The day that a foster family arrived to take one boy away and he unraveled in a cacophony of desperate screams and wails. The unforgettable, lost look that a nun named Sister Mary Joseph observed on the face of the deaf girl, E., when she discovered that the beautifully wrapped present that Archbishop John Joseph Krol handed to her at the annual Christmas party at a downtown hotel was in fact just a decoration—an empty box.[7]

It was from the bald, bespectacled Archbishop Krol, who had succeeded John Cardinal O'Hara in 1961 as Philadelphia's top prelate, that Stanley Plotkin got a green light to study his new RA 27/3 rubella virus in the children living at St. Vincent's Home.[8]

In his letter requesting Krol's permission, Plotkin did not explain to the archbishop that he had captured the vaccine virus from one aborted fetus and grown it in cells from another. Krol was passionately antiabortion. In 1973 he would call the Supreme Court's *Roe v. Wade* decision striking down state criminal abortion laws "an unspeakable tragedy for this nation that sets in motion developments which are terrifying to contemplate."[9]

In 1964 the archbishop gave the rubella study the go-ahead.

"The subjects were 31 healthy, normal children, aged 14 to 29 months (average, 21 months), from an orphanage supervised by the Archdiocese of Philadelphia," Plotkin would write in the paper that resulted. "Permission to include the children in this study was obtained from parents or guardians."[10]

As for U.S. government clearance, Plotkin was uncertain whether he even needed it. Products that were clearly vaccines did need formal approval to be tested in humans, from Roderick Murray's DBS, in the form of a piece of paper called an "Investigational Exemption for a New Drug." To win this approval, a product must have passed muster in lab and animal tests specified by the DBS.

In May 1964 Plotkin wrote to the DBS, calling RA 27/3 "non-vaccine material" that he intended for a trial "with the object of producing experimental rubella." Would he need clearance from the DBS to run the trial, he asked? And if so, what safety tests would the division require in order for RA 27/3 to win that clearance? Could he receive a quick response? "We are anxious to begin tests with this material as soon as possible."[11]

The answer came back within a week. "The dividing line between this kind of activity and actual development of a [vaccine] product is not clear," wrote Murray. Many labs that were early in the vaccine-making process did file their proposed trial procedures, called protocols, with the DBS, since at any time their work could "take on aspects of the development of a vaccine," he added. But no, formally, Plotkin did not need to jump through those hoops. Whatever he did, "you would, of course, be guided by ethical considerations."[12] "You should protect yourself," Murray suggested, by making sure that the proposed project was reviewed ahead of time by an "objective group of advisers." Finally, Plotkin should follow the recommendations of the NIH expert committee that had pronounced on the human diploid cells in the pages of *Science* the previous year.[13]

That committee was the one that Murray had put together within a month of the publication of Hayflick and Moorhead's 1961 paper describing the new, normal lab-dish fetal cells and their potential for vaccine making.

The committee had been exceedingly cautious in its report published in *Science* in January 1963, warning that the human diploid cells might harbor hidden viruses that could cause cancer or otherwise harm vaccinees. It had urged that new vaccines made in them should first be put into one or two adult volunteers, who should then be watched carefully for at least six months in case the vaccine was infected with hepatitis or another virus with a long incubation period. Murray added that Plotkin needn't worry about these recommendations if "you already have data from other studies with diploid cells to meet these [committee] suggestions."[14]

Plotkin considered that he did have that data. It was just one year earlier that the quiet, determined Yugoslavian vaccinologist, Drago Ikić, had immunized five thousand Croatian preschoolers against polio with a vaccine made using WI-38 cells—to no apparent ill effect.

As for summoning an "objective group of advisers" to preapprove the study, Plotkin didn't do that. But it is probable, he wrote in 2016, that he consulted with Hilleman, Merck's vaccine chief, and with Robert Weibel, a pediatrician at the University of Pennsylvania who worked closely with Merck running vaccine trials.[15]

When Plotkin wrote to the DBS, he stated that he was using "non-vaccine material." This was because he didn't want Murray and his staff defining the RA 27/3 virus as a vaccine, giving them license to interfere with his trial, require lab and animal testing, and generally slow him down.[16] In truth, he deeply hoped that RA 27/3 was a vaccine in the making. But he also knew that the virus was only very mildly weakened at this point. He had grown it through just four passages on WI-38 cells, hardly enough generations adapting to lab life for it to lose its punch for human beings. Rather, it would likely give the toddlers full-fledged rubella. This would prove beyond a doubt that the agent that he had captured from the kidneys of fetus 27 was indeed the rubella virus. Not that he doubted this. But others might.

The trial would also allow him to study the spread of rubella infection from child to child and the immunity produced in the toddlers by the experimental infection, Plotkin wrote that spring to his colleague, Theodore Ingalls.[17] Ingalls was a well-known epidemiologist at the Henry Phipps Institute (part of the University of Pennsylvania) who was supporting Plotkin's work by funding the salary of a lab technician.

Plotkin was not the first scientist to set out to cause rubella in human beings. In 1938 a pair of Japanese researchers, Sadataka Tasaka and Y. Hiro, produced rubella by injecting children with nasal washings from people with the active disease, demonstrating that rubella was almost certainly caused by a virus.[18] In 1953 Saul Krugman, a pediatrician at New York University, gave one- to six-year-old children rubella by injecting them with blood drawn from infected children.[19] And in 1962 John Sever, the director of virology at the NIH's National Institute of Neurological Diseases and Blindness, produced rubella in male prisoners by injecting virus captured from a sick patient into their noses, using swabs and medicine droppers.[20]

As with other similar studies in this era, these rubella-inducing studies did not elicit protests. And by the time Plotkin set about infecting children, the country was in the midst of an unprecedented rubella epidemic. Just as the clarion call of endangered soldiers on the front lines had made medical researchers willing to use people in unimaginable ways during World War II, in 1964 the urgency of the quest for a rubella vaccine in the face of a devastating epidemic consumed Plotkin and his contemporaries, causing them to lay aside any queasiness about testing their vaccines on powerless, institutionalized children. They also worked with this fact easing their minds: German measles—unlike polio and unlike the hepatitis with which the DBS's Murray had purposely infected prisoners—is generally a mild disease if one is not a fetus. Its one documented, fatal complication—the brain inflammation called encephalitis—was rare in adults and extremely rare in children. So, Plotkin says, he did not worry as he launched the first human trial of RA 27/3.[21]

St. Vincent's Home for Children had an ideal layout for Plotkin's purposes. The two self-contained wings meant that the children in this first study—one-year-olds who lived on East 2 and two-year-olds who inhabited West 2—could be kept isolated, both from the children on the other wing of their floor and from the older children downstairs. But within East 2 and West 2 the toddlers would continue to eat, sleep, and play together.

"They slept in one of two common sleeping areas each measuring approximately 256 [square feet] and containing eight cribs," Plotkin reported.[22] If the RA 27/3 virus was infectious enough to spread from vaccinated toddlers to the unvaccinated controls within each group, the close and self-contained quarters of East 2 and West 2 were the perfect place to observe it.

Such spread of the virus was a top-of-mind concern, since small children were potential targets for a rubella vaccine, and if vaccinated kids spread the vaccine virus to their pregnant mothers, the results could be disastrous. Before the trial began, Plotkin tested the blood of all the female caregivers on East 2 and West 2—nuns and laywomen alike—for antirubella antibodies. All were already immune to rubella, he wrote in the paper that resulted from this first study at St. Vincent's Home.[23]

Plotkin planned to test two methods of vaccine administration: injection under the skin and intranasal administration, which meant squirting liquid vaccine into one nostril with a dropper "while the child was forcibly restrained." Since in naturally occurring infection the virus colonizes the nose and throat before invading the blood, it stood to reason that intranasal vaccination might be feasible. What's more, it might be cheaper, since it would take less skill to administer. And it would almost certainly cause less crying.

Plotkin ran safety tests before he took RA 27/3 to St. Vincent's Home. He tested the vaccine for the presence of bacteria, fungi, and the ubiquitous *Mycoplasma* and found it clean.[24] He also injected it into animals, in the abdomen and in the brain: twenty baby mice, twenty adult mice, and six guinea pigs. He injected ten rabbits subcutaneously. All the animals stayed healthy for six weeks.[25]

By the autumn of 1964, Plotkin felt ready to put his vaccine in the toddlers at St. Vincent's Home. He ran the trial there in two parts, each six weeks long, separated by an interval of ten weeks, which included a break for Christmas. Children received an injection in the shoulder or a squirt in the nose. Some received a higher dose of the virus and some a lower dose. Other children received nothing, serving as controls.

During each trial all the children, vaccinees and controls alike, were examined every day for fever, rashes, swollen lymph nodes, and other signs of illness and had throat swabs taken every three or four days. Their blood was drawn at the beginning and the end of the trial.

Every few days Plotkin drove the three miles down Woodland Avenue to the Home for Children, where he examined the toddlers and collected throat swabs. On the days that he wasn't there, others took temperatures and checked for rashes. They may have included Sister Mary Joseph, who was a trained nurse, and the archdiocese's staff doctor, Horst Agerty.

Virtually all of the small inhabitants of East 2 and West 2 participated in the trial. Almost none of them had antirubella antibodies before the study began.[26]

Plotkin learned several important things from the trial. From the toddlers who had no rubella antibodies before the trial began he learned that his virus, given as an injection, caused rubella. Nine of the eleven antibody-lacking children who received injections developed a rubella rash. Most had swollen glands and half ran a fever of up to 103 degrees. All of them showed rubella living in their throats for about a week before and a week after the rash appeared. And their rubella spread. Four unvaccinated children were infected by the kids who had received the shots in the shoulder. Children who got the higher-dose injection got sicker, and they were the only ones who spread the virus to unvaccinated children.[27]

RA 27/3 seemed to have an immunizing effect. All eleven of the children who lacked antibodies before the trial began and who received injections developed antibodies. And these antibodies were apparently protective: three of the toddlers who got jabs in the shoulder during the first trial were injected again on day one of the second trial. This time they didn't get rubella. The antibodies generated by the first injection had shut down the virus when it was injected a second time.

The intranasal vaccine, by contrast, didn't cause illness or spread it—but neither did it generate antibodies in nine of the eleven toddlers who received it.

Plotkin submitted his paper to the *American Journal of Diseases of Children* in early June 1965, and it was published that October. His coauthors—included largely as a courtesy, as Plotkin himself had done most of the work—were Ingalls, the University of Pennsylvania physician who had funded his lab technician's salary, and David Cornfeld, a well-liked senior pediatrician at the Children's Hospital of Philadelphia, who, Plotkin says, smoothed his entrée to St. Vincent's.

There are tables in the 1965 paper and diagrams that lay out how the RA 27/3 virus was developed. And there is one full-page photo. It is a close-up of a child who received a lower-dose injection. It shows his or her naked back and the back of a chubby left arm. There is a head of soft brown curls just visible at the top of the photo. In the middle of the child's back there is a heart-shaped patch of white skin. The rest of the skin is covered by a bright red rash.

The experiment had succeeded but also failed, Plotkin wrote. RA 27/3, when injected, generated protective antibodies but caused disease, and spread it too. And while the disease it produced was "a mild illness" with no complications and no severe symptoms, "only an agent unable to spread to contacts and, therefore, unable to endanger pregnant women" would make an acceptable vaccine, he concluded.

Plotkin made clear his intention to keep moving forward. The results, he wrote, were a baseline. From here he would further weaken the rubella virus, with the goal of inventing a workable vaccine.

In Plotkin's view the work couldn't happen quickly enough. The threat of the next epidemic hovered five or six years off, if history was a guide. In the meantime the heartbreaking results of the one that had just ended seemed to be everywhere.

Steve and Mary Wenzler were all about family. The newlyweds joked that they would have nine children. They had the energy. And they had the time, back in the early 1960s. Both were schoolteachers in their early twenties, settling into life in Toms River, New Jersey, an Atlantic shore town fifty-six miles due east of Philadelphia.[28]

The jobs were what had drawn the Wenzlers to Toms River from the busy suburbs of Trenton, where they had grown up. Thanks to Toms River Chemical, a huge Ciba-Geigy plant that opened in 1952, the town of seventeen thousand was growing so quickly that school construction could hardly keep up. So Steve—officially Stephen Joseph Wenzler III—had taken a job as band director at Toms River High School, and Mary was teaching music at Central Regional Junior High School in nearby Bayville.

During their first year of teaching, the Wenzlers also began working on the family project. In the autumn of 1963 Mary had a miscarriage. That was common. She wasn't worried.

Then, one morning in March 1964, Mary found a bulletin from the school nurse in her mailbox in the junior high school office. It announced that an epidemic of German measles was hitting the school. She recalls the bulletin warning that a big uptick in student absences could be expected, if teachers hadn't noticed it already. Any student who was feverish or had a rash should be sent home. Any teacher with symptoms should not come to work.

This was the first that Mary had heard of an epidemic. It wouldn't have mattered, except that her period was slightly late and she had a gut feeling that she was pregnant. She knew what that felt like, after the miscarriage. She also knew, she does not remember how, that German measles during pregnancy was not a good thing. Some 125 seventh- and eighth-graders trooped through her music classes every day. It seemed highly likely that she had been exposed.

That evening Mary and Steve discussed what to do. She hadn't yet connected with a doctor in Toms River. They decided that she should see Vincent Pica, her trusted family doctor from her former life in Hamilton Square, a Trenton suburb.

A day or two before her appointment with Pica, Mary developed a mild fever and a light red rash on her face and called in for a substitute. Worry gnawed at her as she drove the thirty-six miles to Pica's office in Hamilton Square.

As Pica heard her story and examined her, Mary saw worry in his eyes too. He gave her a shot in her rump. It was gamma globulin, an injection of pooled antibodies from blood donors. The aim was to confer the same kind of ready-made immune protection that the horse antibodies conferred when given immediately to people bitten by rabid animals—in this case to protect the fetus from infection by killing off the rubella virus in a woman's bloodstream before it penetrated to her womb. But with rabies there was a vaccine that was also given, to kick the victim's own immune system into high gear, producing its own antibodies. With rubella there was no vaccine to give along with the gamma globulin.

Whether gamma globulin alone was actually effective against rubella was painfully uncertain. And there was no way to answer the question. Researchers couldn't run a prospective, controlled clinical trial that would deny the gamma globulin treatment to some pregnant women and give it to others. Pregnant women would not put up with that. They were not orphans or prisoners who could be easily manipulated. So the scientists had to draw what conclusions they could from case histories.[29]

The British physician J. C. McDonald examined the medical records from nearly thirteen thousand pregnant women given the injection between 1954 and 1961 after being exposed to rubella. The evidence suggested that gamma globulin effectively prevented rubella in these women, he wrote. But it was not conclusive. What was more, it was well known that the virus frequently invaded the blood without causing visible sickness. Therefore, preventing manifest illness in the mother might not be synonymous with protecting the fetus.[30]

So when Mary Wenzler visited her longtime family doctor in the spring of 1964, the state of the art was this: physicians gave gamma globulin because it couldn't hurt and it might help. And patients hoped.

After she missed a second period, Mary made an appointment with Alfred Pietrangelo, a local family doctor who delivered babies. He confirmed her pregnancy. She told him she had had rubella and the gamma globulin injection. Unprompted, he said that he didn't do abortions and that if she was interested in ending her pregnancy, she would have to find a different doctor. She told him that she didn't believe in abortions either. She said she wanted him to see her through this pregnancy and deliver the baby.

On Tuesday, November 24, 1964, at 3:33 p.m., after an uneventful labor, Stephen Joseph Wenzler IV was born at Toms River Community Hospital. Baby Stephen weighed five pounds, thirteen ounces and was nineteen inches long. (This was less than the 10th percentile for weight and about the 17th percentile for length.) The delivery-room staff soon noticed something that they did not tell Stephen's parents: there were milky white opacities where the black pupils of the baby's eyes should have been. They didn't give Mary the baby to hold.

Mary still hadn't seen baby Stephen when, that evening, as she lay flat in a shared hospital room, a thin, young ophthalmologist came to see her. He stood by the bed looking down at her and told her that her baby had cataracts blinding both eyes.

Mary's exhaustion didn't blunt the shock. "What do we do?" she finally managed to say.

"If I were you, I would get pregnant again, and then it won't bother you so much," he said.

Fifty years later Mary doesn't remember the rest of the conversation. She does remember the feeling of hopelessness that overwhelmed her and the kind nurse with a European accent who, after the ophthalmologist had gone, comforted her while she cried.[31]

Mary spent Thanksgiving Day 1964 in the hospital. She and Steve took baby Stephen home that weekend and set him up in a bassinet beside their bed. He nursed and slept and cried when he was hungry. When his eyes were open, they gazed emptily into space.

Mary tried to do normal mother things. She dressed Stephen in a blue velvet vest with his name embroidered on it and matching blue short pants—a gift from her sister-in-law. She wrote in Stephen's baby book: "I smiled when I was one week old." "I grasped at a toy at three months." "I cut my first tooth at ten months." Then she stopped writing entries.

Mary did something else soon after Stephen was born. She called her family doctor, Pica, the one who had given her the gamma-globulin injection. He had decided to leave general practice and happened to be completing specialty training in ophthalmology at the Newark Eye and Ear Hospital. He wanted Stephen's eyesight to be assessed by his boss, Richard Stern, just as soon as possible.

On February 1, 1965, Richard Stern, a tall, big-boned surgeon who despite his massive hands somehow performed microsurgery on infants, made a half-centimeter slit in one of ten-week-old Stephen's eyes and inserted a tiny, spoon-like instrument, which he used to scoop out the liquefied, milky substance that composed the cataract. The next month he did the same thing to the other eye.

Cataract surgery on rubella-stricken infants in the 1960s had a dismal success rate and was fraught with postoperative complications that weren't necessarily the surgeons' fault.[32] First, pediatric ophthalmology didn't yet exist as a subspecialty. Surgeons like Stern were adult ophthalmologists trying to adapt their methods.[33]

Second, they were almost always dealing with undersized eyes.[34] Rubella stunts the growth of the eyeball in the womb, just as it stunts other organs. That made the surgery technically tough.

What's more, the operative techniques then available were primitive by today's standards.[35] Even such basics as operating microscopes were only just becoming available. (Stephen's surgeon did have a microscope. That wasn't a given. A nascent group of pediatric ophthalmologists in Washington, DC, were reduced to borrowing a microscope from their ear, nose, and throat colleagues.)[36]

Finally, in many cases the surgery itself stimulated a chronic, vicious cycle of inflammation that affected other parts of the eye, including the iris, the cornea, and the retina. Many infants and children ended up needing additional surgeries, and some in the end had to have their entire eye or eyes removed. It was a brutal, depressing business.[37]

So it was with great trepidation that several days after his first surgery Mary and Steve watched as the bandages were removed from Stephen's eye. They had tied colored balloons on the rail of his hospital crib. Their baby brightened, waved his arms, and turned his head toward the balloons. He looked like he was trying to *focus* on something for the first time in his life.

The Wenzlers recall that when it came time for them to pay their bill, Stern charged them just $10 beyond what their insurer paid. "You are going to have enough challenges as it is," he told them. They would consist, he said, not so much in getting Stephen used to the world as in getting the world used to Stephen.

Stern sent the Wenzlers to an expert who crafted a tiny pair of glasses with ultrathick plastic lenses and round frames of muted gold. Stephen began wearing them everywhere.

Mary recalls that, at some point when he was a child, Stephen's sight measured 10/200 with his glasses on. This meant that he had to be ten feet away to perceive an object that a perfectly sighted person could see clearly at a distance of two hundred feet.

It was also during his first months of life that Mary began to suspect that Stephen couldn't hear. He didn't react when she came into a room. He didn't look up when she clapped her hands or banged the lid of a pot. He didn't turn his head at the sound of her voice. He reacted to her only when she touched his crib or touched him.

At a checkup with Dr. Pietrangelo, the Toms River family doctor who had delivered Stephen, Mary asked if her son might be deaf. Pietrangelo laid Stephen on his back on an examining table. He banged his hand, hard, on the table beside the baby. Stephen looked toward the noise.

There's no problem with his hearing, Pietrangelo told Mary, handing him back to her. Later Pietrangelo called Steve Wenzler to tell him that he had found a heart murmur while listening to Stephen's chest. The infant should be booked for a cardiac catheterization.

The Wenzlers did not go back to Pietrangelo. They asked Pica's advice on where to take their son for heart care. That July cardiologists at Deborah Hospital in Browns Mills, a tiny town in the center of the state, inserted a catheter into a vessel in eight-month-old Stephen's groin and snaked it up into his heart. Injecting dye, they saw the classic signs of a heart malformation that is common in congenital rubella. It was first flagged by Norman Gregg, the Australian physician who discovered congenital rubella syndrome; it was found in autopsies of three of his tiny patients and reported to him by other doctors.[38]

Patent ductus arteriosus (PDA) results when a fetal blood vessel—the ductus arteriosus, which links two major arteries leaving the heart—fails to close in the first few days of an infant's life, as it should. ("Patent" means "open" in medical parlance.) In the fetus, which is not yet breathing on its own, the ductus allows blood to bypass a high-pressure circuit through the fluid-filled lungs. But in the newly born infant, now breathing for itself, an open ductus results in an aberrant circuit in which some of the freshly oxygenated blood that should be pumped out to the body, where it's needed, instead is diverted to the lungs. The upshot is that the infant's heart has to work extra hard to keep the body supplied with oxygenated blood. With time, and in severe cases that are not surgically repaired, the extra work can push the little heart into failure, with blood backing up into the lungs and slowly, in effect, drowning the child. Babies with PDA have symptoms like feeding poorly, shortness of breath, a racing heart, and poor growth.

The Wenzlers don't remember that baby Stephen showed any of these signs. Mary does recall that he was slow to hit developmental milestones and that when he did so, he did so in his own way. He never crawled. Instead he lay on his back and propelled himself around the floor by pushing with his legs—possibly because this position allowed him to see better what little he could. Finally he started sitting up. And then, after his PDA was surgically repaired when he was not quite two years old, he began growing like a weed.

During his initial cardiac catheterization, the cardiologists also found that Stephen had multiple narrowings of branches of his pulmonary artery, the big vessel that carries blood from the right ventricle of the heart to the lungs for oxygenation. Along with PDA, this problem, called branch pulmonary artery stenosis, is the most common heart defect in children with congenital rubella.[39] Severe branch pulmonary artery stenosis requires surgeons to insert a balloon catheter and sometimes a stent to clear the blockages. When this fails, a heart-lung transplant may be the only option. In Stephen's case the narrowings did not cause symptoms, and the doctors adopted a wait-and-watch approach, as they still do today.

Just before Stephen's first heart procedure, his parents' life got even more complicated. Leonard, a brother for Stephen, was born in April 1966, when Stephen was seventeen months old. Mary was sick with worry during the pregnancy—irrationally, she told herself. But Leonard was born perfectly healthy.

The Wenzlers don't remember, because the time was so busy and so chaotic and so traumatic, when it was that they got Stephen to a specialist who confirmed Mary's suspicion that their son was profoundly deaf. What is clear is that Stephen was wearing hearing aids by the time he was three years old and that he heard, or perceived through vibrations, a very minimal degree of sound.

In September of 1968, when he was not quite four years old, Stephen Wenzler enrolled in the Beachwood Nursery School in Toms River. The preschool's director, Mrs. D'Arienzo, had no special experience with disabled children, but she said she was willing to try including Stephen with her gaggle of able-bodied three-, four-, and five-year-olds.

Mrs. D'Arienzo soon came to like Stephen a lot. He was very bright, she told Mary, and just wanted to learn and do everything that the other kids did. In this Stephen was lucky; one third or more of children with congenital rubella are intellectually disabled, the virus having invaded their brains in the womb, damaging blood vessels, starving the brain of oxygen, and often stunting its growth. Nearly one in four children have abnormally small heads. In up to 7 percent of children with congenital rubella, the brain damage causes autism.[40]

Stephen didn't fuss when Mary dropped him off for preschool—a good sign in a boy who didn't hesitate to let his displeasure be known. He brought home drawings, which Mary posted on the refrigerator. Perhaps most important, Mrs. D'Arienzo taught Stephen discipline: that he needed to follow the rules of the school just like everybody else.

Photos of Stephen in this era show a grinning brown-haired boy wearing thick glasses with heavy black temples, his chin tilted up as he tries to focus through the lower half of his bifocal lenses. He is thin, and his blue gray eyes are magnified by the thick lenses. In his ears there are hearing aids bigger than quarters and ten times as thick. Cords hang down from them, leading to a receiver he wears strapped to his chest.

Mary Wenzler has kept many photos. Here is Stephen as a baby, chin dimpled and cheeks like porcelain, tiny glasses perched on his nose, a faraway look in his eyes that says he is not focusing on the camera. Here are photos of Stephen with his brother, Leonard, in a child's wagon; here are the brothers riding on their dad's back in the backyard; here they are ringing a big black bell at a firehouse. Here is Stephen, four years old, in a bed at the Children's Hospital of Philadelphia, where C. Everett Koop, the pediatric surgeon who will go on to become President Ronald Reagan's surgeon general, has just repaired Stephen's hernia and undescended testicle—both additional manifestations of congenital rubella.[41] Here, one year later, is Stephen in a red gown with Mrs. D'Arienzo, clutching his diploma as he graduates from the Beachwood Nursery School.

What is striking in many of these early photos is the liveliness, even impishness, that projects from Stephen's eyes and from his grin—how very much like a typical boy he seems. Looking at them, it is hard to imagine that he inhabits a world of near darkness and near silence.

After a long and painful struggle, the Wenzlers found Stephen a place at the Perkins School for the Blind in Watertown, Massachusetts, whose most famous graduate, Anne Sullivan, taught Helen Keller to read and write. Stephen arrived there at age seven and remained there through high school.

He grew up to be an exceedingly bright and gifted man who would live a life of unrealized potential, frustrated ambition, and loneliness. Some of this would be the result of his own failings. But most would be attributable to a society that could not see what he might have been.

CHAPTER THIRTEEN

The Devils We Know

Philadelphia and Hamburg, Pennsylvania, 1965–67
Marburg, Germany, August–November 1967

*In retrospect, it is amazing that cells [freshly harvested from monkey kidneys]
were thought to be the safest choices. Considering all the microbial agents to
which animals are exposed, the subsequent events were predictable.*

—Stanley Plotkin, 1996[1]

As he finished the trial on the toddlers at St. Vincent's Home for Children early in 1965, Plotkin knew that he needed to weaken the RA 27/3 virus further so that it would stop causing rubella in vaccinees—but not so much that it would no longer cause a vigorous immune response. And he knew that he needed to waste no time doing it. He was a generous and outward-looking man; a timely vaccine might forestall the next epidemic and with it the heartbreak and the physical suffering he had witnessed at such close range in London and in Philadelphia. He was also competitive. And he was sure that his rivals, like him, were striving to develop a vaccine before anyone else.

Plotkin was correct. His competitors were pushing ahead just as quickly as he was. Maurice Hilleman, the tough Montanan who headed vaccine research at Merck, had already, in January 1965, completed extensive animal safety tests and was injecting the company's experimental rubella vaccine into intellectually disabled children living in institutions in and around Philadelphia, in northern Pennsylvania, and in Delaware. (The Pennsylvania Association for Retarded Children gave the in-state trials its blessing.)[2] For drug companies like Merck, beating the competition to a rubella vaccine would mean a big new market for a product that would be much in demand. It was not clear whether the CDC would recommend that a rubella vaccine be given to women of childbearing age or to small children as part of routine childhood vaccinations. But either group—the approximately 3.6 million children born annually in the mid-1960s or the roughly 39 million girls and women then of childbearing age—amounted to a huge number of customers.

This fact was not lost on other drug firms. Across the Atlantic in Genval,

Belgium, virologists at a company called Recherche et Industrie Thérapeutiques were hard at work developing their own rubella vaccine, with a virus captured from the urine of a ten-year-old girl who was ill with rubella. Soon a St. Joseph, Missouri, company named Philips Roxane, a subsidiary of Philips Electronics and Pharmaceutical Industries Corporation, would begin developing yet another competing vaccine.

And then there were the Division of Biologics Standards scientists. In the DBS's utilitarian redbrick building perched on a rise in the middle of the NIH campus in Bethesda, Maryland, Paul Parkman and his boss, Harry Meyer, were preparing to launch human studies of their "high passage virus," HPV-77 rubella vaccine—made from the virus that Parkman had captured from the throat of a young soldier at Fort Dix, New Jersey, and then weakened by growing through seventy-seven sequential cultures of African green monkey kidney cells.

Unlike Plotkin, Parkman and Meyer had obtained formal permission from the DBS to experiment with the HPV-77 vaccine in human beings. This permission, called an Investigational Exemption for a New Drug, was new in the regulatory world. It had been instituted as part of a 1962 tightening of prescription drug–approval laws that was spurred by a fresh tragedy.

In a horrible episode in the late 1950s, 1960, and 1961, the drug thalidomide was widely prescribed—or available over the counter—as a sedative and for morning sickness in pregnant women in Europe, Canada, Australia, Japan, and many other countries. As a result, thousands of babies were born with missing or flipperlike limbs and other birth defects. One U.S. regulator, a woman named Frances Kelsey who was a drug inspector at the Food and Drug Administration, single-handedly prevented thalidomide from being approved for market in the United States. It was, however, taken by hundreds of pregnant American women in company-sponsored drug trials. The congressional reaction to the thalidomide tragedy included a new legal requirement that drugs be proven effective as well as safe before they could be marketed.[3,4]

The DBS did not handle approvals for drugs like thalidomide; they were medicines and were separately regulated by the FDA. But it did regulate vaccines. Just one DBS employee—a chubby, good-natured, cigar-smoking physician named Joe O'Malley—handled, with the help of his assistant, all of the applications for permission to test new vaccines in people.[5]

Roderick Murray, the taciturn, inscrutable DBS director, briefly considered moving Parkman and Meyer to the NIH's National Institute of Allergy and Infectious Diseases, to get them out of the DBS and avoid the appearance of a conflict of interest.[6] But the pair didn't want to be slowed down by a move. And Murray didn't want to slow them down. With the pain and suffering of the 1964 rubella epidemic now in full public view in the form of thousands of damaged newborns, the DBS was under tremendous pressure to produce and approve a rubella vaccine before the next epidemic arrived.[7]

In the United States rubella spiked with regularity, every six to nine years. Before the record-breaking 1964 epidemic, the last substantial outbreaks had been in 1958 and 1952. Now 1970 began to loom as the year when another epidemic might well descend—the year by which a rubella vaccine needed to be in use. Murray decided that it would be enough to erect a "paper barrier"—the Investigational Exemption for a New Drug—to shield Parkman and Meyer from allegations of conflict as they went about developing their rubella vaccine in the very division that was the gatekeeper for the U.S. vaccine market.[8]

In October 1965 Parkman and Meyer launched the human first trial of the HPV-77 vaccine, at the Arkansas Children's Colony, a state school for the intellectually disabled in rural Arkansas.[9] There seven hundred students lived in small groups in widely scattered cottages, allowing them to be isolated for vaccine studies. "Children were selected for vaccination only after their parents or legal guardians had been fully acquainted with all the details of the project and had given written permission," Parkman, Meyer, and their colleague Theodore Panos wrote in the resulting publication.[10] (Panos was the chairman of pediatrics at the University of Arkansas School of Medicine, where Meyer had gone to medical school; it was through him that Parkman and Meyer arranged for the trial to be held at the Arkansas institution.)[11]

Decades later Parkman recalled that he and Meyer had also sought approval from the NIH Medical Board Committee, made up of officials of the several NIH institutes and of the Clinical Center, the huge research hospital on the NIH campus. The board could be consulted when NIH scientists were undertaking "any nonstandard, potentially hazardous procedure."[12] However, whether to enlist the board was left to the discretion of the scientists undertaking a study. Parkman and Meyer's proposed trial was unusual in that it was being conducted off campus, not at the Clinical Center. What was more, Parkman remembered, "It was sort of an iffy thing. You could start a rubella epidemic, you know? Things could go wrong. These were retarded children in an institution and that might not be the best thing. Maybe the parents wouldn't understand—a whole lot of things could go wrong. So for the medical board . . . it was a little bit of a struggle. I think we got approval by a narrow margin."[13]

At the Arkansas Children's Colony Parkman and Meyer injected the

HPV-77 vaccine into thirty-four children living in four different cottages and left thirty antibody-lacking children unvaccinated. The unvaccinated children shared the vaccinees' living quarters and would serve as vulnerable contacts; if the vaccine virus was contagious, they could well catch the disease.

In stark contrast to Plotkin's study at St. Vincent's, none of the vaccinated children developed rubella, with its rash, swollen lymph nodes, and fever. The researchers did detect the rubella virus in throat swabs from two thirds of the vaccinated children—meaning that they might have spread the disease even though they didn't get sick. But that did not happen. None of the unvaccinated, antibody-lacking children living with the vaccinees got rubella. What was more, 94 percent of the vaccinated children developed antirubella antibodies, the sought-after result.[14] It was an auspicious launch for the HPV-77 vaccine.

In the meantime, back in his lab at the Wistar Institute, Plotkin had set about weakening his rubella vaccine virus, RA 27/3, in two ways. He had already grown it through four passages on Hayflick's WI-38 cells before the trial at St. Vincent's. Now he grew it through twenty-one more, assuming that the more adapted it became to life in lab bottles, the less virulent it would be in the human body. In addition, he tried a technique that he knew had successfully weakened a different virus, polio: incubating the virus at slightly cooler temperatures than he had done initially.

Plotkin grew the virus at three different temperatures as he passaged it. One set of bottles he kept at 95 degrees Fahrenheit, the same temperature at which he had grown the virus before taking it to St. Vincent's Home. With the other, as he passed the virus-laden solution from one bottle to the next, he dropped the temperature first to 91.4 degrees and then to a relatively chilly 86 degrees Fahrenheit.

Each time he harvested a batch of fluid from the WI-38 cells to pass into a new bottle of fresh WI-38 cells, Plotkin set aside some that could be used as experimental vaccine and put it through safety testing. By the end of many months of work, he had produced RA 27/3 virus grown at three different temperatures and at many passage levels. He had on hand, for instance, eighth-passage virus grown at 95 degrees; thirteenth-passage virus grown at 91.4 degrees; and twenty-first- and twenty-fifth-passage virus grown at 86 degrees.

One of these many recipes, he hoped, was going to strike the right balance, making the virus weaker but not too weak. He didn't know where the sweet spot was. Plotkin went back to St. Vincent's Home to find out.

In the autumn of 1965, in the spring of 1966, and again early in 1967,

Plotkin ran three more trials involving forty one- to three-year-olds at St. Vincent's Home.

He tried different vaccines in different groups of toddlers, injecting them in the shoulders. Then for six weeks he followed up, although not entirely on his own. Plotkin enlisted the help of two physician colleagues. John Farquhar was an amiable, bespectacled pediatrician with a keen interest in infectious diseases who was director of pediatrics at Philadelphia's Presbyterian Hospital. Michael Katz had been Plotkin's medical school roommate and was, like Plotkin, a scientist at the Wistar. He was an erudite man who as a fourteen-year-old in Nazi-occupied Poland escaped under cover of dark from the Janowska concentration camp in Lwów, in what is now Ukraine. He spent most of the rest of the war under an assumed name, working as a runner for the Polish resistance in Warsaw. He made it to the United States in 1946, at age eighteen. All the rest of his family was killed by the Nazis.

Plotkin, Katz, and Farquhar divided the trial work, getting in their cars and dodging the streetcars out Woodland Avenue to St. Vincent's Home. There they checked the toddlers for rashes and fevers and swollen lymph nodes, swabbed their throats, and checked their noses for virus with cotton-tipped wires that the doctors stuck in the children's nostrils. On the days in between these trips, another doctor, Horst Agerty, who was the archdiocese's physician for St. Vincent's Home, examined the children for signs of illness. Sister Agape, the home's administrator, and Sister Mary Joseph, who had trained as a nurse and worked for years in a remote corner of South Africa, were also on hand—Sister Mary Joseph in a nurse's office complete with examination table on West 2, where a goldfish tank was a particular attraction for the small visitors.[15]

As he worked his way through the roughly sixty-five occupants of St. Vincent's Home, where there was little turnover, it became clear to Plotkin that he was running out of subjects. He would need to go farther afield. In August 1966 he drove ninety miles northwest of Philadelphia to the Hamburg State School and Hospital—a sprawling former TB sanatorium on a hilltop one mile outside the small town of Hamburg. The sanatorium, with its circular driveway and elegant Spanish mission–style main building, had emptied as tuberculosis faded with the advent of effective anti-TB drugs, better diagnostic tools, and improved prevention measures. In 1960 the state of Pennsylvania repurposed it, transferring nearly 1,000 intellectually disabled people there from another state institution—and from a swollen waiting list of 2,700 others whose families were seeking spots for them in state facilities.[16]

While the Pennsylvania legislature renamed the white-walled, three-hundred-acre facility a "school" and "hospital," it was, at the time of Plotkin's studies there, little of either. Few of its residents were taught anything, and while it had a surgical facility to deal with urgent problems, it was primarily a warehouse for people incapable of functioning on their own. The residents lived in wards of thirty or forty beds. They were nominally called "children," but many would age into adulthood at Hamburg and remain there for decades.[17]

Many residents were profoundly disabled. They were bedridden or confined to wheelchairs, diapered, incapable of feeding themselves. Many couldn't speak. One ten-year-old boy who died at Hamburg in 1967 succumbed to what an autopsy described as "convulsions due to mental retardation," which in turn was due to being starved of oxygen by a prolapsed umbilical cord at the time he was born.[18] Another, a quadriplegic ten-year-old, was affected by "severe inanition and dehydration"—meaning essentially that he was wasting away. His death was attributed to "cardiac arrest complicating congenital spastic paralysis."[19]

Cleaning, cooking, nursing, and otherwise caring for the roughly 950 residents employed scores of people from the town. The work could be physically taxing and sometimes dangerous: caregivers were on occasion attacked. Six years later a nearby newspaper, the *Observer-Reporter* of Washington, Pennsylvania, reported that straitjackets, heavy drugging, and confinement in three-by-five-foot steel cribs were used to restrain unruly inmates.[20] The staffing challenges complicated Plotkin's efforts to launch a trial. "We will provide help in holding the children," he wrote to Benjamin Clark, the short, stocky physician with a military bearing who was in charge at Hamburg.[21]

In late summer 1966, in the first of many trials he would run at Hamburg—again with help from Farquhar, the affable Presbyterian Hospital pediatrics chief, and Katz, the Wistar physician/scientist who had escaped the Nazis—Plotkin injected the RA 27/3 vaccine in the shoulders of seven children, aged four to thirteen, whom he described as "moderately to severely retarded." Another seven such children served as unvaccinated contacts and shared close quarters—one floor of one building—with the vaccinees. "Contact between subjects was promoted by placing them together in a playroom for regular periods," Plotkin wrote in the paper describing the trial.[22]

Plotkin began conducting trials at the Hamburg State School and Hospital at a critical juncture in the history of human experimentation in the United States.

In June 1966 Henry Beecher, a doctor who specialized in anesthesia at Harvard Medical School, published an article entitled "Ethics and Clinical Research" in the *New England Journal of Medicine*, a leading U.S. medical journal. It shook the medical establishment.

Beecher outlined, without naming names or institutions, twenty-two human experiments chosen from the medical literature from 1948 through 1965 that, he reported, had risked the health or life of their subjects. "Grave consequences have been suffered as a direct result of experiments described here," the paper began—experiments for which, Beecher wrote, "it must be apparent that [subjects] would not have been available if they had been truly aware of the uses that would be made of them."[23]

The experiments had been conducted by doctors at prestigious hospitals and universities, including the Clinical Center, the huge research hospital at the NIH. They had been published in top medical journals and funded by the NIH, the U.S. military, drug companies, and private foundations. What was more, Beecher had chosen the twenty-two from a far longer list of ethically suspect studies that had not been difficult to compile.[24] The abuses were everywhere.

In one experiment doctors had withheld chloramphenicol, a treatment known to effectively fight typhoid fever, from charity patients with the disease, to see if they died at greater rates than patients who received the drug.[25] (They did.) In another, they withheld penicillin, which was known to be effective against strep throat, from a control group of 109 U.S. airmen with the ailment, knowing that they could develop well-known, potentially life-threatening complications. Three men went on to do so after the bacteria in their throats spread.[26] Children in an institution for the "mentally defective" were purposely infected with the hepatitis A virus. (This institution, it emerged, was New York's Willowbrook State School for the Retarded, another horrible holding pen, and the pediatrician who infected the children was the prominent Saul Krugman, the same man who had earlier infected children with rubella.)[27] Separately twenty-two elderly, senile hospitalized patients were injected with living liver cancer cells and told simply that they were receiving "some cells." Leading this trial was the prominent cancer scientist Chester Southam of Memorial Sloan-Kettering Cancer Center, whose example Hayflick and Moorhead had followed in deciding to inject dying cancer patients with the new human diploid cells in 1960.[28] The list went on.

With the article's publication, the NIH was finally called to account for the lack of informed consent or observance of other ethical constraints in the human experiments that it funded. Beecher's explosive findings had been circulating and had received press coverage since he gave a public talk on them in March 1965, and the NIH had already received at least one letter from a member of Congress, the Chicago Democrat Roman Pucinski, asking how the agency was going to react. So the government was ready.[29]

On July 1, 1966, two weeks after Beecher's article appeared—and exactly one month before Plotkin began his first trial at Hamburg—the U.S. Public Health Service, the NIH's parent agency, published new guidelines under which U.S.-funded researchers like Plotkin would henceforth have to operate in running human experiments.[30] They would have to collect, and document, the informed consent of all participants in the research. They would also need to win ethical approval for their experiments ahead of time from new committees at their institutions—if these hospitals and universities wanted their scientists to continue to receive NIH funding.

Institutions, including the Wistar, got the message quickly. By July 22, three weeks after the government guidelines went into effect, Koprowski had constituted a "committee of associates" to pass ethical judgment on human trials run by Wistar researchers. It comprised four scientists, including Moorhead, Hayflick's coauthor on the 1961 paper that announced the mortality of normal human cells in lab dishes, who was not a physician; and Werner Henle, the eminent virologist at the Children's Hospital of Philadelphia, who himself had run wartime experiments testing an experimental influenza vaccine by infecting with influenza both vaccinated and unvaccinated juvenile offenders, and intellectually disabled residents at a state facility, Pennhurst.[31]

As word of the new rules spread, Plotkin scrambled to respond. On July 25, 1966, he wrote to Clark, the physician in charge at Hamburg:

> National Institutes of Health has requested that in any trial using human subjects there be records of consent. Therefore, I wonder if you could write me a letter simply stating that you, as the person responsible for the children to be used in the first trial, assume the responsibility for them. The letter will be filed until such time as NIH requests it, if ever.
>
> With regard to subsequent groups of children, from whose parents you have agreed to obtain consent, I think it would be desirable to include in the consent form, statements similar to the following: "German Measles in the mother may cause birth defects in the infant, including mental retardation. We are trying to find ways to vaccinate against the

disease. In these trials we wish to inoculate your child with a living vaccine virus. We understand that the German Measles infection, if it takes
in our child, may produce a rash and slight fever. We understand that this
is normally a mild or nonserious illness but that very rarely ill effects have
been attributed to German Measles."[32]

Clark responded two days later with the letter that Plotkin requested, taking responsibility for the children injected in the first trial. They were inoculated at Hamburg on August 1, 1966.[33]

Plotkin could have put off the first trial and sought permission from the
parents of the fourteen children involved. But he did not. He had already labo-
riously arranged the staffing and subjects for the first Hamburg trial. Getting
consent from the subjects' families would have meant more delay. And he was
a man in a hurry.[34]

Two years later, after launching a tenth trial at Hamburg, Plotkin would
obtain another for-the-record letter from Clark, attesting to the fact that written permission had been obtained from the parents of the participants in all ten
of the trials at Hamburg—or from Clark, acting in the place of their parents.[35]

By the spring of 1967 Plotkin had conducted a total of four new studies—
three at St. Vincent's and one at Hamburg. Their results showed that thirty-seven
of thirty-eight vaccinated children had developed antibodies—a 97 percent success rate. Admittedly, the numbers were small, but the percentage boded well.

As for side effects and spread to unvaccinated contacts, the message from
the new studies was this: the higher the passage number and the cooler the incubation temperature, the less likely an experimental vaccine virus was to cause
symptoms of the disease, like a rash, fever, and swollen lymph nodes; the less
likely it was to colonize a child's throat; and the less likely it was to spread to
unvaccinated children living in close quarters with the vaccinees. In the very
last of the new trials, five one-year-olds at St. Vincent's Home received coolly
incubated twenty-fifth-passage virus. Not one of them developed a rash or
swollen lymph nodes. Not one of the toddlers' throat or nose swabs tested positive for rubella virus. The kids weren't contagious, and their blood tests showed
that the vaccine had generated protective levels of antibodies.[36]

Plotkin had hit his target.

As he wrote up his new findings for the *American Journal of Epidemiology*,
Plotkin also returned to his argument for WI-38 cells, which was by now
well rehearsed: that the cells were cleaner and therefore safer than the animal
cells his competitors were using to make their rubella vaccines. He had new

ammunition, and it targeted African green monkeys, the very species that Parkman and Meyer had passaged their rubella vaccine in seventy-seven times and that Hilleman had initially used to grow the rubella vaccine that Merck was now developing. Hilleman had captured the Merck rubella virus, Benoit, from that eight-year-old Philadelphia boy with the same last name, grown it through nineteen passages on the African green monkey kidney cells, and then weakened it further in duck-embryo cell cultures.

African greens were perceived as relatively safe. They did not harbor SV40, the cancer-causing monkey virus that was discovered in 1960 in the rhesus and cynomolgus monkey kidneys then in use to make polio vaccines. That problem had forced the DBS to require polio vaccine makers to switch to using the kidneys of African greens, because they did not naturally harbor the SV40 virus. However, since that change in 1963, more disturbing information had emerged about African green monkeys.

In 1966 scientists at Lederle, a leading maker of the Sabin oral polio vaccine, published a paper in the *Journal of Infectious Diseases*. It concluded that the DBS-required safety tests now in place for African green monkey kidney cells were "thorough and effective" at ensuring the safety of the oral Sabin polio vaccine that had by then virtually replaced the killed, injectable Salk vaccine in the U.S. market.[37]

However, Plotkin saw the Lederle data as saying the opposite—that the green monkeys were a continuing source of concern. The Lederle scientists reported that they had studied 865 polio vaccines that they made using the kidneys of 865 individual green monkeys. They had been forced to reject 309 of the vaccines—36 percent—because of the presence of unwanted viruses.

"There is abundant evidence that primary African green monkey kidney is rife with latent viruses," Plotkin wrote in the new paper reporting on his recent studies at St. Vincent's Home and the Hamburg School—and he pointed to the Lederle study.[38]

Plotkin sent his new paper reporting RA 27/3's promising results to the *American Journal of Epidemiology* in mid-May 1967. Then he waited for the journal's production gears to grind. Early in August, as the paper was in press for the September issue, events in the picturesque university town of Marburg, West Germany, made Plotkin's case for his vaccine, produced in a human diploid cell strain, intensely, immediately relevant.

Marburg is a historic university town nestled on the river Lahn in central Germany and dominated by an eleventh-century hilltop castle. In 1967, it was home to Behringwerke, a vaccine and blood serum producer that was then part of the pharmaceutical company Hoechst. The company made live polio vaccine, and for this it harvested kidneys from African green monkeys.

That summer, probably in late July, Behringwerke received a shipment of African green monkeys from Uganda. The animals (*Chlorocebus aethiops*), also known as grivet monkeys, had black faces framed by striking white tufts of hair, bristly fur, and stomach skin with a tint of blue. They weighed anywhere from seven to seventeen pounds, and this lot of them looked healthy, despite their long journey: they had been sidetracked to London to avoid the Israeli-Arab Six-Day War. Then they were delayed in England due to an airport employees' strike. (Two monkeys escaped an animal house in London where the monkeys were housed during the delay. They were later recaptured and sent on to Germany separately.)[39]

At last the monkeys landed at the Frankfurt airport, and from there they were sent to three destinations: the Torlak Institute in Belgrade, then the capital of Yugoslavia, which produced and safety-tested polio vaccines; the Paul Ehrlich Institute in Frankfurt, which safety-tested polio vaccines; and the Behringwerke facility in Marburg. The monkeys at Marburg and Frankfurt were soon dispatched and their kidneys harvested. They weren't killed at Belgrade until about six weeks later. By then they didn't seem so healthy. The Torlak Institute reported a mortality rate 33 percent above what would normally be expected in the monkeys.[40]

In early and mid-August in Marburg, several Behringwerke employees suddenly began feeling ill. So did a handful of workers at the Paul Ehrlich Institute in Frankfurt. They had headaches behind their foreheads and in their temples. They had muscle aches. They spiked fevers—some over 102 degrees Fahrenheit. They felt generally awful. Still, for the first three or four days, the illness didn't seem too unusual, and those affected simply stayed at home.[41]

By the end of one week, though, they were beginning to vomit and to have severe, even explosive, diarrhea. Most developed a generalized skin rash, and many had conjunctivitis—an inflammation of the membrane lining the eyeball and the inner surface of the eyelid. More worrisome still, some of the sick people just weren't themselves. They weren't thinking or acting normally. Something was going on in their brains, it seemed.

Over the course of several weeks, some two dozen patients were admitted to university hospitals in Marburg and Frankfurt. Many of the afflicted were gravely ill. Some patients' faces grew swollen and unexpressive. One patient became psychotic; others were sullen and negative. Some patients' spleens and livers grew far bigger than they should have been.

The signs from their blood tests were frightening: their disease-fighting white blood cells were sinking dangerously low, as were their blood-clotting platelets. The levels of certain enzymes that rise during liver damage had soared.[42]

The sickest patients began bleeding from every orifice: noses and eyes, mouths and rectums, the puncture wounds where hospital workers drew their blood. Under the patients' skin, pinprick spots of bleeding appeared. So did bigger bruises. Their scrotums and vulvas turned a dark reddish blue. More than one patient collapsed from massive bleeding on the day after being admitted to the hospital. Others bled more slowly. Bleeding was an ominous sign. The people who ended up dying—seven would perish before the episode ended—were those who had bled.[43]

In all, twenty-one Behringwerke workers grew ill. In Frankfurt four employees at the Paul Ehrlich Institute became sick. About one month later at the Belgrade center, a veterinarian fell ill after conducting autopsies on five of the monkeys that had mysteriously died in Belgrade.[44]

It emerged that every sick employee had been in direct contact with blood, organs, or cell cultures from the monkeys. They had killed the monkeys, or dissected their kidneys, or bored holes in their skulls. They had handled monkey kidney cells in culture or washed the glassware afterward.[45] The published proceedings of a conference that was held two years later described some of the cases:

> The monkey-keeper Heinrich P. came back from his holiday on August 13th 1967 and did his job of killing monkeys from the 14th–23rd. The first symptoms appeared on August 21st.
> The laboratory assistant Renate L. broke a test-tube that was to be sterilized, which had contained infected material, on August 28th, and fell ill on September 4th, 1967.[46]

Soon several of the hospitalized patients' caretakers got sick too.

> The nurse Anneliese K. began to work in the ward on August 23rd and became ill on August 30th.
> The doctor Inga H., who was infected by the stitch of a needle on August 22nd, showed the first symptoms of the disease 5 days later on August 27th.[47]

Before the outbreak ended, two doctors, a nurse, and a medical student were infected. A morgue assistant helping to conduct an autopsy on one of the dead became infected when he cut his forearm.

The autopsies of the victims showed the internal devastation that had transpired while they were deteriorating and, finally, bleeding to death. Whatever had infected them was an equal-opportunity pathogen, attacking brain and spleen, pancreas and thyroid and liver, kidney and testes and ovaries and skin, leaving these organs strewn with bloody lesions. Their stomachs and intestines were full of blood.[48]

For the survivors recovery was long and painful. Their hair fell out in clumps, as if they had had radiation therapy for cancer.[49] The agent had a predilection for testicles, and the scrotums of some patients—three quarters of the victims were men—grew inflamed and ballooned up with a raging testicular infection. Later, doctors following the patients discovered that the culprit pathogen could linger for months not only in their testicles but also in their eyes and their semen. The wife of one survivor in Marburg became infected three months after the outbreak; his semen tested positive for the disease agent.[50]

From late August labs inside and outside Germany raced to identify the unknown pathogen in blood and serum samples from the patients and in the liver, spleen, kidney, and brain from one autopsy.[51]

In the United States the production of vaccines in green monkey kidneys all but halted, and the release of existing vaccines was suspended.[52]

On November 20, 1967, three months after the death of the first victim, a scientist named Günter Müller at the Bernhard Nocht Institute in Hamburg, Germany, looked through an electron microscope at blood serum from guinea pigs that had become ill after being injected with blood from the sick human patients. He saw a virus unlike anything anyone had seen. Viruses were often spherical; this one was a long, stringy thing, like a thread. It would eventually be assigned to the same taxonomic family as Ebola virus: filoviruses (from the Latin *filum*, meaning "thread"). While it acted a lot like Ebola when it infected a person, it was not Ebola; it preceded Ebola, at least in human awareness. Following scientific practice, the discoverers named it for the place where it first appeared: Marburg virus.

Years later a writer named Richard Preston dug into the question of where and when the monkeys that caused the Marburg outbreak became sick. He found an English veterinarian who had been responsible in 1967 for clearing the animals as healthy before they were exported from Entebbe Airport in Uganda. In Preston's 1994 book, *The Hot Zone*, the airport veterinarian explained that the animals were exported by an unscrupulous monkey trader. When monkeys that appeared sick were pulled off outgoing flights by the veterinarian, the trader was supposed to euthanize them. Instead he had them crated and taken to a small island in Lake Victoria, where they were released. Later, when he was short of animals for a shipment, the trader would visit the island and select the healthiest-looking animals that could be found to fill out the shipment. It is not certain that the Marburg-infected monkeys came from that island, but the veterinarian thought it was certainly possible.[53]

During this first-recorded Marburg virus outbreak, 7 people died of the 32 who were infected, a case fatality rate of about 22 percent. That rate has been much higher in outbreaks in Africa that have happened since. During the deadliest of these, in Angola in 2004 and 2005, 227 out of 252 infected people died, a case fatality rate of 90 percent.[54]

In 1967 the German outbreak came up repeatedly when, early in November, the DBS convened a big conference on the NIH campus. This was several weeks before the Marburg virus was identified by Müller, the scientist in Hamburg. The conferees were gathered to tackle the increasingly contentious question of which cells were best for making viral vaccines. There could hardly have been a more dramatic backdrop than the unsolved outbreak in Germany and Belgrade.

Five years after the publication of the influential *Science* paper that sowed doubts about Hayflick's human cells, the DBS—in effect, Roderick Murray—remained opposed to their use in vaccine making. And it became clear during the three-day conference that the events in Marburg were not going to change the regulator's position one whit.

Plotkin wasn't at the conference, but Hayflick was there, and by the last day he was brimming with frustration. He accused the DBS and its backers—of whom there were a significant number, some illustrious, in the crowd—of a head-in-the-sand position that amounted to preferring the devil they knew to the one that they didn't.

"The better we know our devil, the happier we are with him," Hayflick held forth. "Consequently, this year when seven people tragically died in situations where monkey kidney was used, we presume that we now know our devil a little better. Next year when, perhaps, other tragedies occur and the following year when tragedies of greater magnitude result, I presume that this will reinforce the contention that the devil we know is being known better and better and is consequently more desirable."[55]

Murray had sat in typical, imperturbable silence throughout the conference. Hayflick's angry comments, however, succeeded in rousing him to a rare, brief speech.

"About this monkey situation," he began. Conference participants had said several times, incorrectly, that the DBS had shut down all vaccines made using monkey kidney cells. This was not true—only African green monkey–based vaccines were undergoing "a temporary halt in release," and this would last only until the DBS sorted out whether the unidentified agent causing havoc in Marburg could be a danger "to the vaccine itself." All signs, he added, suggested that DBS screening requirements for imported monkeys "would probably have picked up this agent" if the infected monkeys had arrived on U.S. shores.[56] There would have been, he implied, no danger of distributing Marburg virus–laden vaccines.

Once the current situation was clarified, Murray concluded, the DBS would decide what to do about the use of African green monkeys going forward. Several months later, Murray's office cleared African green monkey kidney cells for use once more in vaccine production.

The Marburg virus had not invaded the polio vaccine supply as SV40 virus had invaded it a decade earlier, which is to say pervasively and silently, being injected into tens of millions of vaccinees without doing apparent, immediate harm. Rather, Marburg's devastation had been wreaked only on those working directly with infected animals or with their remains—and those in intimate contact with such people. And the results of Marburg infection had been immediately and highly visible. For Murray and those who shared his views, the Marburg events were therefore a blip, if an unfortunate one, and no kind of long-term threat to the vaccine supply. They were not a reason to jettison the use of African green monkey kidneys.

But for Hayflick and Plotkin the Marburg tragedy was yet another piece of evidence in the growing case for making the switch away from dubious monkey kidneys to using human fetal cells to make viral vaccines.

CHAPTER FOURTEEN

Politics and Persuasion

Philadelphia, 1967–68

The silent pressure for conformity exists whenever grants and contracts for research are under the direct control of governments; . . . then . . . no science is immune to the infection of politics and the corruption of power.

—Jacob Bronowski, Polish-British mathematician and science writer, 1971[1]

In November 1967, as the conferees debated at the NIH, Plotkin found himself in need of two things: time and money. The first he could control only by sleeping less—and he was already down to six hours a night. The second was becoming a problem that was just as hard to solve. The grant from the NIH's National Institute of Allergy and Infectious Diseases that had so generously funded Plotkin for the last several years was set to expire on November 30, 1967. The steady support from the private Joseph P. Kennedy, Jr. Foundation was also at an end. Only the last $6,100 of the $180,000 from that Washington-based foundation flowed to Plotkin in 1967.[2] "We are operating on a tight budget, unsupported by a pharmaceutical firm," Plotkin wrote in a letter to Benjamin Clark, the superintendent at Hamburg, as he watched the money running out.[3]

Plotkin's competitors—drug companies with deep pockets, like Merck, and the Parkman-Meyer duo at the DBS, with the resources of the NIH behind them—were far better funded, he was sure. The disparity now threatened to sink him.

Looking for a fresh source of support in the early autumn of 1967, Plotkin applied for a piece of a new pot of NIH money specifically earmarked for developing a rubella vaccine. The funds had been authorized by Congress in 1965, as the damaging reality of the huge rubella epidemic sank in with politicians. The $1 million that Congress allotted (equal to $7.2 million in 2016) was enough to fund work on several vaccine candidates, it seemed to Plotkin.

The purse strings of the new rubella-vaccine fund—and the job of getting a rubella vaccine developed quickly—were held by the Vaccine Development Branch in the NIH's National Institute of Allergy and Infectious Diseases.

In the summer of 1967 Plotkin did all he could to put together a first-rate application, gathering data from collaborators in Paris and Leningrad who had injected children with his RA 27/3 vaccine, tabulating the results of his own human experiments at St. Vincent's Home and at Hamburg, and even traveling to the NIH to run his draft application past John Sever, a senior rubella scientist and an influential member of the Vaccine Development Board, which advised the decision makers in the Vaccine Development Branch as to what projects were worth funding. Plotkin was asking for $60,000, which would see him through one more year of work.

Plotkin took time to write a separate letter to the Vaccine Development Branch, suggesting that it use some of its rubella money to run a head-to-head trial of his vaccine against Parkman and Meyer's HPV-77 vaccine, the one that the DBS duo had grown through scores of passages on African green monkey kidney cells. Such a comparison would show clearly which vaccine generated a better antibody response and which one was less likely to result in the vaccine virus colonizing the nose and throat, with the attendant risk of spread from vaccinees to the unvaccinated.

On November 21, 1967—incidentally, the day after Marburg virus was finally isolated in Germany—scientist Earl Beck, an administrator at the Vaccine Development Branch, wrote to Plotkin. The branch and its board of advisers had considered his application carefully. The answer was no.[4]

"As you know," Beck wrote, "there are already a number of rubella vaccine strains available to the program." That was all Plotkin got by way of explanation. A follow-up letter from Koprowski elicited a lengthier reply from Daniel Mullally, the chief of the Vaccine Development Branch, who waxed enthusiastic about Parkman and Meyer's HPV-77 vaccine, while explaining that the branch didn't have the funds to support every vaccine candidate.[5]

Still, Beck had written to Plotkin that his proposed head-to-head trial of his own RA 27/3 vaccine versus HPV-77 was an "excellent" idea. All that Plotkin needed to do to make it happen was to provide the branch with a complete manufacturing protocol for his vaccine, including records of all the safety tests he had run in the lab and in live animals. The latter tests needed to include injecting the brains of live monkeys and watching them for weeks to see if the vaccine caused neurological symptoms, then sacrificing them and studying their brains under the microscope. And, of course, he would need an Investigational Exemption for a New Drug—the permission that the Division of Biologics Standards alone was poised to bestow. Presumably, he had already filed for one.

Plotkin had not done monkey tests. Nor had he applied to the chubby, cigar-smoking Joe O'Malley in the DBS for an Investigational Exemption for a New Drug.

Plotkin would end up being rescued by industry: specifically, by the Philadelphia-based drug company Smith, Kline & French, with whom he had been corresponding that autumn. He had even driven out the Schuylkill Expressway to the company's labs in Upper Merion Township, where he met with Bob Ferlauto, a sometimes-hot-tempered Sicilian who ran the company's microbiology research and had seen the potential in Plotkin's vaccine.*

On December 1, the day after his NIH funding ran out, Plotkin received a grant of $10,000 from the company.[6] The funding—more than $72,000 in 2016 dollars—amounted to a godsend, if a temporary one. Plotkin hoped that there would be more where it came from.

As Plotkin scrambled for funding in 1967, he was also struggling to push ahead with more trials at the Hamburg State School and Hospital. After running the first trial there in the summer and fall of 1966, he had written to Benjamin Clark, the physician who was the institution's superintendent, thanking him for his "magnificent cooperation" and asking if he could begin another trial at the hospital early in the new year.[7]

"I am very sorry to have to inform you that the Nursing Service at this hospital has adamantly opposed any continuance of the rubella vaccine trials," Clark wrote back, without offering further explanation.[8]

Nonetheless, by January 1967 the nurses had somehow softened—or been required to soften—and Plotkin soon launched another trial.

There was another nurses' revolt in the summer of 1967, when Plotkin proposed still more trials at Hamburg, to begin in the autumn.

In a sharply worded memo the director of nursing, Miss Lois Colley—a tall, smart, take-charge woman—pointed out that the hospital was in the middle of training personnel, converting a hospital wing, and opening two new areas. "It will be impossible to provide staff for another Wistar program," she wrote to Clark. What was more, Plotkin's written contention that the new trial "would involve no more work for nursing personnel other than transfer of residents shows little insight into the nursing care and time involved."[9]

"She might be less resistant to the program by September," Clark wrote

---

*Five decades and several mergers and acquisitions later, Smith, Kline & French has morphed into what today is the giant Britain-based drug company GlaxoSmithKline.

to Plotkin. "But I think we should not plan too much on being able to continue."[10]

Plotkin needed the Hamburg children for experiments he considered crucial. In order to generate production-level volumes of vaccine, companies would need to expand his virus through still more passages in WI-38 cells. He had already taken on the task himself, generating new twenty-sixth-, twenty-seventh-, and thirtieth-passage versions of the RA 27/3 vaccine. He needed to test them in human beings.

In July 1967 Plotkin wrote to Director of Nursing Colley at Hamburg with a promise to bring along a technician next time to handle the throat swabs and the blood draws, "since you apparently feel that the nursing time needed to assist us is a significant drain." He assured her that attendants who worked in the area of the trial (one wing of the second floor of building 1), once they were shown to have antibodies to rubella, could circulate freely both in the trial area and in the rest of the institution with no problem—he wouldn't, in other words, commandeer them and keep them away from their workaday duties. If necessary, he would probably allow older women—he meant women too old to get pregnant— to work in the vaccine trial area too, without having their antibody levels tested.[11]

Colley acquiesced, and the trials continued, with promising results.

Plotkin was, and is, a stubborn man, an attribute that worked to his advantage when, on January 5, 1968, he received an ominous letter. Roderick Murray wrote to forbid Plotkin from sending his vaccine to out-of-state collaborators for human trials until he had conducted safety tests in monkeys, and until he had run a much larger volume of vaccine—the amount required of commercial manufacturers—through the full gamut of required safety tests in the lab and in animals.[12] Murray in fact was asking for safety tests on more vaccine than Plotkin had ever made, and far more than he now had on hand.

"For the Wistar Institute to prepare a lot [of vaccine] which would be large enough to satisfy Dr. Murray's criteria would be prohibitively expensive," Plotkin soon wrote to his longtime collaborator, Theodore Ingalls, who had since moved from the University of Pennsylvania to the Epidemiologic Study Center in Framingham, Massachusetts, and to whom Plotkin had hoped to send his RA 27/3 vaccine so that Ingalls could run trials there. The months of delay while he prepared it, Plotkin wrote, would in effect eliminate his human cell–based vaccine from the race for U.S. licensure.[13]

"You will understand, I am sure, how neatly this puts us in a box," Plotkin wrote in another letter, to Wistar lawyer and board of managers member

Robert Dechert. "The Vaccine Development Board has stated that they would test our material if DBS would approve it. If DBS will not approve it, the Vaccine Development Board need not act."[14]

The upshot, he explained to Dechert, was that Murray was shutting down vaccine development as an option for anyone but big pharmaceutical companies.

In the meantime Plotkin's DBS rivals, Parkman and Meyer, had landed in the vaccine race with a splash, with a pair of articles in the *New England Journal of Medicine* in September 1966 reporting on their success vaccinating the intellectually disabled children at the Arkansas Children's Colony with HPV-77, the rubella vaccine produced in African green monkey kidney cells.[15] Two other DBS scientists, Ruth Kirschstein and Hope Hopps, were also authors on the first of these papers; Hopps would remain heavily involved in rubella vaccine work.

Meyer and Parkman had since hurriedly expanded their HPV-77 trials, sending the virus out to several collaborating teams that promptly ran more human studies in more intellectually disabled institutionalized children. The DBS duo's summary report on these new trials, submitted to the *American Journal of Diseases of Children* in October 1967, as the crisis in Marburg unfolded, and published the following June, reported that their vaccine had successfully generated antibodies in 152 of 159 injected subjects, or 96 percent. What was more, the virus had not spread to any of the 145 contacts living in close quarters with the vaccinees.[16]

Before the Marburg outbreak the plaudits had already begun to pour in for Parkman and Meyer's HPV-77 vaccine, which its backers promptly labeled "the first effective" one.[17] The events in Germany did not dampen the enthusiasm. Parkman was also praised, rightly, for his all-important isolation of the rubella virus with his colleagues at the Walter Reed Army Institute for Research back in 1961. And the Parkman-Meyer duo, with four other NIH colleagues, were justly credited for developing, in 1967, an important blood test that allowed worried women with rashes to learn in hours, rather than days or weeks, whether they had rubella. The new test also sped vaccine research by allowing researchers to test hundreds of blood samples quickly for the presence of antirubella antibodies.[18]

For developing the blood test and the HPV-77 vaccine, Parkman and Meyer collected a $3,000 cash award from the American Academy of Pediatrics. The U.S. Junior Chamber of Commerce named Parkman one of America's Ten Outstanding Young Men of the year 1967.[19] At the award ceremony a letter to Parkman from President Lyndon Johnson was read aloud. "Few men can number themselves among those who directly and measurably advance human welfare, save precious lives and offer new hope to the world," the president wrote. "On behalf of all people who look forward to a healthier world, I offer you my congratulations and best wishes."[20]

As Parkman and Meyer basked in praise, executives at Philips Roxane, the St. Joseph, Missouri–based subsidiary of Philips Electronics and Pharmaceutical Industries Corporation, took note of all the attention to the duo's HPV-77 vaccine. It seemed clear that the vaccine developed within the DBS's own redbrick walls, by the two young scientists with the starlike buzz around them, was bound to be looked on favorably by the man who mattered: Roderick Murray.

So, soon after the first results from the Arkansas Children's Colony study were published in 1966, scientists from Philips Roxane asked Parkman and Meyer to send them the HPV-77 virus. They grew it through eleven more passages in kidney cells from three- to six-month-old puppies. The Missouri company dubbed its rubella vaccine HPV-77/DK-11, denoting the additional eleven passages through dog kidney cells.[21] Then its scientists too began testing their vaccine, working against the epidemic clock that was ticking toward 1970.

At the Merck campus at West Point, the hard-charging Maurice Hilleman had labored through 1965 and 1966 developing several formulations of the company's Benoit strain of rubella virus—the virus captured from the throat of that eight-year-old Philadelphia-area boy with the surname Benoit back in 1962. Hilleman and his team experimented by growing the Benoit virus for varying numbers of passages in African green monkey kidney cells, followed by varying numbers of passages in duck-embryo cells. Like Plotkin, Hilleman was trying to find the sweet spot: a live virus that was weakened but not too weak, a virus that prompted the body to make plentiful antirubella antibodies but didn't cause full-blown disease.

By the summer of 1966, Hilleman's team had created several variations of the Benoit vaccine, which differed from one another by the number of passages they had gone through in African green monkey kidney and then duck-embryo cells. The Merck researchers had safety-tested the varying formulations in lab cultures, in mice, and in monkeys and injected them into intellectually disabled children in and near Philadelphia. One of their Benoit vaccine formulations, version B, looked particularly promising. It generated hefty levels of antibodies without giving the vaccinees rubella. Hilleman felt that he was closing in on his mark.

Hilleman was the last man to shrink from an all-out race to a vaccine. In fact, being a drug-company scientist, it was his job to produce his own vaccine candidate and then to make sure it crossed the finish line first. In his career Hilleman would develop more commercially successful vaccines than any other human being before or since. But in the case of rubella vaccine, he backed down from a fight—not once but twice. In both cases it was women who stared him down.

The first of these women, Mary Lasker, was the widow of advertising magnate Albert Lasker. She was sixty-five years old in early 1966 and a veteran of three decades of political activism. She had jumped into public life in the late 1930s as secretary of the Birth Control Federation of America, which would later become Planned Parenthood. But her consuming focus soon became medical research. She and her husband established the Albert and Mary Lasker Foundation in 1942, a decade before he died of cancer. Its medical research awards quickly became the most prestigious in the country. To this day, their recipients often go on to win Nobel Prizes.

An art historian by training, Lasker had never lifted a test tube or cultured a cell in her life. But she was a philanthropist with an outsized influence on U.S. medical research—a woman used to bending ears and twisting arms at the highest levels of U.S. government and industry. And she was convinced that Parkman and Meyer's HPV-77 was on a glide path to victory in the rubella race, if only because its inventors worked just up the stairs from Roderick Murray in the DBS's building 29. One day in 1966, Hilleman and his boss, Max Tishler, the president of Merck Research Laboratories, met with Lasker, at her bidding, in her elegant apartment on Central Park West.[22]

As Hilleman explained to the writer Paul Offit in 2004, Lasker had summoned the two men to put to them her considered view: that competition from a Merck vaccine candidate would do nothing but slow down DBS approval of *any* rubella vaccine. And in any event, whose vaccine did they think would win approval, theirs or the one developed in-house at the DBS? Lasker had seen the consequences of the recent epidemic, she said pointedly, and she was deeply disturbed. The goal was to get a vaccine licensed fast, before the next outbreak wreaked more devastation. Hilleman was dubious. He told Lasker he didn't think the HPV-77 vaccine would get approved, "because [Parkman and Meyer] didn't make it into a vaccine. It's just a goddamned experiment."[23]

Lasker wasn't convinced. She told Hilleman to go away and think about it. He didn't need to think for long. Lasker was politically powerful. She could make life very unhappy for him if she chose. Grudgingly, Hilleman told Lasker that he would get some HPV-77 vaccine from Parkman and Meyer and put it in kids himself and see what he thought. He vaccinated about twenty children and was appalled by the side effects. "Jesus Christ it was awful. Toxic, toxic, toxic," he recalled.[24] The results of this trial were apparently not published.

So late in 1966 or early in 1967, Hilleman further weakened HPV-77. Instead of growing the virus through more passages in African green monkey kidney cells, he put it through five sequential cultures of duck-embryo cells. Hilleman liked ducks as experimental hosts. Unlike most animals, they were, he would soon write, "remarkably free of infectious diseases and [cancer]."[25] Hilleman named the tweaked vaccine HPV-77/DE 5 to reflect the fact that it had now been grown through seventy-seven passages in African green monkey kidney cells followed by five passages in duck embryos.

In June 1967 Hilleman tested the new Lasker-prompted vaccine in eight intellectually disabled, institutionalized children. It generated only one fifth of the antibody levels that Merck's best Benoit formulation, version B, did.[26] But the number of subjects was so small that the difference could have been due to chance. He needed much bigger numbers if he was going to make a decision—and he needed those numbers from a real-life situation: a situation full of mothers and children and pregnant women; a situation in which, if the vaccine virus was going to spread, that would quickly become evident.

In September and October 1967 Hilleman tested the new duck vaccine in a heavily Irish Philadelphia suburb called Havertown-Springfield. "I tell you something, that was a gutsy day for me because I was extremely worried about transmissibility," Hilleman recalled in 2004.[27] He resigned himself to losing his job if a pregnant woman in the trial gave birth to a baby with any kind of abnormality whatsoever. Because of the background rate of non-rubella-related birth defects, he put the odds of that happening at about one in three hundred, even if the vaccine didn't spread and cause rubella in a single pregnant woman.

He and his team injected 269 youngsters, from babies to five-year-olds. Virtually all developed antirubella antibodies. The researchers also tracked the nonimmune, unvaccinated siblings of the children who were vaccinated. None of those 262 siblings became infected. Nor, crucially, did any of the thirty-four nonimmune mothers whose children were vaccinated, three of whom were pregnant. But the new HPV-77 duck vaccine still underperformed in the children: it generated antibody levels just one third of those produced by Merck's best Benoit vaccine, version B.[28]

Hilleman had a decision to make. It wasn't a pretty choice. The Merck vaccine was better, but the adapted Parkman-Meyer vaccine still delivered "substantial" levels of antibodies, didn't spread to contacts, and didn't give vaccinees rubella, he wrote in the paper that resulted.[29] The difference between the vaccines wasn't, perhaps, big enough to go to the mat for. Yet it would be no small enterprise to switch tracks now. It would involve, among other expensive hassles, building duck houses so as to raise colonies of ducks on the Merck campus.[30] Hilleman was caught between the rock of data showing clearly that Merck's Benoit vaccine, version B, was better and the hard place of Mary Lasker's political clout and adamantine will. Her words came back to Hilleman: "Which vaccine do you think DBS is going to approve?"[31]

Hilleman dropped Merck's Benoit vaccine. Lasker had triumphed. The big drug company would pour its resources into Parkman and Meyer's HPV-77 vaccine, developed in African green monkey kidney cells within the walls of the DBS, grown by Merck a handful more times through cultures of duck embryos, and bound for injection, if Hilleman could make it happen, in millions of Americans just as soon as was humanly possible.

It had been six years since Hayflick began proselytizing for lab-grown human fetal cells as clean, safe, normal, noncancerous microfactories for making viral vaccines. European virologists and vaccine makers had been embracing them for nearly as long. It had been more than five years since Mrs. X's abortion and since Hayflick, with NIH funding, had produced from her fetus's lungs eight hundred ampules of what had become the gold-standard human diploid cell strain: the WI-38 cells. It had been three years since the 1964–65 rubella epidemic devastated tens of thousands of American lives and since Plotkin had wrested something good from that tragedy: a rubella vaccine made using the WI-38 cells. And still there appeared to be not the slightest indication that his vaccine, or any vaccine made using the WI-38 cells, would see the light of day in the United States. Not if it required Roderick Murray's approval.

CHAPTER FIFTEEN

The Great Escape

Philadelphia, 1967–June 1968

When Dr. Leonard Hayflick left Philadelphia for California in 1968 he had something with him few travelers carry in their luggage—frozen human cells.
—*Philadelphia Evening Bulletin*, April 4, 1976[1]

If the WI-38 cells were ignored in the United States, abroad they were increasingly in use. In 1967 the Yugoslavian republics of Croatia and Slovenia began the first large-scale, routine use anywhere of a WI-38 cell–produced vaccine—the oral polio vaccine that the tall, soft-spoken Drago Ikić, the vaccine chief at the Institute of Immunology in Zagreb, had championed from practically the moment that the WI-38 cells were launched, in 1962. Yugoslavian authorities licensed that polio vaccine for the whole of the country in 1968, when they also licensed a measles vaccine made using WI-38 cells "for massive use."[2]

Scientists at Britain's Burroughs Wellcome had asked Plotkin to send them his RA 27/3 rubella virus in 1966 and were working on developing a vaccine using the WI-38 cells. France's Institut Mérieux obtained the RA 27/3 virus from Plotkin and began making an experimental rubella vaccine in 1967—the same year that it began commercial development of Koprowski and Wiktor's rabies vaccine, also made using WI-38 cells. And at a research and development center near the tiny town of Sandwich, on England's southeastern coast, vaccine makers at Pfizer were laying plans for a polio vaccine made using WI-38 cells.

The WI-38 cells were also being put to work in labs on the front lines of public health. Britain's Medical Research Council, the rough equivalent of the U.S. National Institutes of Health, had already been supplying WI-38 cells for diagnostic purposes to public health labs in England and Wales for several years. In February 1967 the World Health Organization began paying the MRC to expand the cells' reach by providing them to disease detectives on four continents. Every two weeks, on Wednesday mornings, bottles of WI-38 cells, bathed in medium, were airlifted from London to Dakar and

Montevideo, Hong Kong and Cairo and Port of Spain. They traveled well, seemingly not bothered by the changes in ambient temperature.

At their destinations scientists used the WI-38 cells to identify viruses that were landing children in hospitals with serious, sometimes deadly, respiratory infections. Then they constructed a map of what viruses were at work where. The far-flung users reported back that the WI-38 cells detected a broad range of viruses—a bigger spectrum than the monkey kidney cells that were supplied by the same program. They reported the WI-38s were particularly useful for picking up cold-causing rhinoviruses and two more-dangerous viruses: respiratory syncytial virus (RSV) and herpes simplex.[3]

(Independent of the MRC's effort, Hayflick acted as a roving supplier and ambassador for the cells, training dozens of scientists in how to grow them. Whenever he boarded an airplane, he would carry, wrapped in both arms, a liquid nitrogen refrigerator that looked like nothing so much as a one-hundred-pound bomb, minus the fins. It was eighteen inches high and packed with ampules of WI-38. Sitting in his economy-class seat, he kept it on the floor between his legs. It had to be kept upright or the gaseous nitrogen inside it might leak out, sending up a dense white cloud of vapor that would doubtless have terrified his fellow passengers.)

It was a sign of the esteem in which Hayflick's WI-38 cells were held that the British vaccine authorities—namely, the silver-haired Frank Perkins, the UK's top vaccine regulator, and three of his colleagues at the Medical Research Council in London—decided, perhaps as a matter of national pride, to derive their own analogous normal, noncancerous human diploid cells. They wanted to produce a cell strain that was, if not the UK's rival to WI-38, then its complement to it.

In September 1966 Perkins's colleagues at the Medical Research Council, J. P. Jacobs, C. M. Jones, and J. P. Baille, received a fourteen-week-old male fetus following an abortion in a twenty-seven-year-old woman carried out, they later reported, "for psychiatric reasons." From the lungs of that fetus they derived a cell line of typical, spindly fibroblasts. They named the line MRC-5, after the Medical Research Council, and published their work in *Nature* in 1970. The time lag allowed them to report that the woman had remained healthy and cancer free for three years following the abortion.

The *Nature* paper described the healthy, noncancerous characteristics of MRC-5 cells, along with their vigorous growth in lab bottles, their normal chromosomes, and their susceptibility to infection with a host of human viruses. The scientists wrote that they had frozen a big supply of the new British human diploid cells: 481 ampules. Their paper opened with an homage to WI-38 and closed with the same. "Our studies indicate that by presently accepted criteria, MRC-5 cells—in common with WI-38 cells of similar origin—have normal characteristics and so could be used for the same purposes as WI-38 cells."[4]

Although neither Hayflick nor the British scientists could have imagined why it would become the case, that last fact—that MRC-5 cells could be used for all the same purposes as WI-38 cells—would become important.

As their prominence expanded, Hayflick was only too happy to promote the WI-38 cells. HUMAN CELLS GIVEN ROLE IN VACCINES, the *New York Times* proclaimed—hyperbolically—after Hayflick spoke at a vaccine conference in 1966. The article quoted Hayflick's views—and only Hayflick's views—as he explained that his cells were cheaper, cleaner, and safer than the animal cells used in vaccine making. "Today, a specialist in the growth of human cells in the laboratory said that research on the cells used in vaccine production had been seriously neglected," the *Times* explained.[5]

A few months later, in 1967, after he spoke to an American Cancer Society seminar for science writers, Hayflick was in the *Times* again, promoting his cells as an alternative to disease-ridden monkey kidney cells.[6]

That same year Hayflick ran out of patience with Koprowski and his institute. The disconnect between his contributions and his treatment by the Wistar's boss had become more than he was willing to tolerate. Nine years after Koprowski hired him, Hayflick remained stuck as an associate member of the institute, in sharp contrast to a coterie of his other Wistar colleagues who were full members and yet whose contributions were not, to his mind, any greater than his own. As an associate, not only did Hayflick make less money than full members, but his salary was not guaranteed by Koprowski should he fail to win NIH funding for his work.

Never mind that he was now a father of five who would have appreciated the job security. What galled him was Koprowski's implicit refusal to credit his accomplishments: his discovery of the *Mycoplasma* that was the cause of walking pneumonia; his nearly simultaneous recognition that normal cells aged in lab dishes, and the huge scientific questions that it opened; his production of the WI-38 cells that could—that *deserved* to—supplant archaic methods of making antiviral vaccines; his tireless efforts to make that happen.

He was at bottom just a showy date for Koprowski, and a cheap one at that: early in 1967 he had just won a three-year renewal of the sought-after NIH

Career Development award that had already fully funded his salary since 1962. What was more, his contract with the NIH's National Cancer Institute—the big one to produce, store, and distribute his human diploid cells—had brought the Wistar more than $120,000 every year for five years now and was moving into a sixth year. That contract, among all of the contracts and awards that the institute's scientists brought in, was consistently one of the largest.[7]

Fed up, Hayflick began looking around. He was offered the chairmanship of the department of microbiology at the University of Vermont in Burlington. He also applied for a position as a full professor of medical microbiology at Stanford University in Palo Alto, California. For the Stanford job he had help in high places: one of his recommenders was Albert Sabin, whose live oral polio vaccine, by then being used in most countries that vaccinated, had vaulted its inventor into the most prestigious ranks of U.S. medical science.

In August 1967 Sabin wrote to Sidney Raffel, a leading immunologist who was heading Stanford's search for a new professor of medical microbiology: "In my judgment, he is a very reliable investigator who has exhibited considerable intelligence and originality in the work that he has done. I have a high regard for Dr. Hayflick because his contributions over the years have been sound and trustworthy."[8]

Not long afterward, Hayflick was offered the Stanford job.

To Hayflick, the difference between the positions at Vermont and Stanford was akin to the difference between coaching for the Phillies and playing first base for the Yankees. His choice was easily made. In the autumn of 1967 Hayflick told Koprowski that he would be taking a professorship at Stanford beginning on July 1, 1968.

It seems likely, given his apparently low regard for Hayflick's capabilities, that there was one thing, and only one thing, that concerned Koprowski about Hayflick's impending departure: the fate of the hundreds of ampules of WI-38 cells that were still stored in liquid nitrogen in the Wistar Institute's basement, under Hayflick's watchful eye.

Koprowski had had designs on the WI-38 cells from the beginning. Nancy Pleibel, a lab technician who began working for Hayflick in 1963 or 1964, recalls that more than once as she learned the ropes, Koprowski turned up in the lab within a day or two of Hayflick leaving on a trip. Charmingly, smilingly, the Wistar czar would ask her for an ampule of WI-38 cells. Politely but firmly, she would refuse, explaining that only her boss could hand out WI-38 ampules. After a while Koprowski stopped asking her.[9]

Minutes from the Wistar's board of managers meetings in the early and mid-1960s make clear that Koprowski tried repeatedly to cash in on the human diploid cells. The Wistar sought payment not only from Norden, a Missouri company that was interested in using WI-38 to develop the nascent rabies vaccine, but also from Pfizer for the use of Hayflick's cells to make a measles vaccine, and from Wyeth, another Philadelphia-based drugmaker that by 1965 had used the WI-38 cells to make an adenovirus vaccine to protect U.S. Army recruits during basic training.[10]*

Koprowski's attempts to turn a profit with the WI-38 cells were far from successful. By 1965 the board of managers had appointed "a special committee of lawyers and scientists to deal with problems" selling the Hayflick cells to industry.[11] The only support that the institute landed, according to budget documents from 1965 to 1967, was $5,000 in each of those years from Norden.[12]

Today it seems incredible that an institution like the Wistar, full of sophisticated scientists, was so at sea when it came to profiting from unique and desirable cells produced under its roof. But in that era living things, like WI-38 cells, could not be patented. It would take a landmark Supreme Court decision in 1980 to change that. (Nonetheless, it appears that Koprowski or Hayflick did take steps to patent the cells, in 1966; a patent attorney they enlisted told them it was too late to even try as Hayflick's discovery had been published in 1961 and the one-year window of opportunity for filing for a patent had closed.)[13]

However, what *could* be patented was a method of using the cells to produce a novel vaccine. Koprowski had already applied, back in 1964, for such a patent for the new, improved rabies vaccine that he and Wiktor were developing using the WI-38 cells. Soon the Wistar would apply for a patent on Plotkin's method of making the RA 27/3 rubella vaccine.

If and when the rabies and rubella vaccine patents were granted, Koprowski's access to at least some of the original ampules of WI-38 carefully frozen in the Wistar basement might be vital. Vaccine companies would want original ampules full of the youngest cells—cells that could be expanded exponentially into a nearly endless supply: original ampules that Hayflick had frozen in the

*This new adenovirus vaccine replaced the vaccine made using monkey cells that was fed to about 100,000 members of the U.S. military between 1955 and 1961 and was dropped when it was discovered to be widely contaminated with the silent monkey virus, SV40.

summer of 1962, after their cells had divided just eight times; original ampules that were, Koprowski hoped, so much cellular gold.

By the autumn of 1967 Hayflick vaguely suspected, without having the corroborating evidence, that Koprowski intended the WI-38 cells to serve as something more than vaccine factories deployed for the good of mankind; that his boss hoped to turn any vaccines made with the cells into plentiful sources of cash, boosting the Wistar's income and freeing Koprowski from the odious fund-raising duties that he considered beneath him and that had been the bane of his existence since he took the helm of the institute a decade earlier.

Hayflick's instincts were right, and for understandable reasons: as the year 1967 drew to a close, a financial vise was tightening on Koprowski.

While the Wistar had remained solvent under its high-living boss, it had never been exactly flush with funds, especially after Koprowski blew through $271,506 to fund the major renovations that were completed in 1959. By the mid-1960s his struggle to find cash that wasn't tied to specific grants was becoming acute. An operating loss of $12,000 in 1965 grew to some $23,000 in 1966, causing Koprowski to defer repairs. The windows in the seventy-three-year-old building needed replacing. So did the roof. The air-conditioning in some labs required new, high-end filters. The public toilets and the sewer system both needed overhauls.[14]

By late 1967, when Hayflick announced his impending departure, Koprowski was facing an expected 1968 deficit of $65,460—$469,000 in 2016 dollars—and board of managers members were calling for long-term solutions. Adding to the tightening financial screws was the fact that the NIH had recently launched a "high-priority audit" of the Wistar's grants. The agency felt that Koprowski's methods of claiming overhead costs—an important source of general-purpose funds for the institute—from individual scientists' government grants were "in conflict" with agency regulations. The NIH aimed to apply the findings of the audit both retroactively and prospectively—meaning that Koprowski could soon owe the agency some unknown, and possibly frightening, amount of cash.[15]

In the autumn of 1967, when the NIH's National Cancer Institute learned that Hayflick would be moving to Stanford in July 1968, officials there decided to end Hayflick's contract to produce, store, and distribute human diploid cells to any qualified researchers who needed them. The cancer institute had been paying the Wistar hundreds of thousands of dollars for Hayflick to produce and distribute the cells ever since the contract was launched in February 1962, soon after Hayflick's paper announcing his human diploid cell strains to the world sent demand for them soaring. Now, institute officials set January 1, 1968, as the end date. The timing seemed right, and not only because of Hayflick's impending move. The sense at the government cancer research institute was that the demand for the WI-38 cells had been sated.[16] Those scientists who wanted them, it seemed, had them by now, more than five years after Hayflick had first produced them. They were being used widely and had already been cited in scores of papers.

On January 18, 1968, several men traveled from Bethesda to the Wistar Institute to sort out the physical disposition of the WI-38 cells now that the contract had ended. Koprowski summoned Hayflick to meet with them. One was Charles Boone, a tall, thin, edgy MD/PhD who since the previous May had been overseeing the human diploid cell contract for the NIH's cancer institute. Also present were John E. Shannon and Marvin Macy, two senior scientists from the American Type Culture Collection. The independent, nonprofit ATCC, as it was popularly known, was housed in a boxy brick building in Rockville, Maryland, six miles from the NIH. It was the country's highest-profile cell bank, and was often the repository that biologists turned to when they needed a certain type of cell for an experiment.

Hayflick joined the group in a small conference room. Also present was Koprowski's deputy director for scientific administration, John D. Ross, who took notes and wrote up minutes of the meeting. According to these minutes, the assembled men agreed that all but twenty of the roughly 375 remaining original ampules of WI-38 cells would be transferred to the ATCC, which would maintain them, deeply frozen, on behalf of the NIH. Hayflick would be permitted to take ten ampules with him to Stanford, and the Wistar itself would be allowed to keep ten ampules.[17]

The group also decided that any use of the 355 precious original ampules being transferred to the ATCC—they were precious because the WI-38 cell populations in them had divided just eight times and so could be expanded into untold billions of cells for vaccine making—"should be totally arrested." They meant that there was to be no more thawing of the ampules, no more planting of these young eighth-generation cells into lab bottles, and no more splitting of those bottles over and over to generate multitudes of cells at higher population doubling levels for scientists to use. Scientists could use the older cells that were already in wide circulation. The remaining 355 original ampules needed to be kept safely frozen at the ATCC until such time as companies began winning licenses from the DBS to make WI-38–based vaccines. Then, carefully, the ampules could be thawed one at a time, expanded through a handful more doublings, and handed out as needed to companies to make vaccines, the minutes specified.[18]

Before the meeting broke up, the group set a date for the transfer of the WI-38 cells from the Wistar Institute to the ATCC: March 1, 1968.

The dry, black-and-white minutes of that January 1968 meeting do not convey the subtext of what transpired that day, according to Hayflick, the only participant, except perhaps Boone, who is still alive. Nor, Hayflick says, does a letter sent from the NIH's Boone to Koprowski four weeks later capture the subtext of that meeting. The letter formalized the decisions made on January 18 as "official policy" of the NIH. It concluded with a flourish that "the WI-38 cells . . . will remain the property of the NCI," referring to the NIH's National Cancer Institute.[19]

Hayflick said in a 2013 interview that he was compelled to attend the meeting and that he had no choice but to concur with the "alleged agreement" transferring the cells to the government, which was pointedly asserting its ownership. "It was not written by me; it's written by five people who are saying, 'You better agree to this.' I have no muscle. What's my muscle? Honor?"[20] He also noted that he didn't sign either the meeting minutes or the follow-up letter from Boone to Koprowski. He did get copies of both.

What Hayflick does not mention is the wording of the 1962 contract between the National Cancer Institute and the Wistar Institute that committed the institute, in the person of Hayflick, to produce, store, and distribute human diploid cells to qualified scientists. That contract, signed several months before he derived the WI-38 cells, addressed explicitly what was to happen when the contract was terminated: "The contractor agrees to transfer title and deliver to the Government, in the manner, at the time and to the extent, if any, directed by the Contracting Officer, all data, information and material which has been developed by the Contractor in connection with the work under this contract."[21]

In the January 1968 meeting at the Wistar, the government's contracting officer, Boone, had directed Hayflick on March 1 to deliver to the government, in the form of Boone and Shannon, all but twenty original ampules of the WI-38 cells.

Hayflick was unhappy after that meeting. He didn't want to give up control of the WI-38 cells, and he especially did not want to cede them to the ATCC, whose past management of cells, in his experience, had been sloppy. So he stewed. But as January rolled into February, he did nothing else. It took a very specific event to precipitate an action on his part that would have profound consequences for his life and his career.

When he relates this event nearly fifty years after the fact, it's clear that Hayflick can see it and feel it as if it had happened yesterday. His voice gets more intense. He is at pains to be clear.

Sometime during his last months at the Wistar, he was working in one of the tiny "sterile" rooms that adjoined his lab. Plotkin squeezed in the door and pulled up the only other chair in the six-by-nine-foot room. The two chatted for a bit, and then Plotkin showed Hayflick the document that had brought him on this visit. It was a letter, on Wistar letterhead, from Koprowski, written to a senior official at Burroughs Wellcome, the British drugmaker. It was not a final document, or a contract, or a bill of sale. But it was definitely conveying the following: that Koprowski was offering to provide to the company ample supplies of WI-38 cells, along with the recipe for making a vaccine with the cells, and the vaccine virus itself, all in exchange for royalties.

Hayflick had suspected, but up until this moment had not known, that Koprowski planned to use the WI-38 cells to make as much money as he could for the Wistar—and to do so without so much as a by-your-leave from Hayflick. He was stunned and sat in silence. He did not think to ask Plotkin for a copy of the letter, or even to ask why he was showing it to him.

Plotkin does not remember this conversation at all, a fact that upsets Hayflick to this day. But in Plotkin's papers there is a single sheet entitled "Chronicle Burroughs-Wellcome Proposed Agreement." It lists, without detailing their contents, eight items of correspondence exchanged between the Wistar Institute and the company during 1968, beginning with a March 6, 1968, letter to Hilary Koprowski from a Dr. Edward at Burroughs Wellcome. Next there is a letter from Koprowski to Dr. Edward, dated March 14, 1968—after which lawyers and bankers take over the correspondence.[22]

It seems probable, especially in light of the events that followed, that the March 14 letter from Koprowski to Burroughs Wellcome is the one that so shocked Hayflick as he sat closeted in a sterile room with Plotkin.

The following month, April 1968, a patent application describing Plotkin's new method of making a rubella vaccine using WI-38 cells arrived at the U.S. Patent and Trademark Office. It listed one inventor, Stanley Plotkin, and assigned Plotkin's ownership rights to the Wistar Institute, meaning that if the patent was granted, the Wistar would be the owner, and if the institute then licensed the vaccine to companies, the royalty money would flow to the Wistar.

As was the case with Koprowski and Wiktor's rabies vaccine, Plotkin's rubella vaccine work had been funded by the NIH, meaning that, under the law in that era, the U.S. government owned Plotkin's invention. But as it had done with the rabies vaccine, the United States would grant a waiver that allowed the Wistar to claim title to Plotkin's process for making a rubella vaccine, patent it, and license it to companies.[23]

The 1968 correspondence between the Wistar and Burroughs Wellcome culminated in Koprowski's traveling to the company's headquarters in London that October. There, on the assumption that Plotkin's rubella vaccine would indeed be patented, he and senior company officials began negotiating the terms under which the Wistar would grant Burroughs Wellcome a license.[24] (Koprowski also at that meeting prodded the firm to think about licensing the rabies vaccine, also made in WI-38 cells, for which the Wistar had won a patent just two months earlier. The company did not end up doing so, although others did.) A few years later, after a patent for Plotkin's rubella vaccine was granted, Burroughs Wellcome negotiated a license with the Wistar Institute and began manufacturing the RA 27/3 rubella vaccine.

Hayflick was profoundly upset after the conversation with Plotkin in his tiny sterile room. He had spent the previous decade deriving the normal human diploid cells, noting that they aged in their lab dishes, and in doing so opening up a new, important field: the study of cellular aging. He had derived enough WI-38 cells to serve vaccine makers into the distant future, and he had worked as hard as was humanly possible to win their acceptance for vaccine making. In the process of all this, he had sometimes been ridiculed, and he had struggled for respect and validation.

Then, as he prepared to leave the Wistar, the letter from Koprowski to Burroughs Wellcome signaled that he was being entirely sidelined by Koprowski in major decision making—and likely profit making—around WI-38. For Hayflick, at this juncture, "to have the vultures descend on what I had struggled so hard to give value to and try to take it for their benefit—I think that an average person would understand why I was, to put it mildly, concerned."[25]

The pill was especially bitter, Hayflick says, because he knew—it was no secret among Wistar scientists—that Koprowski was under tremendous financial pressure. He needed, among other things, to keep paying the salaries of the scientists who were full institute members when they were between, or without, grants.

"I was not a full member of the institute, so that did not apply to me. And here I'm being asked to leave [the WI-38 cells] behind to benefit the full members of the institute, very few of whom could argue that they gave value to those cells."

On or around March 1, when, under the January agreement, the ampules were to have been moved from the Wistar to the ATCC, a specially outfitted station wagon arrived from Maryland, carrying the NIH project officer, Charles Boone, and John Shannon, the ATCC's curator of cell lines. Hayflick turned them away, saying that he wasn't ready to hand over the cells because he hadn't prepared an inventory of them.[26]

Sometime not long after this, when no one was looking, Hayflick visited the Wistar basement. There he packed every single one of the remaining original WI-38 ampules—some 375 frozen vials that composed the largest stock of young WI-38 cells on earth—into one or more portable liquid-nitrogen refrigerators and departed the premises. He left behind not even the ten ampules that Koprowski's institute had been promised in the January agreement. He left behind, in fact, not a single original, low-passage ampule of WI-38.[27]

Hayflick stored the frozen cells temporarily with a friend and colleague, Eugene Rosanoff, a vaccinologist at nearby Wyeth Laboratories who obligingly, from time to time in the next few months, topped off the liquid nitrogen that kept the cells frozen.

Hayflick says that he took the ampules with the intention of keeping them only until title to the cells could be properly sorted out. In his mind, he says, their ownership was in question. He believed that there were several potential stakeholders who might reasonably claim ownership: him and Paul Moorhead; the "estate" of the WI-38 fetus, by which he meant the WI-38 fetus's parents; the Wistar Institute; and, just possibly, the NIH. But he was not going to be so naive as to leave the cells in the NIH's possession while ownership was sorted out. Do that and he was sure he would never see them again.

Moving a family of seven 2,900 miles was no small undertaking. The Hayflicks split the travel. Ruth flew out to the Bay Area with five-year-old Rachel and two-year-old Annie. Hayflick drove the three older children cross-country in their dark green Buick LeSabre—the latest in a series of sedans that Hayflick's father had passed on to the family.

Joel, age eleven, sat up front with his father most of the time, with ten-year-old Deborah and nine-year-old Susan in the backseat. They drove west through Pittsburgh, stopped to see drag races in Joplin, Missouri, and then drove on to Arizona, where they gazed at the world's best-preserved meteor crater and marveled at the immensity of the Grand Canyon. All along the way, Hayflick says, some extra cargo traveled with them. Carefully strapped on the backseat beside his daughters was a liquid-nitrogen refrigerator stuffed with ampules of WI-38.

CHAPTER SIXTEEN

In the Bear Pit

Philadelphia and Bethesda, 1968–70

Science appears calm and triumphant when it is completed; but science in the process of being done is only contradiction and torment, hope and disappointment.
— Pierre Paul Émile Roux, French bacteriologist and developer of
the first effective treatment for diphtheria[1]

As 1968 began, the race to develop a successful rubella vaccine intensified. Maurice Hilleman's Merck team, now committed, thanks to pressure from Mary Lasker, to the HPV-77 duck vaccine, was conducting human trials in the United States and abroad. With all of the huge drug company's resources behind the effort, eighteen thousand children were injected in field trials of the vaccine by early 1969.[2]

The Missouri company Philips Roxane, which had tweaked Parkman and Meyer's HPV-77 virus by growing it in puppy kidney cells, had sent it for human testing to physicians at Georgetown University in Washington, DC; at an inner-city clinic in Nassau, Bahamas, run by the Catholic Church; and at Parke-Davis, a drug firm in Detroit.[3] The Detroit company, which had in-house expertise in running clinical trials, would inject nearly twelve thousand children.[4]

In Genval, Belgium, the company Recherche et Industrie Thérapeutiques, known as RIT, was racing forward with a rubella vaccine made from a virus that its scientists had captured from the urine of a ten-year-old girl with rubella and grown in African green monkey kidney cells and then in kidney cells from three-week-old rabbits.[5] By the end of 1968, the company had injected its experimental vaccine, called Cendehill, in 25,000 people.[6]

All three companies, as they armed themselves with the extensive volume of vaccine, and the attendant monkey tests and other safety data that Plotkin was ill-equipped to provide, were keenly aware that the DBS was under pressure to get a vaccine licensed soon. And all were hoping that their vaccine would cross the finish line first. Whichever company persuaded the U.S.

regulator to issue the first license would be poised to capitalize on the vast U.S. market of worried women and their small children.

Plotkin was now the only independent scientist trying to push forward a vaccine. Despite Smith, Kline & French's $10,000 grant late in 1967, the way forward was anything but clear. Roderick Murray's January 1968 letter forbade him to send his RA 27/3 vaccine to out-of-state colleagues, unless and until he made and safety-tested what for him were daunting amounts of the RA 27/3 vaccine, including injecting the vaccine in the brains of monkeys and observing the animals for weeks on end.

Plotkin could, and did, supply the vaccine to foreign collaborators in countries including Israel, Iran, Japan, and Taiwan. (Murray was using an arcane law to prevent Plotkin from shipping the vaccine between states but could not block him from sending it out of the country.) But even with their help, he couldn't hope to muster the tens of thousands of vaccinees—and the corresponding invaluable data—that the three companies were accruing. Plotkin and his collaborators were looking at vaccinating hundreds of subjects at most in the coming year.

In Philadelphia Plotkin wound up his trials at St. Vincent's Home for Children. The turnover among the one- to five-year-olds at the home wasn't large, and he had by now vaccinated most of the sixty-one children living there in the spring of 1968. That April he wrote a letter of thanks to Sister Agape, the administrator, and Sister Mary Joseph, the trained nurse at St. Vincent's, and offered to vaccinate the handful of remaining children who had not received his vaccine.[7] He also wrote to Archbishop Krol, who had recently been elevated to cardinal.

Dear Cardinal Krol:

During the last 3 years, we have been testing a vaccine against German measles at the St. Vincent's Home for Children. This fact will, of course, be known to you, since permission was obtained before the studies began. What may not be known to you is that the results have been excellent and the development of this vaccine has been immeasurably advanced by these studies done at St. Vincent's. This vaccine will ultimately protect women from the risk of having deformed children as a result of German measles during pregnancy.

I want to express my appreciation to Sister Agape and Sister Mary Joseph of the St. Vincent's Home for their unstinting cooperation and kindness during these studies.

Very sincerely yours,
Stanley A. Plotkin, M.D.[8]

Within days he received a reply from the cardinal, pronouncing himself "delighted" with the news."[9]

In the spring of 1968 Plotkin also launched what he considered a crucial test of his vaccine in which he enlisted fourteen Philadelphia families, mostly from among friends and neighbors in the complex of town houses where he lived. Many of them had connections to the university and were likely well versed in the risks of contracting rubella during pregnancy. They included his own family.[10] By this time the new government rules required him to get his subjects' formal, written consent.

Plotkin wanted to see how his RA 27/3 vaccine functioned in the real world. The children he had vaccinated in institutions certainly lived in close contact with one another, but their situations did not exactly mimic conditions in the outside world. If the vaccine virus was transmissible, the intimacy of family life would surely be a cauldron for its spread. In each of the fourteen families a child lacking rubella antibodies was injected. One of the child's siblings, and the child's mother, both also lacking antibodies, were followed as intimate contacts who would be at risk of contracting the disease if the vaccine virus was capable of spreading. If the nonimmune mothers and siblings had not developed antibodies by the end of the trial seven to nine weeks later, it was a sure thing that the vaccine virus had not spread.

In Plotkin's case his six-year-old son, Michael, was vaccinated. His wife, Helen, had given birth to a second son, Alec, in 1966. Alec was approaching two years old at the time of the trial early in 1968. Neither Helen nor Alec had antirubella antibodies. Asked in a 2015 interview if he had concerns about giving the vaccine to his son, Plotkin responded: "No, no, not at that point. If I had concerns, of course I wouldn't have done it. I felt it was in some sense obligatory that I give it to my own family."[11]

By September 1968 the family study was finished and Plotkin had pulled together the results, along with the data from all of the trials he had conducted at Hamburg since he had published the first trial there one year earlier, in the midst of the Marburg outbreak. His paper on the new Hamburg studies, and on the family study, was published in February 1969 in the *American Journal of Epidemiology*. On the first page he and his coauthors—Farquhar, the genial director of pediatrics at Philadelphia's Presbyterian Hospital; Katz, Plotkin's Wistar colleague and former medical school roommate, who had escaped the Nazis; and another University of Pennsylvania pediatrician, a jokester named Charlie Hertz—thanked several people for their "invaluable help." Among them was Hayflick.[12]

Plotkin had a lot to be thankful for: Nearly all of the children he and his colleagues vaccinated had developed significant levels of antibodies—levels comparable to those of his competitors' vaccines. And none of the children's contacts, either at Hamburg or in the Philadelphia families, including his own, had become infected with rubella. Nor had any of the vaccinees developed signs of clinical rubella, save for a few swollen lymph nodes.

He concluded that all of his latest RA 27/3 virus formulations—the formulations that in practice would be used in manufacturing by companies—generated robust levels of antibodies, were safe for vaccinees, and were noncontagious for their contacts. But the studies had been tiny: at Hamburg they had been conducted on seven, or four, or six children and, in some cases, a similar number of unvaccinated, antibody-lacking children living in the same ward and sharing the same playroom with them. And his collaborators who had tested the RA 27/3 vaccine abroad, while they had produced excellent results—antibodies in 100 percent of vaccinees—had injected a sum total of 163 people.

"Larger studies will have to be done, of course," Plotkin concluded, "to define completely the safety of RA 27/3 or other rubella vaccines."[13]

And how to execute those larger studies, hobbled as he was by the constraints imposed by the DBS's Murray? Fortunately for Plotkin, Bob Ferlauto, the feisty Sicilian who was the director of microbiology research at Smith, Kline & French, had become keenly interested in Plotkin's RA 27/3 vaccine since he first funded Plotkin with $10,000 at the end of 1967.

This might seem mystifying for two reasons. First, Murray's implacable opposition to using WI-38 cells to make vaccines was well known among drugmakers. Smith, Kline & French had no reason to be optimistic that he would change his mind and allow such a vaccine to come to market in the United States. But the Philadelphia company had its eye on a route of administration of RA 27/3 that it hoped might overcome Murray's resistance.

After that first trial in 1965 at St. Vincent's Home, Plotkin hadn't abandoned the idea of giving the vaccine as nose drops. He had continued to include the nasal formulation in his human trials. And it had begun to work. His was the only vaccine that induced immunity when given intranasally, and Smith, Kline & French, in the person of Ferlauto, was hopeful that the DBS would be less averse to a WI-38 cell–manufactured vaccine that wasn't injected and therefore didn't come into such deep and intimate contact with the vaccinee. What was more, children clearly would love to avoid needles, and that could conceivably give the company a market edge. It would later emerge that large amounts of the vaccine needed to be squirted into the nose in order to ensure its effectiveness. Because the big volumes would offset any market edge that it might give companies, the nasal vaccine became a financial nonstarter. But as Ferlauto considered Plotkin's vaccine, this drawback wasn't yet evident.

There is another reason that Smith, Kline & French was interested in Plotkin's vaccine. Early in 1968 the company acquired RIT, the Belgian firm that was pushing hard to bring to market its own injectable rubella vaccine, the Cendehill vaccine. With the purchase of RIT, Smith, Kline & French became the owner of the Cendehill vaccine at the same time as the company was investing in Plotkin's RA 27/3 vaccine. At a time of fast movement in the rubella vaccine race, and with uncertainty about how well any single vaccine would perform once on the market, the company was hedging its bets.

"They are riding two horses as a hedge against an accident which could remove Cendehill from contention," Plotkin wrote in a memo that fall.[14] If the Cendehill vaccine failed to perform, then RA 27/3 could take its place.

Early in 1968 Smith, Kline & French's Ferlauto brought all the resources he could muster to Plotkin's RA 27/3 vaccine. In April he saw to it that the monkey-brain injection tests required by the DBS were contracted out to scientists at RIT in Belgium.

"For Ferlauto. Stop. Monkey safety and neutralized virus controls will be started on Plotkin's vaccine as soon as possible. Stop. Results will be available within six weeks. Stop," read a telegram that arrived at Smith, Kline & French on April 9, 1968, from Constant Huygelen, the head of vaccine development at RIT.[15]

And beginning in January 1968, a smart young Smith, Kline & French patent agent—Alan D. Lourie, who would go on to become one of the country's top intellectual property lawyers and a senior federal judge—was enlisted to get Plotkin's vaccine patented, pronto.[16] The patent had to be applied for quickly, because Plotkin's *American Journal of Epidemiology* publication of September 1967—the one that had announced to the world that he had created an effective vaccine—had set the patent application clock ticking.[17] A patent had to be applied for within a year after an invention was divulged publicly.

On April 1 the company, with government permission, filed for a patent on Plotkin's behalf.

By August Ferlauto was proposing that his company make ramped-up quantities of the vaccine and hurrying to apply to the DBS for what was by then called, as it is today, an Investigational New Drug permit, or IND, which would allow the company to launch human studies.[18] In the meantime Ferlauto wrote to Koprowski, "I would like to strongly recommend" that the Wistar not seek financial support or commercial interest from other companies.[19] He wanted to protect his company's market, in the event that the Plotkin vaccine was licensed.

Ferlauto didn't know that on that score he was already too late. Throughout 1968 Koprowski had been actively talking to London-based Burroughs Wellcome and the family-owned French firm Institut Mérieux in Lyon, where the Wistar's chief was good friends with the owner, Charles Mérieux. The three companies would spend 1969 sparring over who would be licensed the rights to sell Plotkin's RA 27/3 vaccine in what portions of the world.

Plotkin may have garnered important industry support, but an event in the autumn of 1968 made clear just how far out in the political cold Plotkin's vaccine still stood—along with the fetal cells it was grown in.

That October the NIH held an unusual press conference to announce the results of the first trial of rubella vaccine in the midst of a natural outbreak of the disease. In the spring of 1968 an islandwide rubella epidemic had taken hold in Taiwan. The disease had been completely absent there since the previous epidemic ten years earlier. As a result, children under ten years old had never been exposed to the virus and lacked immunity.

Researchers from the University of Washington, from the U.S. Naval Medical Research Unit 2 in Taipei, and from the Municipal Health Bureau in Kaohsiung, a seaport city in southern Taiwan, seized the opportunity to test Merck's newly tweaked, duck-based vaccine—and, alongside it, Plotkin's RA 27/3 vaccine. In late April and early May the scientists vaccinated hundreds of first to fourth graders at four schools in Kaohsiung. In the subsequent months less than 1 percent of the children who received injections contracted rubella. Among unvaccinated children the attack rate varied from 13 percent to 28 percent, depending upon the school.[20]

Daniel Mullally, the chief of the Vaccine Development Branch at the NIH's National Institute of Allergy and Infectious Diseases, delivered the good news to the press conference that October. (It was Mullally who one year earlier had sung the praises of Parkman and Meyer's HPV-77 vaccine in a letter to Koprowski explaining why his branch was refusing to fund Plotkin's work on RA 27/3.) Both the *New York Times* and the NIH's in-house newspaper, the *NIH Record*, covered the story on their front pages.[21,22] The HPV-77 vaccine, invented on the NIH campus by Parkman and Meyer and tweaked slightly by industry, had shown itself to be remarkably effective in the Taiwanese epidemic, the articles reported.

A reader of either of these articles would have had no way of knowing that Plotkin's vaccine too had been injected in the Taiwanese children, and that it had performed slightly better than Merck's. One child out of 187 injected with the Merck vaccine had contracted rubella. One child of 198 injected with Plotkin's RA 27/3 had gotten the disease.[23]

Either Mullally did not mention Plotkin's vaccine at the NIH press conference or he gave it very short shrift. Either way, the result was the same. Neither the *New York Times* nor the *NIH Record* mentioned Plotkin's RA 27/3 vaccine. It was as if it didn't exist.

On a chilly evening in mid-February 1969, Plotkin attended an evening reception at Governor's House, a hotel in Bethesda, Maryland, that catered to visitors and scientists from the nearby NIH. The hosts were Lewis Thomas and Saul Krugman, two leading lights of U.S. medicine and the dean and chair, respectively, of pediatrics at New York University Medical School. Their reception opened the International Conference on Rubella Immunization. Held in the largest venue on the NIH campus—the auditorium in the imposing Clinical Center, the NIH's research hospital—the three-day meeting was attended by everyone who was anyone in the world of rubella vaccinology. More than four hundred people from all over the world turned up.

The meeting was charged with energy and expectation: virtually everyone present knew that the DBS was preparing within several months to license the one or more rubella vaccines it judged to be fit for the U.S. market. There was equally a sense that licensure could come none too soon, with the new epidemic that everyone feared expected as soon as 1970.

Murray himself had invited Plotkin to the conference in a letter months earlier. "Great strides are being taken in research directed to the control of rubella," the DBS director had written in what passed for enthusiasm in his buttoned-down parlance. "Your participation will do much to contribute to the value of the conference."[24]

Onstage at the huge meeting were the accumulated data for four vaccines. There was Plotkin's RA 27/3, grown in WI-38 cells. There was Parkman and Meyer's HPV-77, in its adaptations by Merck in duck embryo cells and by Missouri-based Philips Roxane, which had grown it in dog kidney cells. And there was Smith, Kline & French's Cendehill vaccine, developed several years earlier in Belgium by virologists at RIT, the subsidiary of the Philadelphia drugmaker. They had captured their rubella virus from the urine of a ten-year-old girl.[25]

Plotkin's vaccine was the only one that had not been in intimate contact with African green monkey kidney cells. It was also by far the underdog when it came to institutional support. More than fifty thousand people had received one of the four experimental rubella vaccines; of these, just five hundred people had received Plotkin's injected vaccine.[26,27]

There were several surprises in store for the conferees.

First, when Hilleman stood to speak, he explained that while both Merck's Benoit vaccine, the original vaccine that the company had developed, and its HPV-77 duck vaccine were "acceptable" for use in children, Merck had dropped the former in favor of the latter. Even though the HPV-77–based vaccine generated levels of antibodies in vaccinees several times lower than Merck's Benoit vaccine, it was in the "best national interest" to concentrate on a single virus when it came to getting a vaccine licensed as quickly as possible, Hilleman explained cryptically.[28] He did not address why that single vaccine should not be the Merck Benoit vaccine that generated superior antibody levels. And he did not mention a powerful philanthropist named Mary Lasker.

Then there was unsettling news from new studies of the HPV-77 vaccine candidates in adult women. Women who received the Merck and Philips Roxane vaccines were getting joint pain and inflammation—arthralgia and arthritis in medical parlance. Not just a few women. Lots of them.[29] Hilleman's group reported that in a trial of Merck's HPV-77 duck vaccine, twenty of thirty-five women (57 percent) developed a rash, joint pain, joint swelling, or a combination of these. The pain and swelling afflicted their fingers, their big toes, their ankles, their wrists, and their knees. Two women were in enough pain to need steroids. The fluid sucked out of one woman's swollen knee was found to contain live rubella virus.[30]

The dog vaccine made by Philips Roxane was just as bad: fourteen of twenty-five women (56 percent) developed joint pain, swelling, or both. In half of them it was distressing enough that they sought out their doctors.[31]

The Merck paper on the women's joint problems presented at the conference would be published several months later and would conclude by noting that tests of Merck's HPV-77 vaccine, further weakened by more passages through duck embryos, were in progress. It would be "desirable to eliminate" the joint pain and swelling, the authors conceded. Nonetheless, they put it on the record that "the vaccinated women accepted their vaccine-related illness and were reassured by their [new] immunity against rubella."[32]

Ostensibly the adult women's problems should not have been a deal breaker. The experts at the CDC who made vaccine recommendations were expected to target the vaccine at young children, on the theory that if they did not contract the disease, they would not pass it on to their pregnant mothers. In all the clinical trials so far, vaccinated children had rarely been bothered by joint pain or swelling. And, in fact, universal childhood vaccination is what the CDC did end up recommending, just two months later.[33]

At the time, experts were divided on whether it made the most sense to vaccinate children or women of childbearing age—after first ensuring that they weren't pregnant, because it wasn't clear if a weakened vaccine virus could damage the fetus.*[34,35,36]

The United Kingdom would opt to vaccinate schoolgirls just before puberty. (After serious rubella epidemics followed in the UK in 1978, 1979, and 1983, the country in 1988 began immunizing preschoolers of both sexes.)[37]

In the winter of 1969, with licensure of one or more rubella vaccines imminent, every physician in the packed auditorium at the NIH knew one thing with certainty. Even if the United States chose to routinely vaccinate small children rather than their mothers, their women patients with pregnancy plans would be at their doors demanding the vaccine the moment it was licensed. And they, as these women's doctors, were going to be hard-pressed to refuse them. So the physicians may have taken note when, during his turn at the podium, Plotkin touched on his findings with his own vaccine in adult women.

Plotkin had decided to test the RA 27/3 vaccine's effects in grown women the previous year, as soon as he heard by the grapevine about the arthritis problems with the Merck and Philips Roxane vaccines. By the time of the NIH conference, he had vaccinated sixty-one student nurses in Philadelphia hospitals; not one had developed joint pain or swelling, he told his audience.[38] What

---

*Regarding the risk of vaccination to pregnant women, over the next two decades, through 1989, the CDC would follow to term the pregnancies of 305 rubella-susceptible women who were inadvertently vaccinated within three months before or after conception. None of the women gave birth to rubella-affected infants. Today the CDC advises that women who inadvertently receive a rubella vaccine while pregnant should be counseled that there is a theoretical risk of up to 2.6 percent that the fetus will be affected. It adds that several additional studies have shown no cases of congenital rubella in about 1,000 infants of susceptible women who were vaccinated while they were pregnant, or just before conceiving.

was more, in preparation for the conference, he had gathered data on an additional eighty-six adult women vaccinated with RA 27/3 by his foreign collaborators. Again, none had developed joint symptoms.

Plotkin concluded his talk by touching on what he saw as his vaccine's other advantages: it was as good at generating protective antibodies as his rivals', but, unlike theirs, it was made in human cells, without the risk of lurking animal viruses. And it could, unlike its competitors, generate antibodies when given as a squirt in the nose.

Plotkin didn't spell out his question as he wound up his talk, but to him it seemed to hang in the air. Why, with the disadvantages of an HPV-77–based vaccine—its joint pain, its arthritis, its rashes, its long lab-dish exposure to African green monkey kidney cells—would the DBS choose to license it, as it was almost certainly going to do in the coming months?

When Roderick Murray, the DBS chief, stood up to speak on the last day of the conference, it was clear that Plotkin's unspoken question wasn't bothering him. His talk did not mention Plotkin's vaccine or Hayflick's human cells. He did point out that the other candidate vaccines had had intimate contact with monkey kidney cells. As to the safety concerns this provoked, producing a rubella vaccine indisputably free of silent monkey viruses was simply unaffordable, he announced.

"The production of a [animal virus] free vaccine suitable for general use appears at this time to be infeasible from the economic standpoint," Murray told the audience. The best he could suggest was careful, long-term follow-up of the vaccinees from the field trials for any untoward effects.[39]

Plotkin's RA 27/3 vaccine, made using Hayflick's virus-free human diploid cells, seemed once again simply not to exist.

The single session from the conference that participants remember vividly fifty years later came on its last afternoon, during a discussion that began with a question about the origins of the WI-38 cells used to grow Plotkin's vaccine. The questioner, Dr. Kevin McCarthy from the University of Liverpool, in England, worried aloud that bits of the WI-38 cells' DNA might be present in the vaccine and then become incorporated in the DNA of vaccinated people. McCarthy asked for information about the siblings of the WI-38 fetus. Were they normal? And what was the reason for the abortion? His implication was that there would be more reason to worry if the fetus had been aborted because it was diseased or abnormal, or if its parents had other diseased children.[40]

Plotkin offered the following confident response.

"This fetus was chosen by Professor Sven Gard, specifically for this purpose," he told the audience. "Both parents are known and, unfortunately for the story, they are married to each other, still alive and well, and living in Stockholm, presumably. The abortion was done because they felt they had too many children. There were no familial diseases in the history of either parent, and no history of cancer specifically in the families; that is, the maternal or paternal sides."[41]

(In fact, Mr. and Mrs. X were no longer married at this point and did not live in Stockholm. But this is beside the point, scientifically.)

Plotkin's reassuring comments immediately prompted a response from the man who was arguably the most powerful person in the room. There were few people as respected—and feared—in U.S. virology as Albert Sabin, who had been known for his piercing intellect and first-rate science even before his live polio vaccine was licensed in the United States in 1961. Since then he had taken on near-godlike proportions in many eyes.

At age sixty-two Sabin was old enough to be Plotkin's father. His receding hair was snowy white. He had thin lips, a thin mustache, and an intimidating gaze. He spoke with a self-assurance born from defeating innumerable lesser men. (He had been on the New York University debating team as an undergraduate, and, it was said, he had never lost a debate.) Hilleman recalled Sabin later as a "mean goddamn bastard" who "went from one field to another, always sneaking in."[42] "He always thought he was more important than anybody" was the DBS scientist Bernice Eddy's recollection.[43] In the 1960s Sabin's influence on the Division of Biologics Standards was enormous.

Sabin took the floor to make two points. First, he challenged the repeated claim by WI-38 defenders that the cells were "fully characterized," with its implication that they did not, and thus never would, show any nasty tendency to become cancerous; that they did not, and never would, be found to harbor hidden "passenger" viruses.

"There is no full characterization for any cell line, because for everything we do, there is always some hypothetical something for which we cannot test," Sabin argued. Sure, Hayflick and Plotkin and other WI-38 backers had scrutinized WI-38 cultures with their electron microscopes for years and never found an unwanted virus. That didn't mean such a virus mightn't be there. They could keep looking until doomsday and never be sure.[44]

Sabin also objected to critics who had labeled him as reacting from emotion rather than reason. His audience need only consider that viruses had been found that caused leukemia in mice. And of course everyone was aware of the now-famous virus named after Peyton Rous that caused sarcomas—nasty tumors of connective tissue—in chickens. There was an urgent hunt on, Sabin continued—one that the NIH's National Cancer Institute and Sabin himself were part of—to discover if a comparable human leukemia virus existed. If it did, there was every possibility that it was hidden in WI-38 cells. What was more, it was known from studying cells from these species in culture that leukemia-causing viruses didn't always reveal themselves in early passages in the lab. They were there, nonetheless, invisible, waiting. So, Sabin concluded, "reasonable people can disagree" about whether WI-38 critics were an emotional bunch reacting from their hearts and not their heads.[45]

As Sabin held forth, Plotkin had begun making notes in his crabbed handwriting on a sheet of lined paper. "Pure theology," he wrote. "No factual basis." Plotkin was not a religious man, but at that particular moment a phrase from the first book of Samuel in the Hebrew Scriptures came, unbidden, into his mind: "The Lord hath delivered him into my hands."

When Sabin surrendered the floor, Plotkin stood and walked toward the microphone at the front of the room. On the dais Murray tried to shut Plotkin down, pleading limited time for debate.

"Let him talk!" someone shouted from the audience.

Murray acquiesced.

"As I recall from polio days," Plotkin began, "debating with Dr. Sabin is very much like getting into a bear pit. One does not come out in exactly the same shape as one went in."

But it was Plotkin who then bared tooth and claw.

Sabin had made the point that leukemia viruses sometimes did not show up in early passages of infected animal cells. Well, he should be aware that, using electron microscopy, viruses had not been found in young, middle-aged, or old WI-38 cells—up through and including the point of cell death after fifty or so divisions. Nor had tests using antiviral antibodies to home in on viruses hidden in WI-38 cells ever been positive—not even tests using the blood serum of leukemia patients, whose antibodies presumably would have sought out any lurking "leukemia virus." Nor had exhaustive chromosomal studies raised concerns. Nor had the cells *ever* turned cancerous.

"It has always been curious to me," Plotkin continued, "that the same people who worry about WI-38 do not worry about the unknowns in other tissues." How many lots of monkey kidneys that were used in vaccine production were studied to ensure normal chromosome numbers, he asked? How many were studied to see if the cells became cancerous with time?

"Dr. Parkman," Plotkin went on, "indicated that the monkey kidney cells used in the first seventy-six passages of the HPV-77 strain were not subjected to detailed tests for silent monkey viruses. But this apparently does not disturb anyone.

"What we are dealing with here is theology," Plotkin concluded. "And, you see, in theology it is very hard to disprove the existence of things. One cannot disprove the existence of ghosts. But that is not, to my mind, a basis for making intellectual decisions."[46]

As Plotkin left the microphone, the auditorium filled with what he would recall decades later as "a thunderous ovation."[47]

There was one person who wasn't there clapping. Hayflick had left the conference early, upset, according to Plotkin, by the sidelining of his cells.

Four months later Murray's DBS published the official manufacturing standards that first allowed the licensing of a rubella vaccine in the United States. The guidelines specified that the vaccine must be propagated on duck embryo or dog kidney cells, effectively anointing Merck and Philips Roxane as the only two companies that could seek a license.[48] Merck's vaccine was licensed by the DBS the same month. The decision was no surprise to the company, which had a flock of Pekin ducks, complete with pond, waiting at its West Point campus, and 600,000 vaccine doses ready for release on the June day that it won FDA approval.[49] The Philips Roxane vaccine followed it to market in December 1969. A few months later the DBS expanded its guidelines to allow rubella vaccine to be made using rabbit kidney cells, allowing Smith, Kline & French's injected Cendehill vaccine to be licensed early in 1970. Because it did not produce joint symptoms as often as its competitors, it became the vaccine of choice for doctors confronted with adult women asking to be vaccinated.[50]

It quickly became apparent that the vaccine produced in puppy cells by Philips Roxane was unacceptable. Not only did it cause arthritis and joint pain in women, but children, who rarely got arthritis after vaccination, began getting it after they received the vaccine, which went by the trade name Rubelogen. The children suffered arthritis in their knees, wrists, and hands. Some had carpal tunnel syndrome. The inflammation could last for weeks, and in nearly 30 percent of children, and probably more, it recurred.[51] Other kids came down with a strange, painful nerve syndrome that woke them at night with hand, arm, and wrist pain and in the morning with pain behind their knees that caused them to hobble around in what was nicknamed the baseball "catcher's crouch." This syndrome too could persist for months.[52]

Within months state agencies began returning the dog-kidney vaccine, unused, to the company and refusing to accept new shipments.[53] In September 1970, nine months after it was licensed, federal government agencies began refusing to buy it.[54] The company withdrew the vaccine from the market in 1972.

And what of Smith, Kline & French's enthusiastic backing of Plotkin's vaccine? Two events early in 1970 torpedoed the collaboration. One was the DBS's licensing of the company's Cendehill vaccine. The other was the resignation, just weeks before the Cendehill vaccine was licensed, of Plotkin's patron at the company, Ferlauto, who suddenly retired to Puerto Rico.

"There [is] probably no one at SK&F now that would really push the program," Koprowski's deputy, Norton, wrote in a memo to his boss and to Plotkin, after taking a phone call from the company with the news of Ferlauto's departure.[55] He was right. Several months later the company dropped development of Plotkin's vaccine. Norton called Plotkin, who was at a meeting in Zagreb, to break the news.

In the United States Plotkin's vaccine had hit what seemed to be a final, insurmountable wall. The huge U.S. vaccine market had become the property of Merck and of Smith, Kline & French and their respective duck embryo and rabbit kidney cell–propagated vaccines.[56]

Events in Europe couldn't have been more different. Shortly after he returned from Zagreb, Plotkin wrote to Hayflick at Stanford, letting him know that the meeting participants had done some back-of-the-envelope math and concluded that about four million people worldwide had been vaccinated with WI-38–based vaccines. He added: "You will be glad to know that [WI-38] rubella vaccine is now licensed in France and Yugoslavia [and] is on application in the UK."[57] The UK would license the vaccine on December 29, 1970, and Burroughs Wellcome would quickly bring it to market. Separately, Pfizer was preparing to apply for licenses in the UK and the United States to market polio vaccine made in Hayflick's human diploid cells. On the other side of the world, the Australian government's vaccine maker, the Commonwealth Serum Laboratories, was busy outfitting the facility where Plotkin's rubella vaccine would be made; Hayflick would travel to Melbourne in March 1971 to advise the Australian scientists. In the country where the ophthalmologist Norman Gregg had first recognized congenital rubella, it was the WI-38–propagated vaccine that the government would deploy to eliminate it.

CHAPTER SEVENTEEN

Cell Wars

Stanford, California, June 1968–December 1969

The entire history of our development of the human diploid cell strains is a litany of failures, disappointments, misunderstandings, and, finally, law suits. Yet, it is not without humor.

Leonard Hayflick, 1984[1]

When Hayflick arrived at Stanford in the summer of 1968, he had neither house nor lab. The family rented a temporary home near the campus fraternity houses. There they waited for fifteen months while a new, boxy house of yellow ocher was built just for them on a quiet enclave called Mears Court, in the comfortable neighborhood that students ironically call the Faculty Ghetto. It had five bedrooms, and even a small study for Hayflick. There were almond trees in the yard and a magnolia with huge white blossoms that was spectacular in the spring.

As for the precious trove of original WI-38 cells, Hayflick recalls leaving them, while he waited for his lab to get up and running, with a highly qualified friend and colleague, Walter Nelson-Rees, who helped run the cell-culture facility at the Naval Biological Laboratory across the San Francisco Bay in Oakland. (Hayflick also remembers that a Nelson-Rees lab technician thawed and expanded several of the WI-38 ampules without permission, just out of curiosity, setting Hayflick's teeth on edge when he learned of the loss.)

After the cold, gray winters and humid summers of Philadelphia, Stanford made for a welcome change. Eleven-year-old Joel admired the limpid light. Ruth, who had been looking forward to living in California, found that the bucolic campus, with its red tile roofs and sandstone masonry, its fragrant eucalyptus and its legions of tanned, bike-riding undergraduates, didn't disappoint. The faculty wives were friendly. She learned that, as a spouse, she could attend classes. She enrolled in a biology course and met a friend who, like her, was an artist. Soon she was grabbing what time she could as a mother of five young children to attend the Palo Alto Art Club and to use the printmaking press in her new friend's garage.

For an ambitious biologist like Hayflick, Stanford could hardly fail to be a draw. Joshua Lederberg had founded the Department of Genetics a decade earlier, just after winning a Nobel Prize for discovering how bacteria transfer genes. Arthur Kornberg in the Department of Biochemistry had won a Nobel in 1959 for elucidating how DNA is built. His biochemistry colleague Paul Berg, who would soon take over from Kornberg as chair of the department, would win a Nobel in 1980 for being the first to splice genes from different organisms together. The trio of "Bergs" made Stanford a mecca at a time when the frontiers of biology were expanding in thrilling ways.

Admittedly, Stanford's Department of Medical Microbiology, where Hayflick became a professor, was not on a par with the Bergs' dominions. Even before accepting the job, Hayflick knew that the department was a backwater, a fact reflected by its quarters in a remnant of the old museum that had been two thirds destroyed in the huge 1906 earthquake. But that left plenty of room for him to blaze a path of his own.

The department didn't have room for Hayflick's lab in the turn-of-the-century building that it shared with the anatomy department. Instead Hayflick was given a Butler building: a rectangular, prefabricated structure of about a thousand square feet. It stood on a patch of gravel that also served as a parking lot, between the medical microbiology building and the back of the university museum. Across the gravel patch, up against the anatomy building, was a kennel of narcoleptic dogs kept by the pioneering sleep researcher William Dement. His staff regularly let the small black and white dogs out to run in the parking lot, where they would sometimes fall asleep in midstride. (Hayflick also parked the family car here during Saturday football games, to avoid paying for parking near the Stanford Stadium across the campus, his son Joel recalls.)

The Butler building was an empty shell, and the Hayflick lab had to be built from scratch, which took months. In the meantime hollow doors served as lab benches and Hayflick created the "walls" of an office for himself by stacking cardboard packing boxes. Within a couple of years Hayflick found NIH funds to help him expand the lab: a double-wide trailer was positioned at one end, and a door was built connecting the two buildings.

After the addition the lab—dubbed "the Cancer Virus Building" for the NIH program that funded it—had a corner room for two secretaries, an office for him next to that, a warm room for incubating cells, a large liquid-nitrogen freezer, and enough common lab space to accommodate technicians, graduate students, and visiting colleagues.

The Hayflick lab was a hive of activity occupied by, among others, an excitable junior technician from Iran whose true desire was to be a banker and who drove his green Volkswagen Beetle around the campus like a maniac; and Eric Stanbridge, a blue-eyed, thin-lipped PhD student who had already worked under Hayflick at the Wistar. Woodring "Woody" Wright, a brainy Harvard graduate with a red beard and earnest brown eyes, would arrive in 1970 to complete his PhD in Hayflick's lab, at the same time as he charged through an MD at the medical school. The mainstay of Hayflick's support staff was his highly capable technician, Nancy Pleibel, a tall, athletic twenty-eight-year-old who had studied chemistry at Penn State and resisted her parents' urgings to become a gym teacher. She had wanted to be a lab technician since she was twelve years old, and after she joined Hayflick's Wistar staff in 1963 or 1964, she advanced to become his chief technician and followed him from Philadelphia to Stanford.

The lab was nothing special, and its subpar air-conditioning meant that it could get uncomfortably hot during the summers. But it was Hayflick's, right down to the custom-labeled glassware, which announced, in big, brown, permanent uppercase letters, one word: HAYFLICK.

During the hot, humid East Coast summer of 1968, the Wistar Institute's scientific administrator, a PhD microbiologist named Robert Roosa, was in Toronto, Canada, attending a tissue-culture meeting with his Wistar colleague Vince Cristofalo, a rotund, phlegmatic native Philadelphian with six daughters who had nonetheless found time to become absorbed by studying aging in Hayflick's fetal cells. Roosa was sought out and summoned to the telephone. On the line was Hilary Koprowski's secretary. She told him and Cristofalo to drop everything and fly back to Philadelphia immediately. The Wistar's chief needed to speak with them on a matter more urgent than anything taking place in Toronto.

At the Wistar Roosa found Koprowski "rushing around" his office, beside himself. The precious, young, original WI-38 cells were gone. Hayflick had made off with every last ampule. Koprowski was furious. Pacing back and forth, he asked Roosa and Cristofalo whom they thought the WI-38 cells belonged to. "He was excited," Roosa recalled. "We were just speculating on how much those cells were worth. To Hilary, they were worth a lot."[2]

Before long, Koprowski sent Cristofalo to Stanford to try to secure, at a minimum, the ten ampules of cells that belonged to the Wistar under the January agreement. Hayflick recalls someone—he doesn't remember who—from the Wistar arriving in his Stanford lab to demand the cells. He remembers feeling annoyed and sending the person back to Philadelphia empty-handed, claiming that he needed lead time and he wouldn't just hand over the cells when the request was sprung on him.[3]

Cristofalo, who died in 2006, recalled the following episode in an interview with Roger Vaughan, who published a biography of Koprowski in 2000. Vaughan writes:

> Koprowski sent Cristofalo out to get Wistar's share back. He flew to California only to be stonewalled by Hayflick. "We had words," Cristofalo says. Koprowski had a session on the phone with Hayflick, and told Cristofalo the deal was all set. Again Cristofalo flew to California. "Hayflick asked me if I had a container. I told him yes, liquid nitrogen. He asked me if it had been tested. He said unless it tested good for seventy-two hours, he wouldn't let me have the cells. I told him it was brand new, and that we were talking about a ten-hour trip, maximum. Again he refused. Again we had words."[4]

Hayflick's departure with the cells became known to those who relied on them very soon after he left Philadelphia. Koprowski was "enraged," recalls Plotkin, who wrote to Hayflick after he had been on the job at Stanford for precisely nine days.[5] "This is to remind you of our discussion concerning WI-38. As soon as you have reconstituted the ampule, we would appreciate receiving a culture. Similarly, we would like to receive a culture from each subsequent ampule. I hope things are going well in California."[6]

Other WI-38 cells were not by any means unavailable at this point. Scores of scientists across the country and around the world had them growing in incubators or tucked away in freezers. The ATCC, the well-known cell repository in Rockville, Maryland, already had in its possession about one hundred ampules of cells that had doubled fifteen times, which were useful for research.[7] But it was the original eighth- and ninth-passage ampules of WI-38 that were golden to vaccine makers, because the cells were so young and so immensely, exponentially expandable. It was hundreds of ampules of these original, oh-so-young cells that Hayflick now watched over at Stanford like a jealous mother hen.

Hayflick was soon just as busy at Stanford as he had been at the Wistar. He continued his research on *Mycoplasma*, the tiny organisms that he had studied as a PhD student. The microbes were the subject of about half of the scientific papers and articles that he published during the late 1960s. Most of his other publications continued his campaign to get WI-38 accepted for making viral vaccines.

When Hayflick had left the Wistar in mid-1968, it had seemed to the experts at NIH's cancer institute that the scientific need for the WI-38 cells had been sated.[8] So the agency had let lapse his contract to provide the cells to scientists. But one year later Donald Murphy, an enterprising project officer in a different part of the NIH, realized that the demand for the cells hadn't waned among scientists in the active and growing field of cellular aging. The WI-38 cells were perfect vehicles for examining the aging process—a field that Hayflick and Moorhead had thrown open with their landmark 1961 paper noting that normal cells aged in the lab.

So in mid-1969 the NIH awarded a new contract to Stanford, with Hayflick as principal investigator and Murphy as the project officer. Hayflick was hired again to produce, study, store, and distribute the cells on the NIH's behalf, this time to biologists doing research on aging.[9] (Murphy did not know that Hayflick had taken the cells to Stanford against the wishes of the NIH's cancer institute. It was a case of one arm of the NIH not knowing what the other was doing.)

But in the meantime, during his first year at Stanford, from July 1968 through June 1969, Hayflick kept distributing the WI-38 cells as he saw fit. He sent twenty-seven original eighth-passage ampules to the Medical Research Council in London, three to the USSR, and five or ten to Drago Ikić, the vaccine master at the Institute of Immunology in Zagreb.[10] He also thawed eleven of the eighth-passage ampules and expanded their cell populations until they had doubled between nine and twelve times, then sent the resulting cells, in bottles, bathed in medium, to scientists. Keeping up with the brisk demand for these dozens of "starter cultures" had him thawing an original eighth-passage ampule every four to six weeks, as he had already been doing for several years.[11]

Hayflick simply ignored the January 1968 agreement and the February 1968 follow-up letter from the NIH to Koprowski, stating that the agency owned the cells and that it was now "official policy" that no more eighth-passage ampules should be thawed and expanded.

Since he didn't have NIH contract support in this first year that he was at Stanford, Hayflick began charging $15 to offset the costs of growing, preparing, and shipping the cells—less, he noted to a meeting of cell-culture experts, than the $25 that the nonprofit ATCC was charging for older, fifteenth-passage WI-38 starter cultures.[12] "These funds," he told his colleagues, "are used to defer [sic] charges for postage, glass-ware, packaging and culturing these cells for distribution."

In November 1968 he established a fund at the university called "Medical Microbiology—Culturing Expense," and deposited the money there.[13] He would eventually, he told himself, get the lawyers to sort out who owned the cells. In the meantime the funds could simply accumulate.

Once the new NIH contract to distribute the WI-38 cells for aging research became operative, in June 1969, Hayflick began shipping the cells for free to researchers studying aging. The contract covered those costs. But he continued to charge costs when others asked for the cells. He made no exceptions for researchers at the NIH, where his paying customers included the Division of Research Services and scientists at the agency's environmental health institute.[14] (Both groups of NIH purchasers would later explain that they thought Hayflick owned the cells and had no idea that the NIH claimed title to them.)[15] Hayflick also charged his former Wistar Institute colleagues, writing to Plotkin at one point in 1969 to ask why he had not received the $15 payment for two cultures he had shipped over the past several months. Was it simple negligence, or had the institute adopted a policy of not paying him? Plotkin soon wrote back to say that he had nudged the appropriate party to make the payment.[16]

After Wistar scientist Vince Cristofalo failed to retrieve the WI-38 cells from Hayflick at Stanford, another Koprowski emissary finally succeeded. On January 3, 1969, Pavel Koldovsky, a Czech physician who was then a visiting scientist at the Wistar, collected ten original eighth-passage ampules from Hayflick and returned them to the Philadelphia institute.[17]

Nonetheless, it seems that Koprowski kept up the pressure on Hayflick to return more of the hundreds of remaining original WI-38 ampules, and that Hayflick responded by trying to wrest from the Wistar's boss a written assurance that, if he did, Koprowski wouldn't turn around and sell the cells: Plotkin wrote to Hayflick in March 1969 telling Hayflick, "Naturally everyone disclaims the intention to sell WI-38," but that if he wanted something in writing, he would have to get it from the board of managers.[18]

It seems highly unlikely that Hayflick ever got that written statement, given Koprowski's active efforts in 1969 to negotiate license agreements for Plotkin's RA 27/3 rubella vaccine, made using the cells, with the Institut Mérieux in France, with Smith, Kline & French in Philadelphia, and with Burroughs Wellcome in the UK. What is clear is that, as the spring of 1969 turned into summer and Hayflick entered his second year at Stanford, he was taking definite satisfaction in holding out on Koprowski.

At one point Plotkin wrote to inform Hayflick of moves by the Philadelphia drugmaker Wyeth to pressure the DBS to accept the WI-38 cells. Wyeth had used the cells to manufacture a new adenovirus vaccine for the U.S. military. Now it wanted to bring that vaccine to the civilian market. The company was getting fed up with Murray's resistance. Wyeth officials were preparing to take their case to Richard Schweiker, a Republican U.S. senator from Pennsylvania.[19] Hayflick responded by complaining about being left in the dark and threatening to tell the senator that the Wistar couldn't either provide or control the precious young ampules.[20]

At some point Koprowski seems to have given up on any serious effort to recover more of the hundreds of remaining ampules of WI-38 cells that Hayflick had at Stanford. Perhaps this was because Hayflick was now so geographically out of reach. Perhaps it was because Koprowski disliked direct conflict, much as he was prone to it. Perhaps it was because several companies already appeared to have adequate supplies of the youngest WI-38 ampules. Or maybe it was because by now Koprowski understood just how obdurate Hayflick could be.

CHAPTER EIGHTEEN

DBS Defeated

Washington, DC, Spring 1972

The DBS has acted in many major areas by simply not acting at all.
—Leonard Hayflick, in testimony to the United States Senate, April 20, 1972[1]

In March 1972 Hayflick saw a decade of relentless proselytizing finally pay off. The Division of Biologics Standards approved for the U.S. market a vaccine made in WI-38 cells, for polio.

Pfizer was the company that had at last wrung a license from Roderick Murray. The huge drug firm was headquartered in midtown Manhattan, but—thanks to the United Kingdom's friendliness to WI-38—it had a facility in Sandwich, England, that was already online to make the new polio vaccine. Initially it would be shipped to the United States from Britain.[2] It was called "Diplovax," named with a nod to Hayflick's "diploid" cells, which Pfizer had put to use pumping out Sabin's polio vaccine virus, now the dominant polio vaccine in most of the world, including the United States.

It was a deeply satisfying moment for Hayflick. The very box in which the new vaccine was packaged proclaimed his accomplishment for the world to see, in bold blue letters on a bright white background: "Propagated in a Human Diploid Cell Strain (WI-38 Hayflick)."

Characteristically, Hayflick promoted the new vaccine in the pages of the *New York Times*, where on March 8, 1972, an article appeared under the headline VACCINE PRODUCED IN HUMAN CELLS.[3] Once again Hayflick was the only person quoted in the article, and once again he reminded readers that "the monkey kidney is a notorious reservoir of unwanted viruses." He went on to forecast that rabies, rubella, and measles vaccines made with WI-38 cells would soon be on the U.S. market.

Why the inscrutable Murray had at last seen fit to approve a human cell–based vaccine is difficult to know. To a reporter from *Science* magazine he claimed that the approval—ten years after WI-38 cells became available—happened quite simply because Pfizer's was the first application the DBS received for a vaccine using the cells.[4] His comment is hard to take at face value, given Murray's power, his resistance to Wyeth's efforts to license the adenovirus vaccine it had made for the U.S. military using WI-38 cells, and his undoubted awareness that no drugmaker aiming at the civilian market would invest the time and resources required to develop a vaccine without at least an informal indication that the DBS would be open to such an application.

In any event, Murray had much more preoccupying him in March 1972 than explaining the reasons for his decade of resistance to WI-38. The previous autumn a seasoned, outspoken U.S. senator had taken a keen interest in Murray's DBS and how it was—or wasn't—performing.

Sixty-two-year-old Abraham Ribicoff, a liberal Democrat from Connecticut, had briefly been secretary of the Department of Health, Education and Welfare, the huge department that housed the NIH, during the administration of his longtime friend President John F. Kennedy. A lawyer from humble origins and a former governor of Connecticut, Ribicoff was a tough political street fighter and someone whose scrutiny must have made for severe indigestion at the DBS. He was in a perfect perch from which to train a laserlike eye on Murray's division, because of his chairmanship of a Senate subcommittee overseeing "executive reorganization and government research" as part of the larger Committee on Government Operations.

Ribicoff had been contacted in the summer of 1971 by a whistle-blower, J. Anthony Morris, a veteran DBS scientist who had been a firsthand witness to, among other things, Bernice Eddy's silencing and demotion a decade earlier for finding and pursuing the "substance" in the rhesus and cynomolgus monkey kidney cells used to make polio vaccine that caused cancer in hamsters— the "substance" that turned out to be the silent monkey virus, SV40.

The whistle-blower Morris wasn't always easy to work with, but he was an accomplished scientist who among other things had discovered an important respiratory virus, respiratory syncytial virus, and had demonstrated, after its discovery in polio vaccine, that SV40 could directly infect human beings when squirted into their noses. His complaints about the DBS were lent gravitas by Morris's scientific record. What was more, his lawyer, James S. Turner, was a serious and capable DBS opponent, having recently written *The Chemical Feast*, a critical study of the FDA, for Ralph Nader's Center for Study of Responsive Law.

In October 1971 Ribicoff took the Senate floor to detail Morris and Turner's long and extremely damaging catalog of the DBS's alleged incompetence, paralyzing conservatism, and inherently conflicted position acting as both regulator and developer of some products, notably rubella vaccine.[5] Most disturbing, Morris claimed that the DBS had repeatedly shut down or ignored scientists' research when their findings pointed to safety or effectiveness issues with vaccines already on the market or soon to be approved.

Morris alleged that the DBS knowingly approved watered-down influenza vaccine—a charge confirmed six months later by the nonpartisan Government Accounting Office (now called the Government Accountability Office, the investigative arm of Congress).[6] He highlighted the DBS's slowness to insist that companies stop using potentially SV40-infected cells from rhesus and cynomolgus monkeys to make polio and adenovirus vaccines in the early 1960s. He charged that the suppression of Eddy's findings on SV40 was part of a pattern in which the division regularly shut down or ignored scientific research that challenged DBS decisions or views on the safety of vaccines—including the Merck rubella vaccine.

At one point in 1969, Morris alleged, a DBS scientist found foreign viruslike particles in lab cultures of duck embryo tissue—precisely the kind of tissue that Merck used to make its rubella vaccine. The scientist was told to abandon the research because the particles were "biologically inactive." Murray was heard to comment: "We must be very careful because if we were to reveal viral contamination this would cause a severe financial loss to the producer."[7]

Morris himself was removed from a position heading influenza vaccine regulation in 1966, when he concluded from a clinical study he had just conducted that the vaccine protected only about 20 percent of vaccinees and told his bosses he wanted to explore the reasons why.

Ribicoff's highly public release of the charges from Morris and Turner made it immediately clear that the DBS was in deep political trouble.

To Hayflick the scrutiny was beyond overdue. And when Ribicoff invited him to be one of the witnesses during four full days of subcommittee hearings the following spring on Capitol Hill, he accepted. Hayflick was the third among the seventeen witnesses who appeared before the Ribicoff panel over four days in April and May 1972. He testified on the first day.

Hayflick held nothing back. He lauded the "clear superiority" of his "absolutely clean" WI-38 cells over monkey kidney cells, telling the senators: "Each monkey is a universe unto itself and may carry harmful viruses in its cells, unlike WI-38. . . . The scope of this problem can be appreciated if one realizes that each lot of [polio] vaccine may require the sacrifice of several hundred monkeys whose kidneys are a veritable storehouse for the most dangerous kinds of contaminating viruses. In fact, monkey kidney is in this sense the 'dirtiest' organ known."

Hayflick lamented that the DBS staff had "dug in their heels" after SV40 was found contaminating polio vaccine, by refusing to offer companies even the slightest indication that the agency would license a vaccine that used WI-38 cells. He assailed the division for its "slothful," "ultraconservative" positions, which had led to the Sabin polio vaccine being fed to fifteen million Russians before it touched American lips. He deplored the situation in which Murray sat as judge and jury on a rubella vaccine developed by his own scientists. He called for panels of expert but external advisers to weigh in on licensing decisions. He bemoaned the finality of Murray's decisions, for which there was no mechanism of appeal. He noted that there was no limit to Murray's now-seventeen-year tenure. And he called for Murray's replacement with someone from outside the DBS.[8] (It was already public knowledge that the NIH had convened a search committee to look for an immediate replacement for Murray.)[9]

Nowhere in his comments was Hayflick sarcastic or glib. He was, on the contrary, deeply sincere. But nowhere was he the least bit politic, either. In this he was classically Hayflick: unable to resist a podium and, once on it, unable or unwilling to pull any punches.

"I think you have made a devastating case against DBS and its present policies and practices," the chairman of the full Senate committee, Senator Fred R. Harris, told Hayflick the moment he concluded his remarks.[10]

It seems certain that the attendant humiliation was deeply felt at the DBS. In one memo written six days after Hayflick's testimony—a memo that became part of polio-vaccine litigation and was unearthed in the book *The Virus and the Vaccine*—a member of the public relations department at Lederle Laboratories, then the dominant U.S. maker of polio vaccine, wrote: "Received a call from DBS Information Office informing us that they have finally decided to take strong action in opposing Dr. Hayflick's allegations concerning monkey tissue vaccines. Apparently the CDC is involved in this counter move."[11]

Hayflick likely aggravated feelings at the NIH further with a hard-hitting attack on DBS scientists that he published the next month in *Science* under the headline HUMAN VIRUS VACCINES: WHY MONKEY CELLS?[12]

The contrast of Hayflick's in-your-face testimony with Plotkin's response to the Senate hearings is striking and aptly crystallizes the difference between the two men. It also reflects the fact that the stubborn Plotkin still hoped that his rubella vaccine would one day win approval from U.S. regulators. So he had no wish to offend them. When Ribicoff asked him to testify, Plotkin kept his head

down, replying to the senator with a succinct letter that explained: "I have decided against testifying in person because this whole history has been rather
unpleasant for me and I do not wish to become actively involved in the problem again." Instead, he wrote, because he felt compelled to do so as a citizen, he
would summarize in writing "the facts as I know them."

Plotkin then related, in pale, measured language, the chronology of his attempts to develop a rubella vaccine in the face of DBS resistance and of his being driven as a result to look for manufacturing interest from English and
French companies. He concluded with the most passive, inoffensive sentence
he could construct: "I believe that the events unfolded above could have been
avoided under different administrative circumstances."[13]

Plotkin, who as a boy had survived pneumococcal pneumonia, influenza,
and asthma, continued, decades later, to be a survivor.

The man who didn't survive was Murray. By the time of the Ribicoff hearings, Murray was "requesting [to] be reassigned" within the NIH, where he
became "special assistant" to the director of the National Institute of Allergy
and Infectious Diseases, until his mandatory retirement sixteen months later.
The man of few words left his position with only a cloaked regret about his
"request": "I do this even though I have strong feelings of loyalty to my staff."[14]

With Murray out, it was the NIH deputy director for science, a kidney
physiologist named Robert Berliner, who was left trying to explain the DBS's
failings to the senators. He was one of a bevy of officials who accompanied
Elliot Richardson, then the Secretary of Health, Education and Welfare, to the
hearing.

By the time it happened, the outcome of the Senate subcommittee's investigation surprised no one. In July 1972 the DBS's 258 employees were transferred to the Food and Drug Administration, and the division—which would
nonetheless remain physically on the NIH campus until 2014—was renamed
the Bureau of Biologics. Later it became today's Center for Biologics Evaluation and Research at the FDA. In 1972, its new director was familiar to both
Plotkin and Hayflick. He was Harry Meyer, of the Meyer-Parkman duo in the
now-defunct DBS, who had developed the HPV-77 rubella vaccine that Merck
had later made its own.

The Pfizer polio vaccine that the DBS approved in March of 1972 ended up
surviving not a great deal longer on the U.S. market than human diploid cells
in a lab bottle. As documented in the book *The Virus and the Vaccine*, the

company struggled with chronic supply shortfalls, making pediatricians reluctant to rely on it.[15] At the same time a hostile campaign against Pfizer's WI-38–propagated newcomer was launched by Lederle Laboratories, which in 1972 dominated the U.S. live polio vaccine market. (By the early 1970s Sabin's live vaccine had supplanted Salk's killed one in much of the world, including the United States, where companies had stopped making, and pediatricians had stopped giving, the Salk vaccine.) Only one other company, Wyeth, made a live polio vaccine for sale in the United States, and the company withdrew that vaccine in the early 1970s.[16]

Lederle did all it could to sink Diplovax, the new vaccine made in WI-38 cells with Hayflick's name on the box. Among other things, the company targeted an important committee of the American Academy of Pediatrics, the main membership organization for the pediatricians who dispensed polio vaccine. Lederle urged the committee not to recommend, in its annually updated *Red Book* desk reference for children's doctors, that they switch from the Lederle vaccine to Diplovax. In this effort the company sought to enlist help from the DBS.

Ruth Kirschstein oversaw polio-vaccine safety testing for the division and was as close as any employee could be to Murray. *The Virus and the Vaccine* recounts how, as Lederle geared up to lobby the association of pediatricians, the company wrote to the DBS: "In preparation for such a confrontation [with the American Academy of Pediatrics committee] it would be valuable to have the support of the DBS. Could we count on Dr. Kirschstein?"[17]

Whether the late Kirschstein went to bat for Lederle with the pediatricians' group—and, if she did so, how heartily—is unknown. But it seems unlikely that she would have been given pause by any warm feelings toward Hayflick, who for most of the decade that Kirschstein had been charged with overseeing polio-vaccine safety at the DBS had been publicly assailing the monkey kidneys used to make the vaccine.

Pfizer stopped selling Diplovax in the United States in 1976. From 1977 until 2000 Lederle's African green monkey cell–propagated vaccine was the only live polio vaccine available in the United States, in an era when live oral vaccine was used almost exclusively. In 2000 the Centers for Disease Control recommended that pediatricians move back to using a more potent version of Salk's killed polio vaccine, and Lederle's live vaccine was withdrawn from the market.

The killed, injected vaccine was substituted for Sabin's live vaccine because rarely—about once for every million doses administered—the live vaccine virus can mutate into a more toxic form and cause paralysis, especially in babies with compromised immune systems.[18] When polio was a scourge, paralyzing and killing hundreds of thousands of people around the world every year, that was a risk worth taking. But because over the last sixty years vaccines have obliterated most polio on the planet, that risk-benefit ratio has changed. As a result, in 2015 the World Health Organization asked countries around the world to begin phasing out Sabin's live vaccine, beginning in April 2016. The phase-out won't be completed until naturally occurring polio is completely eradicated. In 2015 it was reported in two countries: Afghanistan and Pakistan.[19] In the summer of 2016, several more cases were documented in Nigeria.[20]

CHAPTER NINETEEN

Breakthrough

Philadelphia, Pennsylvania, and New Haven, Connecticut
September 1970–October 1973

Vaccinology, I would say that it's not rocket science. It's a lot harder than rocket science.

—Alan Schmaljohn, virologist at the University of Maryland, 2014[1]

G iven all that went before, one could be forgiven for assuming that nobody in the United States much noticed, or cared, when Smith, Kline & French dropped its development of Stanley Plotkin's rubella vaccine in the autumn of 1970, after the company won approval for its Cendehill vaccine—and after Plotkin's patron, Ferlauto, left the company. Nobody, that is, besides Hilary Koprowski, Plotkin, and the small circle of colleagues who had worked with him to test it.

After all, the DBS-approved rubella vaccines, developed in the cells of African green monkeys and then grown in duck-embryo cells by Merck—the company branded its vaccine Meruvax—and in rabbit kidney cells by Smith, Kline & French, as Cendehill, had been on the U.S. market now for more than a year. Children were being routinely immunized by their pediatricians, and women who wanted the vaccine were getting it—usually the Cendehill vaccine in the case of grown women, because it seemed to cause fewer cases of arthritis than did Merck's vaccine. (The vaccine made in puppy kidney cells by Philips Roxane of St. Joseph, Missouri, was by this point being rejected because of the bad side effects it caused.) The vaccines seemed to be doing their job: the feared rubella epidemic of 1970 had not materialized.

But among the few Americans who cared about the fate of Plotkin's RA 27/3 rubella vaccine was an exceedingly bright, determined woman named Dorothy Horstmann, who had been paying keen attention to the studies of the various rubella vaccines being published in medical journals—and who rarely, if ever, took no for an answer.

In the autumn of 1970 Horstmann was a fifty-nine-year-old pediatrician and vaccine scientist at Yale Medical School in New Haven, Connecticut. She had watched closely the studies that Plotkin had published over the preceding five years. She had read, and continued to read, the papers on the other rubella vaccines. She had coauthored some of her own studies on them. And she was becoming convinced of the superiority of Plotkin's RA 27/3 vaccine. Like Plotkin, Horstmann resembled a dog with a bone when she got onto a cause. Unlike him, she had a very high profile and a long and distinguished track record in vaccinology. She was also known as a woman of unimpeachable integrity.

Horstmann was born in Spokane, Washington, in 1911 and grew up in San Francisco, where as a girl she shadowed a physician who was a family friend while he saw patients at the local hospital. By 1940 she had earned an undergraduate degree from the University of California at Berkeley and an MD from the University of California at San Francisco. Later, at Yale, she completed a specialization in pediatrics, and in 1961 she became the first woman professor at the medical school in New Haven. She had by then played a crucial role in the development of polio vaccines by discovering that polio invaded the central nervous system by way of the bloodstream.[2]

In September 1970, the same month that Smith, Kline & French dropped Plotkin's vaccine, Horstmann's direct gaze—accented by a brunette bob and a toothy grin—stared out at readers from the pages of the *New York Times* under the headline NEW RESEARCH ON RUBELLA CHALLENGES EFFECTIVENESS OF VACCINATION PROGRAM. Recent studies, the article explained, indicated "that many vaccinated persons—perhaps more than half of them—can become reinfected with rubella virus and that the vaccine might not give life-long protection as originally thought."[3]

Horstmann's was one of those recent studies, and it was days away from being published in the *New England Journal of Medicine*.[4] Its findings were alarming. With colleagues from Yale and the University of Hawaii, Horstmann had set out to discover how well vaccination actually protected vaccinees when they were placed in the midst of a natural rubella epidemic. She knew where to find one: every year a group of military recruits known as the all-Hawaiian company arrived at the U.S. Army's Fort Ord, California, for training. And every year at Fort Ord there was a rubella outbreak after these young Hawaiians—who were known to be particularly vulnerable to rubella because of their geographical isolation—arrived and began living in close quarters with others in a state where the virus circulated freely.[5]

So in a study begun in the spring of 1969, the Horstmann team vaccinated the Hawaiian recruits with Smith, Kline & French's Cendehill vaccine two to three months ahead of their departure for Fort Ord. During the outbreak that then ensued at the fort, 80 percent of the vaccinated recruits were reinfected with rubella. This is not to say that they broke out in rashes and nursed fevers, although on rare occasions, in other studies of reinfected vaccinees, this did occur.[6] Rather, spikes in their antibody levels showed that the virus had reentered their bodies and multiplied silently.[7] For the reinfected recruits this hardly mattered. But for pregnant women in contact with them, it might be positively dangerous.

The problem was not that the vaccinations given in Hawaii had failed to stimulate the men's immune systems to develop antibodies. In fact, the men had all developed decent levels of antirubella antibodies after being vaccinated and before shipping out to Fort Ord. Nor was it because the California microbe was some particularly aggressive strain of the virus—a group of men at Fort Ord who were naturally immune to rubella because they had been infected in childhood easily fought it off. Blood tests showed that just 3.4 percent of those men who were immune because of childhood bouts with the disease were reinfected.[8]

The message was that the antibodies generated by the vaccine—as opposed to those generated by natural infection—were somehow failing to prevent reinfection by naturally occurring virus. They were in some way acting differently—they were weaker—than the antibodies generated by natural infection.

Horstmann's study joined others about rubella vaccine then being published in the *New England Journal of Medicine*, the *American Journal of Diseases of Children*, and the *Journal of the American Medical Association*.[9] These studies found that people vaccinated with the Merck and Cendehill vaccines were being reinfected in the open community, in places like first-grade classrooms in Memphis; in other, nonmilitary institutions, like the residential Nazareth Childcare Center in Boston; and during clinical studies in which vaccinated people were deliberately exposed to the virus through nose drops.[10] The sum total of the work raised a question that Horstmann put bluntly in her *New England Journal of Medicine* paper: "If vaccinees can be readily reinfected a few months after successful immunization, what are the long-term prospects for durable protection of the young woman in the childbearing age who was successfully vaccinated at the age of six?"[11]

Ten months before her paper was published, Horstmann had begun a quiet campaign to respond to that question—not with more studies on the vaccines already on the U.S. market but by presenting an alternative.

In late 1969 or early 1970 Horstmann wrote to Plotkin asking for data on his RA 27/3 vaccine.[12] Plotkin was only too happy to supply it and ended up sending Horstmann hundreds of ampules of his rubella vaccine, provided by the two companies that were in the process of winning market approvals for it abroad: the Institut Mérieux in Lyon, France, and Burroughs Wellcome in London.[13]

Horstmann wanted to repeat her study of Hawaiian recruits when they landed at Fort Ord, this time using the RA 27/3 vaccine. She wanted to inject Plotkin's vaccine in Hawaiian recruits who had never been exposed to rubella, watch them ship out to Fort Ord, and see if, once they were living in close quarters with other, stateside, recruits and were exposed to rubella for the first time, they got reinfected at the same high rates as had already been demonstrated in the recruits who had been vaccinated with the Cendehill vaccine. Her hunch was that they wouldn't be. She suspected that Plotkin's vaccine generated better immunity. But she needed to do the study to prove it.

Then Horstmann hit a roadblock. She soon wrote to Plotkin that Colonel Edward Buescher, the deputy director, soon to be commandant, of the Walter Reed Army Institute of Research and as such the one in command of what studies could and could not be done on army recruits, was standing in the way of the new study of RA 27/3 vaccine in Hawaiian recruits "for reasons that do not make sense, either scientifically or administratively."[14] Horstmann couldn't make out his "ulterior motives."

Not one to be turned aside, Horstmann proceeded to find another route to her goal. In December 1970 she launched a study of Plotkin's vaccine among several hundred children in Danbury, Connecticut, a developing commuter town of fifty thousand outside New York City. She partnered with a local doctor, Martin Randolph, whom she described to Plotkin as the leading pediatrician in Danbury; with Warren Andiman, a young infectious disease doctor at Yale; and with a Yale virologist and mother of five named Ann Schluederberg. The group vaccinated children in Randolph's private practice, in a Head Start program, and at a day-care center in Danbury, comparing Plotkin's vaccine, head to head, with the Merck and Cendehill vaccines. Then they watched the young vaccinees' antibody levels over time. Their results would take fully eight years to generate and publish.[15]

Horstmann also launched a smaller, quicker trial with Plotkin's longtime collaborator, Ingalls, at the Boston School for the Deaf. The resulting paper, published early in 1972, showed that the antibody levels in children vaccinated with the Plotkin vaccine in 1968 hadn't significantly declined two years after they were first vaccinated. What was more, the scientists had injected a "challenge" strain of disease-causing rubella into the noses of eighteen of the children who had been successfully vaccinated two years earlier. Just two of them—11 percent—showed a subsequent rise in antibody levels that indicated they had been reinfected.[16] Eleven percent wasn't trivial, but studies showed that even people who had natural rubella immunity from having had the disease in the past were reinfected at rates of 3 percent to 9 percent. And 11 percent certainly didn't come close to the 40 percent to 100 percent reinfection rates that were being found with Merck's duck-embryo vaccine and in Cendehill vaccinees.[17]

Horstmann, as usual, was blunt in her conclusions. Yes, her numbers were small, and larger studies would be needed to confirm them. But the point was that to protect the fetuses of the future, it was crucial that any rubella vaccine deliver long-lasting immunity—and that that immunity be "qualitatively comparable" to the immunity that resulted from natural infection.[18]

By "qualitatively comparable" Horstmann meant that a successful rubella vaccine needed to generate not only robust *levels* of antibodies to rubella virus but also the same *types* of antibodies that are made during natural infection. When the body's immune system recognizes a foreign invader, it generates antibodies to numerous different pieces of that invader, called antigens. It also generates categories of antibodies that differ in how and where they attack the virus. Horstmann pointed to increasing evidence that showed that Plotkin's vaccine mirrored the natural immune response in the *kinds* of antibodies it produced. This, she remarked, was in notable contrast to Merck's HPV-77 duck vaccine and the Cendehill vaccine made by Smith, Kline & French using rabbit kidney cells.[19]

Anyone reading the paper couldn't miss Horstmann's clear conviction that the RA 27/3 vaccine—Plotkin's vaccine made in the WI-38 cells that Hayflick had launched a decade earlier—was simply better than anything then available in the United States.

In the early 1970s Horstmann began calling Maurice Hilleman, the vaccine chief at Merck. Regularly. Persistently. Always with the same message: Plotkin's RA 27/3 vaccine was superior to Merck's. The Plotkin vaccine was better because it mirrored natural rubella infection much more closely. It mirrored natural infection much more closely because it was far less weakened than the other rubella vaccines. It was less weakened because it had been passaged through cell cultures just twenty-five times. Not fifty-one times, like Cendehill. Not seventy-seven times, like the original Parkman-Meyer vaccine. Not eighty-two times, like Merck's vaccine, which took Parkman-Meyer's virus and sent it through five additional passages in duck-embryo cells.

Hilleman took call after call from Horstmann. Finally he caved.

"When she called me enough times, I said, 'Dorothy, for Christ's sake, you're a pain in the ass,'" Hilleman told interviewer Paul Offit thirty years later. "I said, 'I don't know how much damage you're doing in the scientific community, the medical community, with all of your promotion of RA 27/3. But I guess I have to think about it.'"[20]

Hilleman may also have been influenced by a paper with Plotkin as the lead author that appeared in the *Journal of the American Medical Association* in August 1973. It began with Plotkin-esque self-deprecation.

"Another [rubella] vaccine strain may seem like Marshal Ney at Waterloo, arriving after the battle is over," he began. (Plotkin was soon informed that he had gotten his history wrong: it was the Marquis de Grouchy who arrived late at Waterloo.) He conceded that forty million American children had already been vaccinated with rubella vaccines that were, pointedly, not his. But then Plotkin laid out the facts that had accumulated about his vaccine and the others.

People who received his vaccine generated numbers and kinds of antibodies that mimicked natural infection much more closely than the already-marketed vaccines. Similarly, people given his vaccine developed a resistance to reinfection that looked a lot like the strong resistance that occurred following natural infection with rubella. This was in stark contrast to people who received the Cendehill vaccine or Merck's Meruvax, in whom reinfection occurred far more frequently than in RA 27/3 vaccinees. He offered one reason that this might be so: In some 40 percent of trial subjects who received jabs of his RA 27/3 rubella vaccine, the vaccine prompted antibodies to appear in the cells lining the inside of the nose and throat, where rubella first lands and multiplies before it invades the blood. Vaccines made using HPV-77 or the Cendehill virus did not produce any antibodies in the nose and throat.

His rubella vaccine was now licensed, Plotkin wrote, in Israel, Ireland, New Zealand, and Yugoslavia, in addition to Great Britain, France, and a number of other countries. With trademark understatement, he noted, "There are . . . several reasons for dissatisfaction with the vaccines currently licensed in the United States."[21]

For Plotkin 1972 was a year of big ups and downs. In May the patent on his rubella vaccine was granted, after four years and many rounds with the U.S. Patent and Trademark Office. Hilary Koprowski told him that, while the proceeds would flow to the Wistar, he was going, in turn, to direct 15 percent of them to Plotkin personally, as the inventor of the vaccine. Of course, that amounted to nothing in the United States, where Plotkin was doubtful that his vaccine would ever be licensed. But the Institut Mérieux in France and Burroughs Wellcome in the United Kingdom, having licensed the vaccine from the Wistar, were now selling it. There was going to be some significant money coming his way; that was clear.

At the same time Plotkin's marriage was falling apart. Three months after his fortieth birthday in August 1972, he decided to call it quits. He left Helen and with her his sons, Michael and Alec, aged ten and six. He moved into a high-rise near the Wistar and saw his sons on weekends. Leaving his boys was the hardest thing he had ever done. The experience left him feeling, he says, that between death and divorce there was not much to choose.

During the coming half decade Plotkin's personal life would improve dramatically. A weight of depression lifted from his shoulders. He began dating. A while after the separation, Koprowski, ever attuned to such details, noticed Plotkin's new haircut and detected a livelier spirit emanating from his colleague. He declared him "a new man."

In 1976 Plotkin met Susan Lannon, a vivacious medical librarian who shared his love of classical music. The couple were married in 1979 in a ceremony in the anatomical museum on the first floor of the Wistar Institute, with a glass case of pelvic bones serving as the altar.

A photo of Plotkin from the era seems to reveal the new man that Koprowski observed. He is standing in his office at the Children's Hospital of Philadelphia, where he had become in 1969 the director of the Division of Infectious Diseases—an appointment he carried along with his Wistar position. He has one hand in his pocket, the other loosely on his desk. His posture is open to the camera. An unforced smile is on his face. He looks happy and relaxed.

It was in this same office, back in the early autumn of 1973, soon after his "Waterloo" paper came out in the *Journal of the American Medical Association*, that Plotkin took a call out of the blue from Merck's vaccine chief. After he hung up, it took quite a while for Hilleman's words to sink in. A few days later Plotkin wrote a letter to Hayflick that began with the following words (he referred to Hilleman by the first name that those who knew the man tended to use: "Morris"):

October 3, 1973

Dear Len:

I am writing to ask for several things but I would like to mention first of all that the other day I received a telephone call from Morris Hilleman asking me if the RA 27/3 strain is available to Merck. Needless to say I was left open-mouthed! ... Please keep this under your hat for the moment.[22]

Plotkin turned the request over to Koprowski, because the Wistar owned the patent. It was with Koprowski that Hilleman and a senior Merck licensing executive soon met.[23] Koprowski gave them an estimate for the royalties the company would need to pay to the institute. After the requisite legal vetting, the deal was agreed, and RA 27/3 was licensed to Merck. Now all that Hilleman needed was WI-38 cells—the youngest ones, and plenty of them. So Hilleman did the obvious thing. He called Hayflick.

In the mid-1970s Hilleman, with the full resources of Merck behind him, ran Plotkin's WI-38–propagated vaccine through the laboratory- and human-testing hoops needed for it to pass muster with the FDA's new Bureau of Biologics and become a licensed vaccine. By early 1978 he and others, led by the University of Pennsylvania pediatrician Robert Weibel, had vaccinated nearly eight thousand antibody-lacking people with the Plotkin RA 27/3 vaccine in New York City, Philadelphia, and Costa Rica. They had also run head-to-head trials comparing Plotkin's WI-38 cell–propagated vaccine against Merck's HPV-77 duck-embryo vaccine in smaller groups of children and adult women. Hilleman would watch the vaccinees for two years before publishing the results, to make sure that the Plotkin vaccinees' superior antibody levels persisted over time. They did.[24]

The findings were unequivocal, Hilleman and his colleagues wrote in a paper that was published in 1980. The Plotkin vaccine "induced antibody in a larger proportion of individuals and at a substantially higher titer level" than did Merck's HPV-77 duck vaccine. And in those people—98 percent of Plotkin vaccinees developed antibodies—the RA 27/3 vaccine induced the *kinds* of antibodies commonly found after natural infection.

The better responses, the authors added, "were achieved without any important increase in clinical reactions and there was no evidence for contagious spread of the infection." Admittedly, a handful of children who were injected with Plotkin's WI-38–grown vaccine developed rashes and swollen lymph nodes—a handful slightly larger than the smattering of kids who developed rashes and swollen lymph nodes after receiving the Merck duck vaccine. But these were "inconsequential" in terms of bothering the children, the scientists wrote. By contrast, the rate of arthritis in adult women receiving Plotkin's vaccine was "substantially lower" than with Merck's.[25]

Physicians and the FDA hadn't needed to wait until 1980 to see similar results. In 1978 Horstmann and her colleagues published the long-in-the-making paper comparing the Plotkin vaccine head to head with its two licensed U.S.

competitors among children in Danbury, Connecticut. It too showed that the RA 27/3 vaccine generated higher, longer-lasting levels of protective antibodies than Merck's HPV-77 duck-embryo vaccine—or the Cendehill vaccine made by Smith, Kline & French.[26]

On September 15, 1978, the FDA granted Merck a license for its WI-38–based rubella vaccine.[27] Merck's new vaccine soon became the only one on the U.S. market and remains so to this day. Its only other competitor, the Cendehill vaccine, was withdrawn in 1979.

The Plotkin vaccine virus, RA 27/3, is also used in most of the rest of the world.

PART THREE

The WI-38 Wars

CHAPTER TWENTY

Slaughtered Babies and Skylab

Stanford, California
June 1972–September 1973

One lung—well, a pair of lungs. The Children for [sic] God and all those other people saying you have to kill embryos constantly. It's not true. One embryo will last generations.

—Leonard Hayflick, October 3, 2012[1]

In June 1972 an obscure marriage counselor in Brighton, Michigan, named Dr. Forrest Stevenson Jr., published a pamphlet entitled "Women, the Bible and Abortion." It linked abortion to communism, to Hitler, and to selective elimination of the ill and the elderly. It had subheadings that included "Dollar Bills and Dead Babies" and "Why Would a Woman Want an Abortion?"

The only text in the pamphlet that was boxed for special attention concerned "Dr. Leonard Hayflick at Stanford University." It explained that "in another atrocity connected with abortion, vaccines are now being produced from aborted babies."

The pamphlet informed readers of Hayflick's motive—to avoid the "senseless slaying" of animals during vaccine making—and his methods: "The baby is injected with a live virus and kept alive for several hours until his body is filled with the disease. They then slaughter him, his blood is drained out and manufactured into serum."

The next sentence was in uppercase: "YOUR CHILD'S NEXT IMMUNIZATION SHOT MAY BE CANNIBALIZED FROM A SLAUGHTERED BABY."[2]

The political environment in mid-1972 was primed for this "news" about Hayflick to take root. The U.S. Supreme Court had recently heard a first round of arguments in the landmark abortion case *Roe v. Wade*, and the high court justices would hear the case argued again that October. Also that autumn the voters in Stevenson's home state of Michigan were preparing to weigh in on a ballot measure permitting abortion on demand through twenty weeks of pregnancy.

Stevenson's bulletin circulated in churches during the summer. It also found its way into an antiabortion publication called *Capitol Region Life Line*, which reached readers in New York State. That autumn the letters began to arrive in Hayflick's mailbox. They came from Michigan and from Troy, Elmhurst, and Brooklyn, New York. They were from women who wrote in elegant cursive and often signed with their husbands' names. One asked if it was true that live aborted fetuses were being kept alive and vivisected for the sake of vaccines. How could he justify this murder? Another pronounced herself "appalled" that humans were being treated no better than animals, which were horribly abused enough. Another asked if it would soon be the turn of the elderly to contribute their bodies to the cause and inquired how old he was.

Stevenson's pamphlet reached a slightly broader audience on October 26, 1972, as part of a letter to the editor in the *Sentinel*, a newspaper in L'Anse, Michigan, with a circulation of three thousand. One week later Michiganders overwhelmingly rejected the abortion-on-demand ballot measure, with 61 percent voting against it.[3]

By late November Hayflick had enlisted the help of James Siena, the legal adviser to Stanford's president. He wrote to Stevenson, threatening to sue if Stevenson did not publish and widely circulate a retraction.[4] By the following April Siena had secured a signed retraction from Stevenson, whom his lawyer, Arthur F. Barkey, described as "a wonderful, warm person, genuinely interested in the future of our human species." In the same paragraph Barkey warned Hayflick and Stanford off suing. If they did, he cautioned, Stevenson "will be defended to the hilt by the combined resources of the pro-life groups in Michigan."[5]

The retraction, parts of which were published in several newspapers, read as follows:

April 7, 1973

Dear Dr. Hayflick,

On prior occasion, I have published and given circulation to a statement which alleged that a process for producing vaccines from live aborted babies was developed at Stanford University.... I now understand that no such process was developed at Stanford University and that you made no such statement. Instead, I understand that you are the originator of a process whereby vaccines are produced from a culture consisting of a strain of cells originally taken from the lungs of a single surgically aborted female fetus of four months gestation in 1962. I apologize for any embarrassment which these misstatements may have caused to you or to Stanford University. I assure you that I have ceased publication of this statement and I intend not to repeat it on any future occasion.

Sincerely,

Forrest C. Stevenson, Jr.[6]

The retraction came on the heels of the Supreme Court's momentous decision that January in *Roe v. Wade*, legalizing abortion-on-demand in the first trimester of pregnancy and, beginning in the second trimester and until the point when the fetus was capable of meaningful life outside the womb, allowing states to enact abortion-governing regulations only if they were "reasonably" related to preserving and protecting the mother's health.

As the huge implications of the high court's ruling reverberated around the country, the baby-slaying allegations kept getting reprinted every month or two, usually as letters to the editor from local antiabortion groups in publications like the *Mining Journal* in Marquette, Michigan; the *Altamont Enterprise* in Altamont, New York; the *Gazette Times* in Heppner, Oregon; and the *American Way Features: A Balanced, Unique and Time-Saving Weekly Editorial Service for News Media Dedicated to God, Truth, Freedom, Morality, Capitalism and Constitutional Government*, based in Pigeon Forge, Tennessee.

Hayflick fought back with typical tenacity, contacting every publication, no matter how small, enclosing Stevenson's retraction and insisting that the publication correct the record. His manila file folder labeled "Stevenson" grew to be an inch thick.

July 20, 1973

Passaic County Right to Life Newsletter

LIFE LINE

P.O. Box P-48

Clifton, N.J. 07011

Gentlemen:

It has recently come to my attention that your April–May 1973 Newsletter "Life Line," Vol. 1, No. 3 repeats a most vicious falsehood concerning my research activities. . . . I regret sincerely that you did not have the wisdom to determine the veracity of this statement before publishing it.

Had you made an effort to do so you would have learned that the originator and perpetrator of this calumny has fully retracted the monstrous charges that he himself invented and which you have repeated. I enclose a copy of his retraction for your consideration.

Because of your deep interest in fairness, honesty, justice and the right to life of the innocent, I would urge you to publish the attachment in order to clear the good names and reputation of the people and organizations that you have unjustly injured.

I would appreciate having a copy of the issue of Life Line in which you will publish the attached retraction.

Sincerely yours,
Leonard Hayflick, PhD.
Professor[7]

His records do not show that Hayflick received any response.

While Hayflick beat back the brushfires lit by Stevenson, one abortion opponent sought to engage the Roman Catholic Church in protesting the use of WI-38 cells. On May 2, 1973, James Ambrose, a resident of Fallbrook, in northern San Diego County, California, perused an article by United Press International that appeared in the *Los Angeles Times*. The writer reported from Stanford that "living human cells will be orbited in outer space aboard Skylab for 28 days to determine effects of prolonged weightlessness."[8]

The article did not identify WI-38 cells, which were in fact the cells that the National Aeronautics and Space Administration had chosen, and that Hayflick had gamely provided, for the upcoming zero-gravity experiments on Skylab, America's first experimental space station. But it did quote Hayflick saying, "This marks the first time that scientists will be able to assess the effects of prolonged space flights at the cellular level."

A suspicious Ambrose got out his typewriter and wrote to Joseph T. McGucken, the archbishop of San Francisco, whose geographical domain included Stanford. He asked for a "penetrating inquiry." The article didn't mention human fetuses or embryos, he conceded, but "silence can be eloquent," he wrote. "I therefore question the morality of this experiment, and I refer it to Your Excellency for consideration and possible action."[9]

Ambrose didn't get the response he was looking for from McGucken, a liberal who supported Cesar Chavez's efforts to organize farm workers, spoke out against Proposition 14, an anti-fair-housing ballot measure, and clearly wasn't predisposed to make an issue of the scientific use of WI-38 cells. McGucken first wrote to Hayflick asking for the facts and received a prompt response. He thanked Hayflick in another letter, saying that he was confident that Ambrose would be satisfied with Hayflick's explanation that the abortion had happened long ago and far away.

There was more to come, however. Hayflick recalls that he was working in his home office on the Stanford campus on Saturday morning, July 28, 1973, when he took a call from a top medical official at NASA.[10] The second manned mission to Skylab had just launched from Cape Canaveral. Aboard were three astronauts and a bevy of experimental material, including the WI-38 cells. The NASA scientist told Hayflick that he was now contending with questions from reporters, prompted by antiabortion activists who were protesting the cells' presence on the *Saturn 1B* rocket now hurtling toward the space station. He needed information about the WI-38 cells' origins.

Hayflick, in what had begun to feel like a very familiar routine, explained that the cells had been derived from the lungs of a single fetus, legally aborted in Sweden a decade earlier. The senior NASA official said he would take the information back to the reporters. Hayflick didn't hear from him again.

The Skylab study, NASA Experiment Number SO15, went ahead. The cells were packaged in a clunky, airtight black box with white knobs called the "Woodlawn Wanderer Nine." It was a miniaturized, fully automated lab, and during the astronauts' fifty-nine-day mission, the cells grew on glass in a minia- ture chamber inside it, bathed in medium and kept at a comfortable 96.8 de- grees Fahrenheit. Every twelve hours a motorized pump injected fresh medium and spent medium was forced out. Tiny microscopes were focused on the cells and time-lapse photos of them taken every 3.2 minutes, creating a film that would allow scientists to study their rate of multiplication, among other things. In the meantime a set of control WI-38 cells was being put through the same paces back on Earth. In the end the NASA team found no differences between the Earth-bound WI-38 cells and those sent on the space adventure: their divid- ing times, growth curves, chromosomes, and appearances remained the same.[11]

"The zero-gravity experiments produced no detectable effects on WI-38," the resulting paper concluded anticlimactically. The study did generate some gorgeous electron micrographs, and it was published as part of the proceedings of a Skylab symposium in 1974. One of the micrographs was a close-up of a long, tendril-like projection of a WI-38 cell as it prepared to divide. At the distant tip of the tendril, its edges had grown roundish extensions, due to cytoplasmic bubbling that characteristically occurs just as the cell divides. Shot against a deep black background lit only by occasional starlike, glowing dots, the long, odd-shaped tendril itself looked like a strange interstellar body. But it was in fact a cell from the unborn fetus of Mrs. X, suspended 270 miles above the Earth, in the midst of circumnavigating 858 times the strange planet that held Hayflick and Plotkin and Koprowski and Mrs. X and untold numbers of babies and children with congenital rubella.

The retraction from the marriage counselor Forrest Stevenson didn't mollify many of Hayflick's critics. That was clear from responses like one from a Mrs. Raymond Somerville of Troy, New York. Hayflick had sent her a copy of the retraction. She replied that she saw very little distinction between what had in fact happened to produce the WI-38 cells and what Stevenson had described in his pamphlet. The "defenseless human being" that was an unborn child, she wrote, could not give informed consent.[12]

Yawning between Hayflick and his adversaries was an apparently unbridgeable gulf in belief about the moral status of the fetus.

The campaign by the adherents of Stevenson tailed off that summer. Abortion opponents were perhaps preoccupied with bigger issues on the heels of the *Roe v. Wade* decision. It would be the new millennium before vocal religious opposition to WI-38 again emerged. By then the organizers would have become more sophisticated.

Cells, Inc.

Stanford, California, 1971–75

I was convinced that I had done nothing wrong. For someone to believe that I would steal something is completely outside of my personality.

—Leonard Hayflick, October 16, 2012[1]

In the 1970s biology was on the brink of major changes. The days of the medical scientist as a selfless, salaried public servant were about to give way to the commercialization of biology. Beginning in 1980, new laws and court decisions allowed biologists to make serious money on their inventions. The changes blurred the once-sacred boundaries between business and biology, and by the early 1980s they were turning some biological scientists into wealthy men. Like an overeager guest, Hayflick would find himself arriving at this party just a little too early.

In February 1971 Hayflick took the next fateful step on the road that began in 1968, when he packed the WI-38 cells into the family sedan and drove them cross-country from the Wistar Institute to Stanford. Like so many roads that people follow to their detriment, this one was paved with many significant but quiet choices, rather than one great, dramatic moment of decision. Nevertheless, it led to a place nearly unimaginable for a man of Hayflick's growing prominence.

That February Hayflick opened an account at the Great Western Savings and Loan Association in Palo Alto. He named it "Cell Culture Fund." During his first three years at Stanford, until this moment, he had been keeping the money he collected to cover the costs of preparing and shipping WI-38 cultures in a Stanford University account. (He would later tell NIH investigators that he abandoned the university account because it wasn't interest-bearing.)[2] The exception was his shipments of WI-38 cells to researchers studying aging. The costs of these shipments were paid for by his 1969 contract with the NIH, which was still in effect, and he didn't collect money for sending out those cells.

Hayflick had intended, he says, when he left the Wistar Institute with most of the world's youngest WI-38 cells in tow, to find appropriate lawyers to weigh in on who owned the cells.*

He believed the question was in dispute. If he was aware of the wording of the original 1962 contract under which he had derived the WI-38 cells, which stated that title to materials developed under it passed to the government when the contract was terminated—as it was in 1968—he chose to ignore it.[3]

As for the action plan formulated in a Wistar conference room on that January day in 1968, under which virtually all of the ampules were to be transferred to the American Type Culture Collection, to steward them on behalf of the NIH—well, he may have been at the meeting where it was agreed on, but he had been an unwilling participant.[4]

During the next three years, from February 1971 until March 1974, Hayflick deposited in the Cell Culture Fund at the Great Western Savings and Loan a total of $13,349.84 that he collected for preparing and shipping the cells to researchers outside the field of aging.[5] Hayflick also raised his prices beginning in 1972, in line, he says, with what the American Type Culture Collection was then charging.[6] The nonprofit ATCC was the highest-profile cell bank in the country, and its pricing could be taken as a standard. But Hayflick kept bumping up his prices, and by August 1974 he was charging academic scientists $35, nearly twice as much as the ATCC charged, for a starter culture of WI-38 cells, and he was charging commercial firms $250—more than eight times as much as the cell repository charged.[7]

On March 12, 1974, he closed the "Cell Culture Fund" with its accrued $13,000 and change. One week later, on March 19, he incorporated Cell Associates in the state of California, with himself and his wife, Ruth, as sole stockholders.[8] Hayflick opened a corporate account in the company's name at the Great Western Savings and Loan, and beginning in April 1974, he deposited the money he collected for WI-38 cell preparation and shipments in this account. He did not spend any of it, he says, adding that he was simply waiting for the opportunity to enlist a lawyer to sort out whose funds they were—to sort out who owned WI-38.

But Hayflick didn't get around to finding a lawyer; he was busy. By the summer of 1974 he had been at Stanford six years and had established himself

---

*The Medical Research Council in London still had some of the original WI-38 ampules—ampules of cells that had divided just eight or nine times. Hayflick had given 100 of them to the MRC in the fall of 1962 and had sent more ampules to the London agency from time to time.

as an important presence on the faculty. His groundbreaking 1961 paper with Moorhead, describing the Hayflick limit, was being cited by other scientists an average of sixty times a year—a huge number of citations compared with most papers—and Hayflick's influence was growing commensurately. His lab in the parking lot behind the medical microbiology building buzzed with activity. He supervised graduate students, lab technicians, and postdocs, traveled extensively, and was the lead investigator on several NIH research grants and contracts that in the course of the years between 1968 and 1975 brought more than $2 million in NIH money to Stanford. (That's nearly $10 million in 2016 dollars.) At home his youngest, Annie, was nearly nine years old, and Joel, the eldest, was preparing to become a freshman at Stanford.

In the midst of his dizzying schedule, in the late summer of 1974 an invitation came Hayflick's way from the NIH. The new National Institute on Aging was being launched, responding to an explosion of interest in aging research. Would Hayflick consider being interviewed for the position of director?

Today ambitious biologists often found companies to commercialize their inventions. They rub shoulders with venture capitalists and dispense wisdom from their seats on companies' scientific advisory boards. They are seen as paragons of success and are objects of envy, not scorn. Their universities applaud them and do all they can to enable their entrepreneurial efforts.

So it's hard to appreciate what a radical departure it was in 1974 for Hayflick to form a company. In those days biology was pursued as an end in itself: for knowledge, for glory, perhaps, but not as a commercial enterprise. It had been so since at least 1923, when Frederick Banting and Charles Best, the Canadian inventors of the first "biotech" product—insulin extracted from animals—sold their patent rights to the Board of Governors of the University of Toronto, for $1 each, as a means of getting insulin made and out to the diabetics who needed it.[9] In 1955, when the famous broadcast journalist Edward R. Murrow asked who owned the patent on the first vaccine against polio, its inventor, Jonas Salk, famously replied: "There is no patent. Could you patent the sun?"[10]

In 1953 the earnest author of a letter to *Science* captured this ethic and the expectations of biologists that accompanied it. The "true" scientist, the letter writer Frederick J. Hammett declared,

is not properly concerned with hours of work, wages, fame, or fortune. For him an adequate salary is one that provides decent living without frills or furbelows. No true scientist wants more, for possessions distract him from doing his beloved work.[11]

Very little in this attitude had changed when Hayflick founded Cell Associates. This did not deter him.

Hayflick launched the company, he says, not as a moneymaking venture but simply to get a tax advantage: being incorporated would lessen the amount of tax he had to pay on the money he was accumulating.[12] (The $13,350 he had in hand at the time he incorporated would be more than $65,000 in 2016).[13] Yet soon after founding the firm, Hayflick began to act like an entrepreneur. He ordered officials at London's Medical Research Council, where he had sent one hundred ampules of the youngest WI-38 cells in 1962, to stop distributing them without first seeking his approval.[14] And, departing from years of practice in which scientists, both commercial and academic, had come to him if they wanted WI-38 cells, Hayflick began to seek out customers in industry. With these firms he jettisoned any notion of recovering only his costs in preparing and shipping the cells. He began flat out selling them, at prices that brought in $47,543.38 in the fourteen months after he launched Cell Associates.[15] (That's $209,000 in 2016 dollars.)

In June 1974 he wrote to the French firm Institut Mérieux, which was using WI-38 cells to make Plotkin's RA 27/3 rubella vaccine. He declared that it "has recently been decided that a supply agreement shall be made with manufacturers using WI-38 for research, development and production purposes." In this letter Hayflick added that he didn't want to set prices that would put WI-38 cells out of reach and asked the company for its view on what price would be reasonable.[16] In August he wrote to the director of the Institute of Immunology and Virology in Zagreb, on Cell Associates letterhead. The Institute's quiet but determined chief, Drago Ikić, had been the driving force behind Yugoslavia's becoming the first country to license any vaccine made in WI-38 cells—a polio vaccine—in 1968. Hayflick now informed Ikić that young starter cultures of WI-38 could be supplied for $1,500 each. "We are currently supplying starter cultures of WI-38 at the ninth or tenth population doubling level (PDL) to several vaccine manufacturers of poliomyelitis," he wrote in that letter.[17] He later sent the Yugoslavian institute two starter cultures and collected $3,000.[18]

He approached New York–based Pfizer, the maker of the WI-38–propagated polio vaccine Diplovax, to ask if it would like to acquire a large stock of WI-38 cells; the company told him it wasn't interested—perhaps because Lederle was in the process of chasing it out of the U.S. polio vaccine market.[19] But another U.S. vaccine maker was highly interested. In mid-July of 1974, or soon after, Hilleman, the vaccine chief at Merck, called him, Hayflick says.

Hilleman was keenly interested in obtaining ample supplies of young WI-38 cells because, nine months earlier, Dorothy Horstmann had finally worn down the tough Montanan and persuaded him to drop the company's HPV-77 duck-embryo rubella vaccine in favor of Plotkin's RA 27/3 vaccine, produced using WI-38 cells. Hayflick remembers Hilleman telling him during that phone conversation that a Merck lawyer, Donald S. Brooks, had established that Hayflick owned the WI-38 cells. In addition, says Hayflick, Hilleman told him that Leon Jacobs, the associate director for collaborative research at the NIH, had affirmed to Merck that Hayflick owned the cells. (The NIH's Jacobs would vehemently deny ever saying this.[20] Later Merck would tell NIH investigators that it did not proceed with the contract until Hayflick—not the company—attested that he owned the cells and that NIH concurred).[21] Now Hilleman wanted to buy the youngest cells that Hayflick could provide.

Hayflick was impressed. "I don't have to tell you how persuasive [Hilleman's information] was," he said in a 2013 interview. "I'm talking to a giant in the industry. A man who doesn't want to risk his job as a vice president or besmirch the name of Merck. All that is critically important. So he certainly didn't lie to me. Whether this [Merck] lawyer made a mistake in his interpretation is another question. Could he have? Certainly. But I acted on information I had."[22]

On October 23, 1974, Hayflick executed a contract with Merck. In it he attested that Cell Associates had acquired ownership of all of the hundreds of young WI-38 ampules that Hayflick had taken from the basement of the Wistar Institute, and that previously he himself had had full right to and ownership of those ampules.[23] The contract stated that, as president of Cell Associates, Hayflick agreed to consign to Merck one hundred ampules of WI-38 cells at the ninth population doubling level and fifty ampules of cells at the tenth. Each ampule contained millions of cells with dozens of doublings left in them. The company would be able to thaw and expand them, one ampule at a time, until Hilleman's great-great-grandchildren were in their graves, and longer. For practical purposes the deal would assure Merck an unending supply of WI-38.

Under the contract Merck agreed to pay Cell Associates $5,000 for each ampule of cells at the ninth population doubling level and $2,500 for each ampule of cells at the tenth. Hayflick also granted Merck the right of first refusal on the sale of one hundred more of these young ampules. At the outset the contract would bring in $625,000. Fully executed, it would be worth $1 million—$4.8 million in 2016 dollars.

With his sales of the WI-38 cells to companies beginning in the spring of 1974, Hayflick crossed a line that would make it extremely hard for him to defend himself later. Had he simply continued to provide the cells to biologists and recover only the costs of preparation and shipping, he might have made a case that he was acting as a responsible steward, even if he was thumbing his nose at the 1968 decision by the NIH that the American Type Culture Collection should play this role and that the youngest cells should no longer be thawed and sent out. The arguments that he would later muster in his defense were true: that NIH officials became aware almost immediately, in 1968, that he had taken all the cells to Stanford and didn't chase him down to recover them; that he had openly charged costs for shipping the cells—including to NIH scientists—for years, and the agency had uttered not a peep; and that he had reported distributing the cells "outside the contract" in his progress reports to Donald Murphy, the NIH project officer who supervised the contract under which he distributed the cells for free to researchers studying aging.[24]

But none of this would mitigate the public damage when juxtaposed with Hayflick's collecting more than $47,000 from firms in the fourteen months that he sold the WI-38 cells through Cell Associates—or with his eye-popping contract with Merck worth up to $1 million. By amassing that kind of money for the cells and by negotiating the Merck contract, Hayflick made his case a political and personal disaster when the facts became public.

Ironically, just as Hayflick was defying the rules of behavior for biologists by launching Cell Associates, the first, hugely important, biotechnology patents—which would create some very wealthy academic scientists—were being planned a few hundred yards from Hayflick on the Stanford campus. There, at the medical school, an intense thirty-nine-year-old associate professor of medicine named Stanley Cohen was having his arm twisted by Niels Reimers, the forward-looking chief of Stanford's young Office of Technology Licensing.

Stanford, like other universities in the late 1960s and early 1970s, was being forced to come to grips with the slowing growth of government funding for medical research after a quarter century of astonishing gains. Partly as a result, the university was becoming less passive about trying to commercialize inventions made by its scientists. Reimers, a former Stanford engineering major with experience in the electronics industry, was the face of its quest to do better.

In the spring of 1974 Reimers was trying urgently to get Cohen to agree to file a patent application. Cohen was the lead author on a landmark paper published in November 1973.[25] He was proud of the paper and the huge contribution to biology that it was clearly going to make, launching a new technology so accessible that virtually any lab could use it. Cohen and his coauthors had succeeded in cutting and splicing genes from different biological sources into circular pieces of bacterial DNA called plasmids. By inserting the plasmids in bacteria, they could be made to pump out proteins coded for by those genes. Soon dubbed recombinant DNA technology, the invention would launch the biotechnology industry, leading to products like human insulin and clot-dissolving drugs for heart-attack patients.

But Cohen had not thought of patenting the gene-splicing invention that the paper described. Indeed, Reimers, Stanford's tech-transfer boss, had learned of the exciting new paper only six months after its publication, because Stanford's news director, Bob Beyers, sent him a copy of a *New York Times* article about it.[26] The patent clock was ticking; any application needed to be filed by November 1974, that is, within a year of Cohen's publication of the discovery.

Reimers recalled Cohen as a reluctant conscript who, when first approached by Reimers, told him he wasn't interested in patenting the new technology. He saw his invention as something that was meant to go out to the research community broadly, not to become a commercial captive. Cohen, who was a physician as well as a scientist, acquiesced, Reimers remembered, only after Reimers argued that patent protection would entice companies that would license the technology and use it to develop drugs that otherwise would not be made—drugs like penicillin, whose commercial production had been delayed for eleven years for lack of patent protection.[27] In the end the patent application was filed just in time.

Cohen's attitude was typical of the era.[28] Biology was not thought of as a commercial venture, and most biologists looked askance on any colleague who moved in that direction. In 1976 Cohen's equal partner in the revolutionary discovery learned this the hard way. Herbert Boyer, a biologist at the University of California at San Francisco, was the senior author on the recombinant DNA paper. After he partnered with a young venture capitalist named Robert Swanson and the duo used $1,000 of their own money to launch a company, Genentech, to exploit the new gene-splicing technology of which Boyer was a coinventor, he took tremendous heat from his academic colleagues.

Boyer recalled later that "the way the attacks went, I felt like I was just a criminal."[29]

A good number of Hayflick's colleagues, when they learned of his activities, reacted the same way. "In those days, in that environment, when you did research with government support, it was in the public domain. When it came out that Len was selling these cells, a lot of people were appalled," said Cristofalo, Hayflick's former Wistar colleague whom Koprowski had sent to Stanford to try to recover the cells from Hayflick.[30]

Hayflick, on the other hand, felt not the least bit like a criminal. He continued to see himself as providing a valuable service to colleagues—which he no doubt was doing—and to companies, in exchange for what now was clearly fair market value for the WI-38 cells. Vaccine-making firms, with their bottom-line focus, would not be handing him tens of thousands of dollars for tiny ampules of WI-38 if the cells were not precisely that valuable.

But there were major differences between Boyer's enterprise and Hayflick's. Boyer was an inventor on a patent—unlike the WI-38 cells, the Cohen-Boyer gene-splicing techniques could be patented because they were *methods*, not living things. What was more, Cohen and Boyer had assigned their rights as inventors to their universities, which would be the owners of any patents that issued. Whatever portion of the royalties might end up flowing to the men as individuals would be at their universities' discretion and would be aboveboard gains for them under Stanford's royalty-sharing policies. Hayflick, by contrast, was selling unpatented WI-38 cells that the NIH had paid him to develop under a contract that specified that title transferred to the government when the contract was up. His actions were rather akin to those of a horse breeder on staff at a stud farm—a breeder who is paid by a client to develop a world-class racehorse to sire more of the same. The breeder, after developing the prize stallion, moves cross-country with the animal and begins selling its offspring to wealthy racehorse owners, arguing that it is not clear who owns the valuable stud and saying that he will keep the proceeds of the sales until lawyers can figure that out.

On October 17, 1974, Hayflick spent the day in Bethesda, Maryland, at the invitation of the National Institutes of Health, in a preliminary series of discussions about the directorship of the new National Institute on Aging. He was briefed by the agency's deputy director, Ronald Lamont-Havers, a jowly arthritis physician with black horn-rimmed glasses; chatted with staff in the

Office of the NIH Director; then met over a lengthy brown-bag lunch with the search committee charged with finding a director for the new institute.[31]

Hayflick would later insist that, during his meeting with Lamont-Havers on that day, he raised the issue of his possession of the WI-38 cells and asked the NIH deputy director to assign a lawyer to determine who owned them.[32] Lamont-Havers would just as vehemently insist that Hayflick made no such request.[33] From the NIH's point of view, it would emerge, Hayflick was anything but aboveboard with the agency about his sales of the cells to companies—until the most senior NIH officials learned of those sales, almost accidentally.

On Thursday, May 15, 1975, Donald Murphy, the midlevel NIH administrator who had launched and oversaw Hayflick's ongoing contract to supply WI-38 cells for aging research, was attending a meeting at the Institute for Medical Research, a cell repository in Camden, New Jersey. There he learned something that bothered him.[34]

Murphy was a former marine and a PhD marine biologist who grew up in Hawaii and had come to the NIH in 1967 as a grants administrator. Hayflick liked him and considered Murphy a friend as well as a colleague. Murphy had been communicating with Hayflick as the NIH prepared to open the National Institute on Aging on July 1, keeping him abreast of efforts to staff up the institute and sharing concerns that key positions shouldn't be filled before a director—presumably Hayflick—was in place.[35]

Now, at the Camden meeting, Murphy was startled when someone told him that Hayflick was selling very young WI-38 cells to companies for $5,000 per ampule and somewhat older cells for $250. He contacted a commercial firm and learned that it had recently bought ampules of the slightly older cells from Hayflick at the latter price.[36]

Then Murphy called Hayflick and confirmed that what he was hearing was true—that Hayflick was selling the cells through Cell Associates. Soon he was on a hastily arranged conference call with Thomas Malone, the agency's associate director for extramural research—and with Lamont-Havers, who by then was serving as the NIH's acting director during a several-month gap between permanent directors. The matter was pressing, for the next day—Friday, May 16—Hayflick's final round of interviews for the position of director of the National Institute on Aging was scheduled.[37]

Hayflick's first meeting the next morning was with the NIH's incoming permanent director, Donald Fredrickson. An NIH veteran who had directed the agency's heart institute, Fredrickson was finishing a year's stint as the president of the Institute of Medicine before taking the helm at the NIH, and so Hayflick met him at the institute's home—the stately National Academy of Sciences building on the Washington Mall—and Fredrickson drove Hayflick the eleven miles to the biomedical agency in Bethesda. On the way, Hayflick remembers, Fredrickson turned to his passenger and said, "I hear you have a problem."[38]

After Fredrickson dropped off his passenger at the NIH, Hayflick proceeded through an orderly round of interviews with other senior agency figures. One of these had been hastily added to the schedule the previous afternoon. Just before lunch that Friday, Lamont-Havers, the agency's acting director, met with Hayflick. He told Hayflick, Lamont-Havers recalled one year later, that the job offer was a firm one—"contingent upon the resolution of a possible conflict of interest in relationship to the selling of the WI-38 cells."[39]

Lamont-Havers had already decided who was going to investigate Hayflick's activities with the WI-38 cells. It wasn't a lawyer trained in intellectual property law. It was James W. Schriver, who headed the NIH's internal auditing office. Called the Division of Management Survey and Review, it was charged with providing detached appraisals of the agency's management practices and with rooting out the misuse of grant funds and conflicts of interest. In this the office worked closely with the Justice Department and the FBI.[40]

The NIH hired Schriver to head its new auditing division in 1963.[41] Schriver, who died in 1999, was a native of Carlisle, Washington, an Olympic Peninsula town so small it scarcely appears on maps, even today. He earned a bachelor's degree in business administration at the University of Oregon in 1939 and in 1945 went to work for the U.S. government, moving through internal auditing positions of increasing responsibility. When he joined the NIH, he had most recently worked in the U.S. Army Audit Agency and the Department of Agriculture investigating, among other things, food stamp fraud. Schriver would, over time, become the man sought out by whistle-blowers at the NIH and the person whom agency directors turned to with thorny, unpalatable problems. (In 1971, when Roderick Murray and the Division of Biologics Standards were under fire from Senator Abraham Ribicoff, the NIH's then-director Robert Marston asked Schriver to investigate.)[42]

In the unsmiling photos that peer out occasionally from the archived pages of the NIH's in-house newspaper, the *NIH Record*, Schriver looks to be all business. He has a square face and jowls and, at the time of his 1980 retirement, silver hair. The *New York Times* would later report him to be "widely respected for honesty and fairness."[43] In 1972 the NIH's parent department, the Department of Health, Education and Welfare, bestowed on him its Superior Service Honor Award, "in recognition of his high professional standards and leadership." Richard Dugas, who worked for Schriver and flew with him to California to investigate Hayflick, recalls Schriver as "the type when he walked into the room, you knew he was the boss."[44]

Nicholas Wade, then a *Science* reporter, interviewed Schriver in 1976. He recalls him as being "very accountant-like. Quite a grandfatherly type, very slow and patient but very determined. . . . He presented this as just an accounting issue. These [WI-38] vials were valuable property and it was his job to find out what had happened to them."[45]

When Hayflick's chief lab technician, the tall, likable Nancy Pleibel, decided to follow her boss from Philadelphia to Stanford in the summer of 1968, she left behind the only life she had known, including her newly widowed father and all of her friends. But despite the big geographical leap, and the empty shell of a lab that she and her boss confronted that summer, Pleibel soon knew she had made the right decision.

For her the Hayflick lab continued to be a fine place to work. The job was important. She was contributing to bigger things, and they kept getting bigger: Dr. Hayflick was enterprising, constantly willing to take on another grad student, apply for another grant, travel to another conference. And she and her boss had a very good working relationship. He was decent and, although he was quiet, he didn't fail to communicate. He made his expectations very clear, right down to the correct way to take phone messages. When he was put out, he never yelled. He spelled out the problem and then spelled out how it should be fixed so that it would not happen again. She had never sensed an ounce of duplicity in him; she couldn't imagine him playing less than straight with anyone.[46]

So Pleibel wasn't alarmed when, in late May 1975, Hayflick explained to her and the rest of the staff that some investigators from the NIH would be visiting the lab, looking into something to do with contract management. His NIH project officer, Donald Murphy, would be with them. Pleibel and the rest of the lab staff were to cooperate in every way, answer all the NIH investigators' questions, and show them whatever they needed to see.

When Pleibel first laid eyes on Schriver, he struck her as belonging in a corporate office. He wore a suit; he was tall; he moved and spoke like a man from

that world. He had with him two colleagues who also wore suits. Richard Dugas, thirty-seven, was an easygoing former high school football player and self-
described "family man" with a bachelor's degree in economics from Providence
College in his native Rhode Island, who was hired by Schriver in 1966. Chris
Curtin, a bald, trim, dapper man in his early fifties who had taken a two-year
accounting course at Benjamin Franklin University in Washington, DC, had a
dry wit but agonized over his work—over the costs, to people's reputations, of
making mistakes. Schriver had supervised Curtin at the Department of Agriculture before hiring him away to the NIH.[47]

Over the course of the next three consecutive days, Schriver, Dugas, and
Curtin interviewed Hayflick at length and inspected the Hayflick lab. In a
return visit the next month that lasted another three days, they searched freezers, counted ampules, and pored over Cell Associates' invoices and bank statements.[48] And Schriver asked what seemed to Hayflick's technician, Pleibel, to
be endless questions. The questions were unsettling to Pleibel for two reasons.
First, they often betrayed a lack of understanding of how a lab functioned and
what cell culture was all about. There was, for instance, the moment when
Schriver found several ampules of WI-38 lying on the bottom of a liquid-
nitrogen freezer, rather than clipped in place on the long, unwieldy canes that
hung down in the freezers. (A cane is a long, straight metal rod with slots that
hold ampules.)

There was a simple explanation for how they got there. In the process of re-
trieving ampules from a liquid-nitrogen freezer, they occasionally got dropped
and fell to the bottom of the freezer. This was no surprise: to retrieve ampules,
a technician had to wear the equivalent of a welder's mask, in case the ampule
exploded, and big, gray insulated gloves, and then peer down into the freezer as
the liquid nitrogen steamed up in his or her face, and then wield a pair of for-
ceps to extract the ampule from the cane. It was no easy matter, and when the
occasional ampule got dropped, it was simplest to leave it where it had fallen at
the bottom and select another. The fallen ampule could be retrieved at some
future point, when the freezer was drained of liquid nitrogen. Now that future
point had arrived, and Schriver found ampules scattered on the bottom of a
freezer. "Are these being hidden?" Pleibel, in a 2013 interview, recalled Schriver
asking.[49]

Schriver's questions also became increasingly adversarial as the days went
by. It seemed to Pleibel that he was looking for something that he wasn't finding, and that he was becoming convinced that Pleibel and others on the lab
staff were hiding that something. "He was trying to press like an attorney to

get us to say things that we didn't know about or didn't have the information. And he was getting frustrated," Pleibel recalled. "It seemed like he didn't believe what we were saying."

Schriver had arrived in the lab, Pleibel decided, with a preconceived idea that there should be a specific number of frozen ampules of WI-38, and that he was going to account for every single one of them. Confronted with what would emerge as Hayflick's less-than-meticulous record keeping, this accounting turned into an exercise in teeth-grinding frustration.

At one point, Pleibel recalled, Schriver told her that if his team was not able to get the information that they wanted on this trip, there could be a Senate investigation. The lab staff would have to go back to testify in Washington. Pleibel wasn't intimidated. She felt she had nothing to hide.

She also thought that Schriver was disruptive, wandering around the lab, picking things up, asking a question here, a question there. The staff couldn't plan, didn't know when they would be interrupted, worried about maintaining sterility, grew upset. Finally Pleibel asked to meet privately with Murphy, the NIH program officer. They met in the lobby of the Holiday Inn near Stanford's Palm Drive. (Pleibel remembers the setting vividly because the lobby looked out on the hotel pool, and as she spoke with Murphy, a man emerged from his morning swim, stark naked.) She asked Murphy to ask Schriver to tamp it down. Things improved, somewhat. And then, as suddenly as they had appeared, the NIH investigators were gone.

After first being interviewed by Schriver in late May 1975, Hayflick got on a plane. His journey took him to the Queen Elizabeth Hotel in Montreal, Canada, for the Twenty-sixth Annual Meeting of the Tissue Culture Association. With him he carried two bottles. They contained cultures of WI-38 cells—young cells, at just the tenth population doubling level. In Montreal he handed the bottles to officials from Connaught Laboratories, Ltd., of Toronto, Canada's major vaccine maker. The company had recently licensed Plotkin's rubella vaccine from the Wistar and wanted to ensure that it had the WI-38 cells it would need to produce the vaccine.[50] Connaught agreed to pay Hayflick $5,000 for the two cultures.[51]

That Hayflick continued to sell the cells as the government investigators trolled through his lab is testament either to his naïveté or his bullheadedness—or, perhaps, to a sense of obligation to the company; it seems at least possible that the sale was arranged several weeks in advance of the meeting, before the Schriver investigation began. Whatever his motive, Hayflick never collected the $5,000 from Connaught for the cells—the last ones he would pass out in the name of Cell Associates. When officials from the Canadian company asked him several months later why he hadn't billed them, Hayflick said he hadn't had time.[52]

On the same day that he handed off the cells in Montreal, Hayflick called Lamont-Havers, the acting director of the NIH, and told him he was withdrawing as a candidate for the directorship of the National Institute on Aging.[53] Hayflick knew that he was in trouble—but not yet just how much.

During his first interview with Schriver in May 1975, Hayflick told Schriver, according to Schriver's documentation of that meeting, that he hoped the investigation could be resolved without having it discussed with Stanford University officials. Schriver said he thought that would be possible.[54] But by June, after consulting with senior NIH officials, he had changed his mind.[55]

On June 17, 1974, Schriver met with Clayton Rich, Stanford's vice president for medical affairs and the dean of the Stanford School of Medicine. Also present was John J. Schwartz, a Harvard Law School graduate who was assistant vice president and counsel for medical affairs. Schriver briefed the men on what he had learned of Hayflick's activities with WI-38, and then they summoned Hayflick to Rich's office. Schriver's documentation of the meeting states that this was the point at which Hayflick revealed Cell Associates' contract with Merck.[56]

But what Hayflick remembers of this meeting is that, as he entered, Rich was behind his desk and Schwartz was sitting off to one side. The first words that anyone spoke came from Schwartz, who said to Hayflick: "You better get a lawyer."[57]

Schriver, Dugas, and Curtin made several trips in the coming months, tracking down and interviewing Hayflick's commercial customers and other key recipients of the WI-38 cells. Curtin and Dugas visited Connaught Laboratories in Toronto and Pfizer's offices in New York City. Schriver went to the Institute of Immunology in Zagreb and visited officials at the Medical Research Council in London. He interviewed Perkins, the former MRC vaccine regulator, who was by then at the World Health Organization. He spoke with Koprowski at the Wistar Institute. And he called Donald Brooks, the Merck attorney who, according to Hayflick, had established that Hayflick owned the WI-38 cells. Brooks told Schriver that the company had entered the pricey contract with Hayflick only after Hayflick attested that he owned the cells, and that NIH concurred.[58]

In between trips, back at the NIH, Schriver became deeply absorbed in the Hayflick case. He spent hours in building 1 on the NIH campus, meeting with top officials—in particular sixty-year-old Leon Jacobs, the NIH's associate director for collaborative research.

Jacobs, a Brooklyn native with a quick-trigger temper, was known for his work on blindness caused by the parasite *Toxoplasma gondii*. He had been at the NIH for most of the previous four decades. It was perhaps Hayflick's misfortune that in the mid-1960s Jacobs had been the scientific director in the Division of Biologics Standards. Jacobs may well have been offended by Hayflick's 1972 excoriation of the DBS at the U.S. Senate—particularly Hayflick's description of the "undirected and subordinate role" of scientific research in the DBS, his contention that this lack of leadership was suppressing the potential of the DBS's capable biologists, and his suggestion that DBS scientists should be allowed to move on to better opportunities "where they will enjoy a full flowering of their capabilities."[59]

Whether or not Jacobs allowed Hayflick's humiliating Senate testimony to influence his initial attitude toward Hayflick, the senior NIH official developed an animus toward Hayflick as Schriver's investigation continued.

In the summer of 1975, Jacobs had been put in charge of an internal NIH committee intended to get a handle on and then supervise the ongoing use of the WI-38 cells. Jacobs was the man who, according to Merck, had one year earlier told the company that Hayflick owned the trove of young WI-38 cells at Stanford.[60]

Jacobs insisted to Schriver that he had said nothing of the kind but had merely ascertained, when an employee named Fritz Miller phoned him from Merck in July 1974, that there was no agreement between the U.S. government and the Wistar that would have allowed the Wistar to own a patent on the cells, if indeed one existed. It did not, and he told the company that too.[61] Later Jacobs telephoned Brooks at Merck and wrote a memo to the record: "Mr. Brooks informed me that, regarding Merck's dealings with Dr. Hayflick, Merck relied strictly on Dr. Hayflick's representations—and not on any information obtained during a telephone conversation with me or anyone else."[62]

Brooks was not entirely forthcoming with Jacobs. A July, 1974, memo entitled "WI-38 Human Diploid Cells" written by Fritz Miller after his consequential conversation with Jacobs and copied to Maurice Hilleman, Brooks, and other Merck executives reads in part:

"A question was raised whether Dr. Hayflick had rights to the cells to offer them for sale. Dr. Leon Jacobs of the NIH had a search of the records made by counsel and he told me by telephone on 7/11/74 that there is no patent that the NIH knows about. . . . Accordingly, Dr. Jacobs is of the opinion that Dr. Hayflick is free to sell the cells. We should now be able to proceed with the development of a price to offer Dr. Hayflick."[63]

In the summer of 1975 Schriver also consulted with the NIH's legal adviser, thirty-seven-year-old Richard J. Riseberg, another Harvard Law School graduate, who would go on to positions of increasing seniority at the NIH's parent Department of Health, Education and Welfare. Riseberg consulted the original 1962 NIH contract under which Hayflick had developed WI-38—the document which stated that, when the contract ended, the contractor agreed to transfer title to the government of any and all materials developed under it by the contractor. He examined the minutes of the January 1968 meeting where the disposition of the cells was decided. He looked at the letter from the NIH to the Wistar following up that meeting, affirming the arrangements for the cells to be transferred to the American Type Culture Collection on March 1, 1968, and reasserting the agency's ownership of the cells. And he concluded that the WI-38 ampules that Hayflick had taken to Stanford belonged to the NIH.[64]

In July, less than two months after Schriver and his deputies first visited Hayflick's Stanford lab, Riseberg contacted the criminal division of the Department of Justice. Lawyers there took a month looking over the facts of the case, then declined to launch a criminal probe.[65] They told the NIH that the agency would be more likely to get the result it wanted—the return of all of the young WI-38 cells to the NIH, along with all of the money that Hayflick had collected—by launching a civil proceeding; the agency's passive acquiescence in Hayflick's possession of the cells between 1968 and 1975 would make a criminal case difficult to win.[66]

In August 1975, two months into the Schriver investigation, Hayflick, Ruth, and their two eldest daughters took a trip to Israel, where Hayflick attended a scientific conference. He was miserable during the trip. When he returned, it was to a call from Schriver's lieutenant, Dugas, informing him that NIH officials would be taking the WI-38 ampules from his lab beginning on August 19.[67]

Over the next eight days those officials—who had called around to find an airline that would allow them to carry those liquid-nitrogen refrigerators that looked like bombs in the cabin (United would not; TWA would)—collected Hayflick's entire remaining stock of low-passage WI-38 ampules and delivered it to the American Type Culture Collection in nearby Rockville, Maryland.[68] The cells had at last landed at the facility where the NIH had intended them to be transferred fully seven years earlier, when Hayflick left the Wistar Institute.

Soon after the transfer, John Shannon, who headed the ATCC's cell culture department, drafted a memo to the cell bank's director, Dick Donovick. Shannon reported that the federal scientists had delivered to the ATCC 103 ampules of young WI-38 cells. Of these, 46 were ninth-generation cells and 57 were the youngest, eighth-passage cells. These 57 were what remained of the 375 eighth-passage ampules that Hayflick had moved to Stanford from the Wistar in 1968. Since their arrival at the ATCC, Shannon added, two of the eighth-passage ampules had exploded. The rest of the cells, he wrote, "are properly inventoried, racked on canes, and stored in a separate liquid nitrogen refrigerator (MVE Model VPS 3500)." They are kept, he added, "under daily surveillance."[69]

Rocky Passage

Stanford and Oakland, California, 1976–78

Because of the government's decade long cold-shouldering of WI-38's, it is hard not to feel sympathy with Hayflick's sense of irony and outrage that the same government now claims the cells to be its own precious property. Nor is it strange to hear him say that "I felt, and I think I am justified in feeling, that these cells were like my children."

—Nicholas Wade, *Science*, April 9, 1976[1]

O n Sunday, March 28, 1976, the following headline appeared on the front page of the *New York Times*: INVESTIGATOR SAYS SCIENTIST SOLD CELL SPECIMENS OWNED BY U.S.[2]

Reporters for the newspaper and several other publications had used the Freedom of Information Act (FOIA) to obtain the fourteen-page, single-spaced report that James Schriver and his deputies completed late in January 1976.[3]

Hayflick had been given until April 1 to rebut the Schriver report and its conclusions, but when in mid-March reporters invoked FOIA to access the document—and a court challenge from Hayflick failed to stop the agency from doing so—the NIH released it without his rebuttal.[4]

The contents, and the media coverage of them, were damning.

Hayflick, Schriver reported, had "sold cell cultures of human diploid cell strains, the property of the United States Government." He had charged most of the costs associated with the sales to NIH research agreements and deposited most of the receipts in his private bank account and later in the bank account of Cell Associates, a company in which he and his wife were the sole shareholders. A total of $67,482.33—just over 1 percent of it interest earnings—had accumulated since 1968. (This is about $286,000 in 2016 dollars.)

The report recommended that the government recover the money and that it consider banning Hayflick from participating in NIH grants and contracts.

There was much else in the Schriver report. A good chunk of it documented the investigators' fruitless efforts to square Hayflick's records of the WI-38

cells he had thawed, expanded, and shipped with the number of WI-38 ampules actually on hand in the lab. Hayflick calculated that he had brought 375
eighth-passage ampules with him to Stanford; Schriver arrived at essentially
the same figure: 379. There any agreement ended. Schriver calculated that
Hayflick could not account for at least 207 of the ampules. Hayflick contended
that all of these ampules either were sent to the Medical Research Council in
the UK, were lost or exploded at the Wistar, were cracked or exploded during
transfer from the Wistar to Stanford, or were found to be contaminated or
dead in 1968 and 1969.

Schriver was not convinced, and his frustration oozes from the pages of the
report.

"Almost every situation concerning the accountability for these cells has
produced unexplainable differences," he wrote. "No record made available to
us fully accounts for the 8th passage ampules."

Nor, Schriver wrote, had Hayflick stewarded the cells responsibly. They
were shipped out at lower passage levels than necessary, when expanding the
cells to higher levels and refreezing those that were in excess of current demands would have preserved the supply of low-passage ampules. (Hayflick
would argue convincingly later that neither scientists nor companies wanted
middle-aged cells; they didn't live long enough to be useful.)[5]

The report also rang alarm bells about "a very high level of contamination"
in the WI-38 ampules. It said that Hayflick had since 1968 used antibiotics to
treat a score of eighth-passage ampules that were discovered upon thawing to
be contaminated with bacteria. These were then shipped, after antibiotic treatment, to scientists studying aging and to paying customers without notifying
them, or the NIH, of the contamination and the "cleaning up" of the cells with
antibiotics. (Hayflick later vigorously disputed that such cells were ever knowingly shipped to companies.)[6]

Contamination was not uncommon in cell culture generally, and not surprising either, given the circumstances under which the ampules were created
on that midsummer day in 1962. And contamination and antibiotic treatment
didn't spoil the cells for use in research. For vaccine makers and regulators,
however, it was a matter of dispute whether "cleaned-up" cells should be used
for vaccine making. Virtually all companies treated their cells with antibiotics
at some point in the vaccine-making process, but many or all began with sterile
cells never treated with antibiotics. The idea was to use antibiotics in such cells
to *prevent* bacterial contamination, not to treat it. All the same, at an NIH

conference one year later, at least one speaker argued that cleaned-up cells could be used for vaccine making without risk.[7]

The Schriver report lamented that Hayflick had not notified the DBS (later the FDA) of the contamination "of a cell line which was in the process of being studied for use in preparing vaccines for human use." From Hayflick's point of view, this was rich. Why would it have made any sense for him to notify of contamination an agency that, during most of the years he was shipping the cells, was dead set against their use for vaccine making? Later he would write of both the former DBS chief Roderick Murray and Harry Meyer, the man who replaced him in 1972: "Neither cared about our work, and both fought our efforts to have [WI-38] cells used for vaccine preparation."[8]

Arguably the most damaging section of the report, though, detailed Hayflick's sales of the cells to companies between March 1974 and the Montreal meeting in June 1975—including the inking of the million-dollar Merck contract.

When the Schriver report became public, Hayflick released a one-page statement to reporters. It read, in part:

> I am shocked and grieved that NIH has released a report which is errone-ous and incomplete, without affording me the opportunity to complete the written rebuttal which they requested be filed by April 1, 1976. I had been assured by NIH that I would be permitted to submit my rebuttal before NIH completed its determination of the facts and consequences of those facts.... I absolutely deny any wrongdoing... and I urge my scientific colleagues around the world who have long relied upon my integrity and that of my work to regard with great caution the statements in the report.
>
> In view of the fact that I have been denied due process by the NIH, I am seeking due process in the courts. I have on March 25, 1976, filed suit in the U.S. District Court in San Francisco against the Department of Health, Education and Welfare and the NIH.... I am confident that I will be totally vindicated by the judicial process.[9]

The appearance of Schriver's findings on the front page of the *New York Times* was difficult enough for an acutely image-conscious man who had so often appeared in that newspaper's pages in a glowing light. But the truly dev-astating article—the one that addressed his own community and the one that Hayflick still speaks of bitterly forty years later—appeared in *Science* magazine

under the headline HAYFLICK'S TRAGEDY: THE RISE AND FALL OF A HU-
MAN CELL LINE.[10]

It was written by Washington-based staff reporter Nicholas Wade, who
had rushed to California, interviewed Stanford officials and Hayflick (with his
lawyers present), and compiled a 3,500-word article in five exhausting days.[11]
The article delved into every aspect of the Schriver report.

Wade pointed out that NIH officials had not exactly avidly pursued the
cells they were now calling a "national resource" in the pages of the *New York
Times*.[12] After making the failed trip to Philadelphia to collect the cells in the
winter of 1968, several months before Hayflick left the Wistar, Wade reported,
Boone, then the NIH project officer, called Hayflick at Stanford in the fall of
1968 to demand the cells' transfer to the ATCC. Hayflick agreed to do so but
didn't follow through, Schriver reported. (Hayflick told Schriver that he had
no memory of ever having such a conversation with Boone.)[13]

After that, "things sort of disintegrated," NIH associate director Leon Jacobs conceded to Wade. "We are not completely faultless in this."

Wade wrote that Hayflick's not-yet-public rebuttal might totally exonerate
him and quoted Hayflick's press statement urging "my scientific colleagues . . .
to regard with great caution the statements in the [Schriver] report."

But Wade's reporting made it clear that the rebuttal was going to need to
surmount some high hurdles.

The facts in the *Science* story were made worse by Hayflick's efforts to explain that he had supplied all of the WI-38 cells to companies, and could continue to do so, by using only the ten original eighth-passage ampules that he
was permitted to keep for himself when he left the Wistar.

Wade described Hayflick's explanation that he could fulfill the Merck contract with far fewer ampules than Schriver believed by growing the cells in
larger-than-usual laboratory bottles. In essence, he meant that he could put the
roughly two million cells from one eighth-passage ampule in a bottle with a
surface area large enough to allow the population to double several times
over—say, to 25 million cells, he told Schriver—before it reached confluence.[14]
Then the cells could be placed into what Hayflick estimated would be fifteen
ampules for Merck. According to Hayflick, the resulting cells would still be, by
scientific convention, ninth-population-doubling-level cells, because they had
been "passed" into new bottles just nine times.

"It is hard to find anyone who agrees with Hayflick's calculations," Wade
wrote, quoting unnamed experts who called Hayflick's reasoning "terribly feeble" and said it "would be laughed out of court." Merck's Hilleman, whom

Wade also interviewed, was less vituperative, but his message was the same. He said simply that one would expect to derive one hundred ninth-passage ampules from fifty eighth-passage ampules.

Wade also reported that it appeared that supplies of pristine, low-passage WI-38 ampules were alarmingly small. The NIH had recovered from Hayflick's lab a total of about one hundred eighth- and ninth-passage ampules. But the ATCC, now in possession of the ampules, had opened nine of them and discovered that six of these were contaminated with bacteria. If all of the one hundred or so ampules at the cell repository were contaminated in equal proportion, there were in fact far fewer ampules suitable for vaccine making than anyone had believed.

"The real tragedy for Hayflick is not what the NIH inquiry or Stanford has done to him but what he has apparently done to the future of WI-38's," Wade concluded. "It now appears that there are sufficient stocks only for the next several years."

The article ended by quoting the NIH's Jacobs saying, duplicitously, that he doubted that the agency would choose to bar Hayflick from obtaining NIH grants, as Schriver recommended, and that "in the long run . . . things will straighten out for this guy."

But its most powerful quote undoubtedly came from Hayflick's old Wistar colleague Stanley Plotkin. Wade wrote: "According to Plotkin, many people warned [Hayflick] about the sales, but he was not open to any kind of remonstrance. 'I think that in the really classical Greek sense it was a tragedy, because it is a man who at the height of his powers brought about his own downfall,' Plotkin says."[15]

By the time the *New York Times* and *Science* articles appeared, the Schriver report was not news in the Hayflick family. Hayflick's son, Joel, studying that winter at Stanford's campus at the country estate of Cliveden in England, picked up the *Stanford Daily* of March 3, 1976, to discover the banner headline PROF IN ALLEGED FUND MISUSE. The article reported that Hayflick had resigned the previous week and that "Hayflick allegedly used Stanford laboratories to produce, store and sell cell cultures in his own commercial enterprises."[16]

"The university was concerned that its laboratories were inappropriately used in the performance of a contract to store, produce, and distribute a strain of human cells, some of which Hayflick sold on his own behalf," a Stanford press release announced on February 27, the day that Hayflick's resignation became public.

Hayflick, in a press release composed by the university and quoted by the *Daily*, maintained that he had abided by Stanford's rules governing contracting. He told the *Daily* that "it was all done in good conscience" and that he resigned voluntarily, on principle, because "the university impugned my motives."

Schwartz, the medical school counsel whose first-ever words to Hayflick had been "You better get a lawyer," responded in the *Daily*: "Our position as to motives is that the funds were improperly deposited in his personal account."

Schwartz would also, with a nod from the NIH, pass the Schriver report on to the district attorney—the local criminal prosecutor in Santa Clara county.[17] The DA ended up electing not to pursue the case, although not before telephoning Fadlo Mousalam, the lawyer whom Hayflick had hired the previous summer, and causing Hayflick to sweat.[18]

Hayflick had forgone, with his unforced resignation, a chance for an adjudication of his case before the Stanford faculty's advisory board, essentially a jury of his peers.[19] The new, highly capable lawyer whom he hired several weeks later—the young William Fenwick, who would go on to become a Silicon Valley superlawyer—rued that decision mightily.[20]

On the Stanford faculty there was a conspicuous silence from many quarters and some vocal support. He remembers that Henry Kaplan, a powerful Stanford Medical School radiologist, lobbied both the medical school dean, Clayton Rich, and Stanford president Richard Lyman on his behalf, to no avail. Kaplan looked at him at one point and, worried about his heart, sent him to a cardiologist colleague for a checkup. (Hayflick was pronounced fit.)

On the other side of the country, Harry Schwartz and Noel Buterbaugh, both senior executives at the laboratory supplier Microbiological Associates and old friends, hired Hayflick as a consultant and began flying him out to Maryland regularly to advise them on cell-culture technologies.

Zhores Medvedev, a Soviet dissident and scientist, wrote a scathing letter to *Science*. (In 1972 Hayflick had organized a group that faced down Soviet authorities and forced them to release Medvedev when he was arrested at an aging conference in Kiev.) The "tragedy" alluded to in *Science*'s headline, Medvedev wrote to the magazine, "lies in the surprisingly tendentious approach of *Science*" to the Hayflick matter—and in the fact "that an outstanding scientist can be forced to resign his professorship at a university without discussion of his case by his colleagues."[21]

Others simmered angrily out of public view. Merck's Hilleman recalled decades later: "I was asked to be a principal witness against him and I said that if there was an intent to convict him, I would make a campaign on my part that two top-level government officials would spend time in jail with him." He added: "He should have been celebrated as a scientific hero instead of being persecuted."[22]

As in many controversies in which one side is more passionately motivated than the other, Hayflick's backers, Hilleman's circumspection notwithstanding, were outspoken, while his detractors—those who were "appalled," as Hayflick's former Wistar colleague Cristofalo put it years later—were largely silent. The fall of a colleague, however much they disagreed with his actions, was not pretty to behold, and it's likely that many were not motivated to add to his humiliation by slinging arrows publicly.

One indication of their feelings, however, is apparent in this letter from Albert Sabin, the polio vaccine inventor who had recommended Hayflick for the job at Stanford. Asked several years later to sign a letter to *Science* in support of Hayflick, Sabin wrote to the letter's organizer:

> I consider myself a friend of Len Hayflick, but I am not in sympathy with the type of activity, especially his association for profit with Cell Associates, Inc., which led to the dispute with the NIH. I hope very much that you have not used and will not use my name in the letter that you propose to have published in *Science* and elsewhere.
>
> Sincerely yours,
> Albert B. Sabin, M.D.[23]

Several weeks after his resignation, Hayflick hired the sharp young Palo Alto intellectual property lawyer Fenwick, whose testimony before legislators had helped shape a new law, the U.S. Privacy Act of 1974. Fenwick was then with a Palo Alto firm called Davis, Stafford, Kellman and Fenwick. Today he is an emeritus partner at Fenwick & West, a highly influential law firm with more than three hundred lawyers that is a Silicon Valley institution. The firm has represented companies including Facebook, Apple, and Google. Fenwick himself first represented the nineteen-year-old Steve Jobs when he was founding Apple Computer.

Fenwick recalled in a 2012 interview that he was convinced that Hayflick genuinely believed he had done nothing wrong.[24] He would not have taken on the case with very little prospect of compensation at his then-$75-an-hour rate had he not believed in the man, he said. Fenwick was ready to contest the NIH's claim to ownership of the cells. He also thought he could make a strong case that the NIH had violated the new Privacy Act by releasing the Schriver report.

There ensued a long legal battle in which Hayflick, represented by Fenwick, sued the U.S. government for ownership of every WI-38 ampule taken from his lab by the NIH, and any progeny they might produce. He also claimed the accrued $67,000, as well as damages for defamation.[25] The government in turn countersued for recovery of the money.[26]

The legal proceedings would stretch to five years. In the meantime, in March 1976, Hayflick suddenly had no job, no lab, and a consuming bevy of legal troubles. He had never been clinically depressed, but he fell into a deep depression. Not a big man to begin with, he lost ten or fifteen pounds. For some reason, he could eat Big Macs when nothing else appealed. He became a regular at a McDonald's on El Camino Real, the commercial thoroughfare that runs along the eastern edge of the Stanford campus.

At home there was the matter of keeping his family afloat. He began collecting unemployment compensation, which brought in $104 every other week. The Hayflick family, with two children in college, a high school senior, and ten- and thirteen-year-old girls, began living on savings, and on the fraction of his former income that Hayflick brought in from consulting jobs.

Hayflick found some relative peace during his frequent trips to Maryland to consult for Microbiological Associates, the laboratory supplier. The facility in rural Walkersville was surrounded by cornfields and farms. The quiet was a tonic, "because there was a point before that time that whenever the phone rang it sounded like a bomb going off because so many phone calls held bad news."[27]

At one point, appearing anguished and aggrieved, Hayflick sent a letter of nearly two and a half single-spaced pages to Plotkin, attaching his eleven-page court complaint and a copy of his press statement.

Plotkin's reply was one paragraph long.

Dear Len,

 I was very glad to hear from you and I wish you well in your legal endeavors. However, my major wish for you is that you have the opportunity to start doing science again at which you are so good. The main advantage of having a profession is to some extent one can submerge one's other problems in one's work, something which I have learned from my own experience.

With warm regards,
Sincerely yours,
Stanley A. Plotkin, M.D.[28]

It was late summer, five months after the Schriver report was released, when the NIH made Hayflick's sixty-five-page rebuttal publicly available to anyone willing to pay $11 for copying costs.[29] (On September 1 Schriver completed his point-by-point rebuttal of Hayflick's rebuttal. It was nearly twice as long because it reprinted Hayflick's arguments and responded to each of them.)[30]

There was little if any news coverage of Hayflick's long retort, beyond a mention by Wade in a *Science* article, noting that Hayflick's rebuttal was now available and that it called Schriver's report "incomplete, inaccurate, and accusatory without proper cause."[31]

Hayflick made many points in his rebuttal. He contended that it was he, Hayflick, who first asked the NIH to investigate the legal status of the cells. He also insisted that his decision, after the Schriver investigation was launched, to remove his name from consideration for the directorship of the National Institute on Aging was made because of many concerns, including the impact that the new job would have on his time for research and the salary cut he would have to take. (Schriver had reported that a conflict clearly existed between Hayflick's activities as president of Cell Associates and the duties of the NIA director. "During our initial discussion with Dr. Hayflick he apparently recognized this and decided to withdraw his name from consideration," Schriver wrote.[32])

Hayflick also, in a stretch, assailed the Schriver report's silence on the fact that Hayflick had filed suit against the government claiming ownership of the WI-38 cells.[33] The Schriver report was completed in January, but Hayflick argued that the NIH could have amended the report after he sued the agency on March 25. This was the very day that, after a court declined to intervene to block its release, the NIH released the Schriver report to reporters under the Freedom of Information Act.

Hayflick complained that the report blithely declared the cells "the property of the United States government." But "the lawsuit puts in issue [that] very question." He did not broach the subject of how an argument could be made for his ownership of the cells.

In his rebuttal Schriver came back at Hayflick on every issue that he raised. The NIH investigator wrote bluntly: "The lawsuit obviously could not have been mentioned in the report because litigation was started after the report was written."

Hayflick had argued that at the pivotal January 1968 meeting where participants including him decided on the disposition of the cells, no decision was reached as to who owned them. Schriver shot back that such a decision wasn't needed—their ownership by the NIH was implicit in the decision made at that meeting: that all but a score of the ampules should go to the ATCC, to steward them on behalf of the NIH.

Hayflick wrote that he "had never received a request from NIH or anyone else to return the cells to ATCC." Schriver retorted that Boone, the rangy pathologist who was the NIH contract officer, had asked him to do just that in the autumn of 1968.

Hayflick decried the report for imputing sinister motives to his less-than-perfect record keeping of the inventories of WI-38. Schriver shot back that the NIH had the right to expect minimum standards, and noted that when NIH scientists, accompanied by an FDA colleague, Hope Hopps, took an inventory of the WI-38 ampules in Hayflick's lab before taking them away, they "expressed astonishment at the slovenly nature of the records and the slip-shod manner in which the ampules were stored."

As for his stewardship of the cells, Hayflick contended, he had filled requests for scientists and institutions "outside the contract" with cells he had already thawed for researchers studying aging, which "would have been otherwise wasted or destroyed."

Schriver retorted that those statements "are unsupportable and can only be regarded as self-serving declarations." Hayflick's own records showed, Schriver wrote, that Hayflick had shipped some 1,700 low-passage cultures while at Stanford and that, of these, nearly 1,200 had been shipped to paying customers. What was more, paying customers had received preferential treatment, frequently receiving younger cells than the researchers studying aging whom Hayflick supplied under the contract. "It is obvious that the interest of the government was not protected."

To explain how he could meet the terms of the Merck contract—to supply the firm with one hundred or more ninth-population-doubling-level ampules—by using just a handful of eighth-passage ampules, Hayflick again argued that he could do so by planting the cells in bigger-than-usual bottles, and giving them more room to expand, generating dozens of ampules' worth of cells before they were harvested at the ninth passage. He also wrote that population-doubling estimations were so imprecise that it was meaningless to attribute significance to differences in population doubling levels of less than ten.

Schriver, in his rebuttal, turned to Hopps of the FDA to respond. She wrote of Hayflick's explanation of how he could generate enough cells to fulfill the Merck contract: "We cannot agree with Dr. Hayflick's comments. Indeed, it is improbable that, on the basis of the number of [eighth-passage] ampules now known to exist, one could derive such a large number of ampules at the 9th PDL. It is possible that by manipulation he could have 9th passage cells, but not cells at the 9th population doubling."

Hayflick had the upper hand in the rebuttal only when he challenged the NIH's passivity and tacit acquiescence in his activities.

He noted that he had repeatedly reported, in his progress reports to the agency, that he was supplying the cells to off-contract scientists. In the face of this, he reminded readers, he had received repeated commendations for his management of the contract delivering cells to researchers studying aging. "At no time were Dr. Hayflick's decisions or policies in this regard ever questioned by anyone," he insisted. Failing to let the report's readers know this, he argued, "unjustifiably distorted" the report.

Hayflick also bemoaned the lack of context in the report:

The report fails to state that Dr. Hayflick pioneered the freezing and preservation of human diploid cells and had them prepared at a time when the state of the art was very primitive. He learned that
1. frozen ampules are very brittle and easily cracked;
2. inexplicably, the bottoms of many ampules dropped out;
3. a portion of ampules exploded upon withdrawal from liquid nitrogen; and
4. ampules, once removed from holders, sometimes are dropped and irretrievably lost.

What was more, he argued, freezers at the Wistar were regularly unlocked and accessible to a variety of people over a six-year period.

But Schriver responded with disbelief to Hayflick's claim that only 139, not 207, ampules were unaccounted for and that all of these 139 had been lost, exploded, cracked, been contaminated, or died. Schriver called the claim "unacceptable." He and his team had found, in the records both at the Wistar and Stanford, numerous instances where it was recorded when an ampule was cracked, exploded, contaminated, or dead. Given that, he wrote, "one is hard pressed to accept such a weak argument that the condition of the 139 unaccounted-for ampules would not have been recorded."

Several items in the dueling rebuttals come down to he-said-he-said arguments, like Hayflick denying that he ever told Schriver that he had behaved "improperly" and Schriver insisting that he had said just that. But on the facts of the sales to companies Hayflick's rebuttal is either weak or silent. Of his transfer of two bottles of young cells to Connaught representatives in Montreal for an agreed-upon price of $5,000, he writes that this information "implies any number of negative conclusions about Dr. Hayflick without serving any legitimate purpose." Therefore, he writes, investigators shouldn't have included it. He does not dispute the $3,000 sale of cells to the Institute of Immunology in Zagreb. As for the Merck contract, Hayflick writes only that it was Merck who approached him, and not vice versa.

"It is immaterial who first approached whom," Schriver responded.

Hayflick closed his rebuttal, which is stiffly couched in the third person, with what can only be termed a lament.

> The report concludes that Dr. Hayflick has acted in contravention of government property rights in the cells. That conclusion is wrong.... But what is so unfair about the report, and so terribly damaging to Dr. Hayflick, is not this wrong conclusion itself, but rather the report's assertion that the issue of ownership has always been beyond dispute and that Dr. Hayflick appropriated government property intentionally and dishonestly with full recognition that he had no colorable right to do so.... It is a cruel irony that the report should fail to mention that there is a significant question as to ownership [of the cells].... The injury to Dr. Hayflick's reputation and career caused by the report is irreparable. He may never fully recover.

Schriver's conclusion was much briefer.

> Dr. Hayflick's comments give NIH no reason to make any changes [in the report]. In fact, much of his material either cannot be corroborated or is in complete disagreement with the facts.

Hayflick had willfully over the course of seven years created the circumstances that had brought his life crashing down around him. But once he had done so, the NIH was not going to show the slightest bit of give in how it dealt with him, especially not after he filed suit against the agency. There was an animus toward him in Bethesda—from Schriver and Jacobs and perhaps from other senior officials—that went beyond the facts of the case and was deeply personal; this would become apparent in the ensuing years.

The dislike also extended to downtown Washington, DC, to the NIH's parent, the Department of Health, Education and Welfare, where senior officials now set out to ensure that "doing science again," as Plotkin had urged him to do, was not going to be easy for Hayflick.

By January 1977 Hayflick had landed a position at a research lab at Children's Hospital Medical Center, across the San Francisco Bay in Oakland. It was a place to hang his hat, but it was certainly no Stanford. What was more, he was not allowed to transfer to Oakland the multiyear NIH grants that he had won at Stanford, like a generous five-year grant to study how normal human cells became cancerous, which had been awarded only weeks before the Schriver report was published.[34] The Oakland hospital was strapped and could in no way support his salary; Hayflick needed to bring in new grant money to fund his science, and he needed to do it quickly.

Just ahead of a March 1, 1977, application deadline, he turned, with trademark tenacity, to the NIH's National Institute on Aging, where an inaugural director, Robert Butler, was finally in place. Hayflick applied for a three-year, $562,000 grant to study aging in laboratory cells. As part of his application, he requested that the NIH provide him with the youngest WI-38 cells—in other words, some of those it had taken from his lab.

In October the advisory council of scientists to the NIH's new aging institute voted to fund Hayflick's grant. Normally, such advisory councils vote to fund groups of grants, each of which has been closely scrutinized by a smaller group of scientific peer reviewers, as Hayflick's had been.[35] But in a highly unusual step, the aging institute's advisers took a separate vote to approve the funding of Hayflick's grant specifically.[36] Their message to NIH administrators couldn't have been clearer: Get off his back and let him do his science again. Then absolutely nothing happened. It would take another twelve months before the NIH informed Hayflick, after he obtained confidential correspondence between the NIH and its parent department under the Freedom of Information Act and asked the NIH to explain it, that he would not see any money before the secretary of health, education and welfare conducted a special review.[37]

Riseberg, the NIH legal adviser, had told the agency in 1977, after it received Hayflick's new grant application, that there was "no sound legal basis" for categorically excluding Hayflick from winning NIH grants.[38] Fredrickson, the NIH director, was of the same mind, and Butler, the inaugural director of the aging institute, agreed.[39,40]

But Leon Jacobs, the NIH associate director, who had been quoted in *Science* saying that he doubted that the NIH would shut down Hayflick's ability to win grants—and who was so aggrieved at Merck's report that he had told the company that Hayflick owned the cells—disagreed, and let Frederickson know. The NIH "could administratively decide not to make the award [to Hayflick], on the basis of questions about Dr. Hayflick's scientific integrity," he wrote to Frederickson several months after Hayflick applied for the funding.[41]

In downtown Washington more powerful players were in Jacobs's corner. In early 1978 Thomas D. Morris, the inspector general for the Department of Health, Education and Welfare, wrote a memo to his Hayflick File, directing Riseberg and others to "develop a paper on . . . Secretary [Joseph] Califano's discretion to withhold a grant award." He wanted another paper on the importance of having NIH reviewers consider any grant applicant from a "managerial and ethical viewpoint." In general, Riseberg and his colleagues were to "shape up the disqualification procedure."[42]

A few months later the inspector general dictated another memo to the Hayflick File. In it he directed an assistant to follow up with Riseberg regarding "the selection of an option by Dr. [Julius] Richmond [the assistant secretary for health] to preclude the award of grant involving Dr. Hayflick as its principal investigator."[43]

Hayflick would finally get his grant money—it's not clear who at the department finally gave up on blocking it—in August 1979, two and one half years after he applied for the grant. It would deliver $562,000 over the coming three years. But Hayflick did little with it, and not only because the NIH refused to provide the young WI-38 cells that Hayflick insisted he needed for his studies.[44] (The agency said that he could obtain older WI-38 cells at cell banks like the ATCC, just like other scientists.)[45] Hayflick's science had all but ground to a halt as he spent time fighting a lawsuit, trying to find a long-term job, and earning what he could from consulting, to keep his family fed. He would rarely publish new, lab-based experiments again.

CHAPTER TWENTY-THREE

The Vaccine Race

Iran and the United States, 1975–81

The sudden attack of a mad, foaming, and enraged animal . . . almost always takes place in daytime in the middle of the village and ends, after a frightful struggle, with the death of the wolf under the sticks and pick-axes of the peasants.

—Marcel Baltazard and Mehdi Ghodssi, director and chief, respectively, of the Rabies Service at the Pasteur Institute of Iran, Tehran, 1954[1]

On October 20, 1975, as James Schriver was laboring over the report that would turn Hayflick's world upside down, a rabid wolf attacked seven people in Aghbulagh, a rural farming village in northwestern Iran. The victims included a fifty-five-year-old man who sustained twenty lacerating face, head, and hand wounds, and a seven-year-old boy with more than twenty-five deep gouges to his head, neck, face, and ears. The mad wolf was at last cornered and killed by the terrified villagers. For the roughly sixty people in the village of mud-walled homes, the attack was a calamity.

The roads were bad, and it was thirty-two hours before the victims, with the wolf carcass in tow, made it to the Pasteur Institute of Iran, a distance of a little more than 250 miles. The institute, which is independent of the famous French institution founded by the rabies vaccine inventor, Louis Pasteur, was a designated World Health Organization center for rabies research. The disease was endemic in both dogs and wolves in Iran, and the Tehran scientists, who still relied on vaccines made in animal brains, had recently been charged with conducting an important clinical trial.[2]

The seven villagers were injected first with blood serum from mules, containing antirabies antibodies produced by Iran's reputable Razi Institute. This was the transient, "passive" immunization that was, and remains, so important to holding the virus in check until victims can be vaccinated and their own immune systems begin to produce antibodies. Within an hour of this first jab, each of the seven villagers was also injected with the trial vaccine. Invented at the Wistar Institute in Philadelphia and produced by the Institut Mérieux in Lyon,

France, it was made using WI-38—Hayflick's human diploid cells. The trial vaccine had already been safety-tested in healthy volunteers, who had produced robust levels of antibodies. Would it work in humans bitten by rabid animals?

The traumatized villagers stayed at the institute in Tehran after receiving the first, "day zero" injection and were again injected on day three. Then they returned to Aghbulagh, where a member of the Pasteur Institute team would travel in order to administer four more injections on days seven, fourteen, thirty, and ninety—if the victims were still alive.

In the meantime a veterinarian at the institute dissected the wolf and tested its brain and saliva for the rabies virus. Both tested positive.

Koprowski and his fellow Polish expatriate Tad Wiktor had begun working to make a rabies vaccine with the WI-38 cells at the Wistar Institute in November of 1962, only a few months after Mrs. X's abortion. They had traveled a long road to arrive at this moment of truth in the waning days of 1975.[3]

In the mid-1960s, after winning the initial struggle to get rabies virus to grow in WI-38 cells, they needed to choose a virus from among the various strains of rabies virus that had been used in making vaccines. They settled on a strain called Pittman Moore, which they obtained from the NIH and which was descended from the original vaccine virus that Louis Pasteur had developed in the 1880s.

In 1965 they developed a "seed" stock of this vaccine, growing it through fifty-two passages in WI-38 cells. They then began handing it out to companies: the Institut Mérieux in Lyon in 1966; the Behringwerke company in Marburg, West Germany, in 1969; and Wyeth in Philadelphia in April 1971. As the companies worked to develop the vaccine, Koprowski, Wiktor, and other scientists refined procedures for concentrating and purifying the vaccine, making it more potent at producing antibodies.[4]

As late as 1966 Koprowski argued for the merits of making the vaccine with a live, weakened virus, rather than a killed one. He believed that in postbite patients, where time was critical, a live virus might well cause a vaccinee to produce more antibodies more quickly with fewer injections.[5] But the dire consequences if even a small amount of disease-causing virus somehow survived in a live vaccine made the approach unpalatable to the companies. They instead zeroed in on different ways of killing the purified vaccine virus. Wyeth used an odorless, colorless chemical called TBP (tri-n-butyl phosphate), which broke the virus into pieces. Company scientists then delivered an additional toxic blow with ß-propiolactone, the same liquid chemical that had been used to kill the virus in earlier vaccine incarnations.[6] Mérieux and the German company Behringwerke, by contrast, used only ß-propiolactone, leaving the dead virus whole.[7] The difference, it would emerge, may have been crucial.

For Koprowski's purposes, something else essential also happened in the late 1960s. In August 1968, not quite four years after it was submitted, the U.S. Patent and Trademark Office granted Koprowski and his colleagues' application for a patent on their new method of making rabies vaccine.[8] Koprowski, Wiktor, Fernandes, and the Wistar were now looking at a reliable royalty stream, presuming that the companies involved could win marketing approval from the regulators in their respective countries.

Finally, on a foggy, drizzly November day in 1971, nine years after they first inoculated a bottle of WI-38 cells with rabies virus, a long-awaited event took place at the Wistar Institute.

Sixteen staffers including Koprowski, Wiktor, and Plotkin became the first human beings injected with the new rabies vaccine. Wyeth provided the vaccine; the Wistar provided the photographer. The volunteers, half of whom had never been vaccinated against rabies, were injected in the forearm or the shoulder. Within days, their antibody levels soared. After another injection more than a month later, the levels in the never-before-vaccinated Wistar staffers exceeded, with one exception, those in people who in other studies had received fourteen doses of the preexisting duck-embryo vaccine—the one that had failed to save the boys in South Dakota and Minnesota and Morocco a few years earlier. In the Wistar volunteers who had been vaccinated against rabies in the past and who thus had preexisting antirabies antibodies, the vaccine had a big booster effect, sending their antibody levels even higher than in those who had never been vaccinated. As for untoward reactions, apart from a few sore arms, there were no complaints.[9]

Koprowski and Wiktor had good cause for optimism as they prepared their results for publication in the *Journal of the American Medical Association*.[10] But it would require four more years of human trials of all three companies' vaccines, in the U.S. and six other countries, confirming the safety of the human-cell vaccine and the fact that it produced much higher antibody levels than existing vaccines—and working out how many doses to give, how many days apart, in order to maximize those levels—before the researchers felt ready to test the vaccine in people bitten by rabid animals, at the Pasteur Institute in Tehran.[11]

In late January 1976, as James Schriver put the finishing touches on his report on Hayflick's activities with WI-38, staff from the institute in Tehran traveled for a final time to the village of Aghbulagh, where they administered the last rabies injections, three months after the wolf attack, to the bite victims. All seven were alive and well.[12]

The Aghbulagh villagers were among forty-five people bitten by rabid wolves and dogs in Iran who were vaccinated by the Pasteur Institute team in 1975 and early 1976, in the first real-world test of the new WI-38 rabies vaccine. All were from rural villages in northwestern Iran. Along with the Aghbulagh villagers, the victims included a ninety-year-old man and a three-year-old boy who didn't receive a first dose of vaccine until six days after a rabid dog bit him deep in the hip and leg. They included seven dog-bitten people, mostly teenage boys, from villages so remote that it took seven and eight days before the first vaccines were administered. All of the animals involved—six dogs and two wolves—were captured, killed, and confirmed through laboratory testing to have been rabid.

Of the forty-five people who were vaccinated, forty-four were alive and well six to twelve months after being attacked, at the time that Koprowski and the Iranian scientists wrote up the results of the trial. Only the ninety-year-old man had died, to all appearances of a heart attack, two months after his last rabies injection and after working a full day and then developing acute chest pain that radiated to his arms. He had no symptoms of rabies.[13]

The trial's spectacular results appeared in December 1976 in the *Journal of the American Medical Association*. "We are convinced that a major breakthrough has been achieved in the postexposure treatment of humans exposed to rabies infection," the authors wrote. "Not only can the number of injections . . . be decreased from between 14 and 21 to 6 or less, but in contrast to the vaccines used in the past and others in current use, the [human diploid cell vaccine] causes virtually no side effects and is highly immunogenic," by which they meant that it was a potent provoker of antibody production.[14]

The French government licensed Mérieux's Institut vaccine, and the company—which has since, through a series of acquisitions, become part of the giant vaccine company Sanofi Pasteur—began selling its vaccine, Imovax, in France in 1978. The U.S. Food and Drug Administration licensed both Imovax and Wyeth's competing vaccine, Wyvac, in 1980. The same year, the CDC pronounced the new human-cell vaccines superior to the existing duck-embryo product, noting that they generated on average ten times more antibodies than the duck vaccine, which soon disappeared from the market.[15]

The vaccine business is risky, and in the worst cases completely unforgiving. Early in 1985, fourteen years after it began developing its WI-38 rabies vaccine but only a few years after bringing it to market, Wyeth was forced to withdraw it. An article in the CDC's *Morbidity and Mortality Weekly Report* explained that in routine studies of the Mérieux and Wyeth vaccines—so-called post-licensure studies, conducted once approved products are on the market—the Wyeth vaccine, Wyvac, had produced subpar levels of antibodies. These levels, which didn't meet the CDC's minimum standards for protection, had shown up in three out of seventeen people who had been vaccinated after being bitten by potentially rabid animals. In fact, one of the three had developed no anti-bodies at all.[16] By contrast, all twenty-two people who had been vaccinated with Mérieux's competing vaccine, Imovax, had developed protective levels of antibodies.

The CDC authors couldn't say why the Wyeth vaccine had failed in these three people—in the past it had consistently prompted protective immune re-sponses. The failure couldn't be attributed to one bad batch of vaccine; several different lots had been involved. Perhaps it was related to age, the authors sug-gested: people's immune responses become less vigorous as they get older, and the poor responders had been some two decades older than the other subjects. Then again, the vaccine's action might have been interfered with when, against CDC-recommended procedures, it was injected into the vast islands of fat in the buttocks of the obese thirty-two-year-old man who developed no antibod-ies at all. (The CDC called for injection into the shoulder.) Possibly the prob-lem was due to intrinsic differences in the two vaccines—in other words, to something inherent to Wyeth's production procedures, in which the virus was broken into pieces in the process of killing it. Or it might have been due to a combination of these factors.[17]

Whatever the reasons, the CDC urged anyone who had been bitten and received the Wyeth vaccine in the previous four months to be revaccinated im-mediately with the vaccine made by Wyeth's competitor, Mérieux.

"We wouldn't want anyone to receive our drug and get rabies," said Wyeth spokesman James Pearce. "It hasn't happened and we don't anticipate it will happen but we want to take preventive measures."[18]

Wyeth executives didn't resurrect the vaccine. In the space of time that it took the company to announce that it was recalling Wyvac, the Mérieux vac-cine became the only rabies vaccine for humans on the U.S. market.

New vaccines against many infectious diseases were still being hotly pursued in the mid-1970s. There was every reason for charging ahead. The approvals of measles and mumps vaccines, in 1963 and 1967 respectively, resulted in plummeting incidence of those diseases and in measles-associated deaths.[19]*

Merck, under Hilleman, pioneered both the measles and mumps vaccines. After the company won approval of its HPV-77 duck rubella vaccine in 1969, it quickly packaged the three vaccines together and in 1971 introduced the first "MMR" vaccine. (The duck vaccine would remain the "R" in Merck's vaccine until Plotkin's replaced it in 1979, at which point the company renamed the vaccine MMR II.)

In the mid-1970s, Merck was also racing to beat its competitors to vaccines against chicken pox and hepatitis A. Both viruses were being grown, on different sides of the world, in human diploid cell strains. In 1974 a Japanese physician and virologist, Michiaki Takahashi of the Research Institute for Microbial Diseases at Osaka University in Japan, published the news that he had successfully vaccinated hospitalized children against chicken pox, which is caused by the varicella zoster virus—VZV for short.[20] The virus is wildly contagious and could and did regularly spread among children on hospital wards, thus the site of his clinical trial. While chicken pox is usually mild, if maddeningly itchy, in ordinary children, it can occasionally be devastating, and even lethal, both to healthy children and to sick kids with diseases like leukemia, whose immune systems are weakened. It can also endanger immunosuppressed adults such as those with HIV/AIDS or transplant recipients who must take immune system–suppressing drugs. When the chicken pox virus, VZV, takes over the body, it ravages the lungs, invades the brain, causes multiple organs to fail, or does all of these things. In the elderly, the same virus causes shingles, the excruciating, blistering skin ailment that also can be fatal.

To make his vaccine, Takahashi had weakened VZV that he captured from virus-filled vesicles on the skin of a sick three-year-old boy with the last name of Oka. This weakening was not easily achieved. The virus is famously finicky about what kinds of cells it will invade. But Takahashi managed it, by growing it first in fibroblasts from an unnamed human fetus, then in fibroblasts from guinea pig embryos—these are the only other kind of cell that VZV will invade—and then in WI-38 cells.[21] In the mid-1970s, as Takahashi pushed ahead in Japan with more

---

*Unlike measles, it's extremely rare for mumps to turn deadly, but it causes permanent deafness in some children and brain inflammation in others. In post-pubertal boys and men mumps can cause an excruciating inflammation of the testicles and, rarely, permanent sterility.

clinical trials of the promising "Oka" vaccine, Merck was struggling to develop its own chicken pox vaccine, using a different strain of VZV that it, too, had grown in WI-38 cells.[22] The company was not meeting with much success.[23] Children in its trials developed antibodies but these soon declined precipitously.[24]

At the same time, on the same Merck campus near Philadelphia, a mild-mannered vaccine scientist named Philip Provost was using WI-38 cells to weaken the hepatitis A virus. Then as now, hepatitis A was often a disease of poverty, spread when people ingested food or water contaminated with virus from the feces of an infected person and abounded where people lacked clean water and basic sanitation. An asymptomatic or mild disease of fever, nausea, and loss of appetite in children, it was much more serious, or fatal, in older adults. Globally hepatitis A was a bane; at home it was a particular problem among gay men. A vaccine could put a huge dent in the damage, and Provost was doing everything he could to make one before some competitor beat him to doing so.

It was into this environment that the news of the NIH's investigation of Hayflick and his WI-38 cells landed like a bomb.

If vaccine makers weren't alarmed by the arresting headline in *Science* in April 1976—HAYFLICK'S TRAGEDY: THE RISE AND FALL OF A HUMAN CELL LINE—they certainly were by the time they reached the second paragraph of Nicholas Wade's long article.

> The cells are increasingly being used for the manufacture of vaccines, as well as for research purposes. . . . Yet it now appears that there are sufficient stocks only for the next several years. Moreover, many of the surviving ampules which NIH authorities decided to remove from Hayflick's laboratories last August are proving to be contaminated with bacteria, a fact which may make them unsuitable for vaccine use and render the supply situation even more acute.[25]

Whether or not the article was accurate in forecasting a supply crisis, its appearance had the effect of spooking vaccine makers. So they turned to the cells that the British had derived using Hayflick's methods: MRC-5, developed at the Medical Research Council in London from the lungs of a fetus aborted in 1966.

At West Point Merck's vaccine chief, Hilleman, ordered Provost to stop using WI-38 cells to develop Merck's hepatitis A vaccine; he was to begin again, using MRC-5 cells.[26] In France the Institut Mérieux stopped making its new rabies vaccine using WI-38 cells and began growing it in MRC-5. And a few years later, in 1981, when Merck gave up trying to make its own chicken pox vaccine and licensed Takahashi's Oka vaccine from the Research Institute for Microbial Diseases at Osaka University, the American company didn't grow its newly acquired vaccine in WI-38. Instead it turned to MRC-5.[27] It would do the same a few years later, when it tweaked the dose of virus in the Oka vaccine to create a vaccine against shingles.

Wade's concluding assessment in the 1976 *Science* article seemed to have become a self-fulfilling prophecy: "The real tragedy for Hayflick is not what the NIH inquiry or Stanford has done to him but what he has done to the future of WI-38s."[28]

The story in the end was not quite so clear-cut. The WI-38 cells remained in use in the United States for three important vaccines. Wyeth had been using WI-38 cells since it obtained them from Hayflick in the early 1960s to make an adenovirus vaccine for the U.S. military, to replace the monkey virus–contaminated vaccine that the Pentagon had used from 1958 until 1963.[29] By the late 1960s more than 250,000 U.S. recruits had received the new, WI-38–propagated adenovirus vaccine in clinical trials. It proved highly effective at fending off the respiratory infections that plagued soldiers in the close quarters of their barracks, and in 1971 the Pentagon began administering the vaccine as oral tablets to every incoming recruit—hundreds of thousands of them annually. When news of the WI-38 debacle broke in March 1976, Wyeth did not change course. Nor did the company stop using WI-38 cells to make its ill-fated rabies vaccine.

However, the markets for the adenovirus and rabies vaccines—about twenty thousand to thirty thousand people received postbite rabies vaccinations in the United States each year—were tiny compared with the customer base that Merck was facing as it pushed ahead with development of Plotkin's rubella vaccine in the mid-1970s. In addition to catch-up vaccines for older children, tens of millions of whom were still being vaccinated, every American toddler received rubella vaccine and would do so open-endedly—more than three million children annually in the mid-1970s. (In 1990 the CDC would add a second dose of the rubella vaccine, just before school entry, to the childhood immunization schedule, instantly doubling Merck's customer base.)[30]

But despite the alarm raised by Wade's article in *Science*, it seems that Hilleman, who was a stickler for getting details and process right, was comfortable that he had enough ampules of young WI-38 cells to vaccinate generations of American children against rubella. If he didn't, he was quickly able to secure them after the Schriver report became public.

In the burst of media that followed the articles in the *New York Times* and in *Science*, Merck officials said that the huge contract with Hayflick, securing the company's right to purchase up to 250 ampules of young WI-38 cells for up to $1 million, was never executed. Rather, a company official said, the firm had purchased only "smaller quantities" of the cells from the Stanford scientist.[31] So it seems likely that, soon after the Schriver investigators descended on Hayflick, Hilleman was able to turn to a different, much cheaper source.

In late 1975, after the government officials got off that TWA flight from California, laden with all of the cells they had removed from Hayflick's lab, and delivered them to the ATCC, experts at the Rockville repository thawed and tested them for contamination. Examining the youngest ampules—those at the eighth passage—they found that the cells in all but 7 of 55 ampules were contaminated with bacteria. But the power of exponential growth meant that those 7 ampules of the youngest WI-38 cells went a long way. The cell culturists at the ATCC expanded the sterile cells inside those 7 ampules until they had doubled a handful more times, producing in the process more than 120 ampules, each one containing enough young cells—cells at the twelfth and thirteenth population-doubling levels—to make hundreds of millions of vaccine doses. The FDA soon certified these as clean and ready for vaccine making.[32] They became available for purchase to vaccine firms for $30 per ampule.[33] But such was the fear generated among vaccine makers by the *Science* article that Merck appears to be the only company that took advantage of that fact.

In 1996 Hayflick used the Freedom of Information Act to obtain an inventory of all of the vaccine-ready WI-38 cells then resident at the ATCC. It shows that the cell repository still had on hand 101 ampules of the young, sterile cells, with another 20 such ampules stored at the Coriell Institute for Medical Research, a biobank in Camden, New Jersey. (Cell banks commonly store a backup supply of cells at another location, to safeguard against losing all the cells in the event of a natural disaster or some other calamity.) These 121 ampules of twelfth- and thirteenth-generation cells were derived from 6 of the 7 original, sterile, eighth-passage ampules removed from Hayflick's lab. But the inventory shows that there are no cells at all available from the expansion of the seventh of those ampules—ampule number 52B. It seems likely that those are the cells that went to Merck in the 1970s, to make Plotkin's rubella vaccine.

In June 2016, another Freedom of Information Act request revealed that in the years between 1996 and 2016, the number of available ampules at the ATCC had dwindled by only eighteen. (At least some and probably all of these eighteen ampules have been set aside for Merck.) Which is to say that, not counting the score of ampules that are safely stored at the New Jersey cell bank, there are still eighty-three vaccine-ready ampules of WI-38 cells, secured by the ATCC's twenty-four-hour surveillance cameras, alarm systems, and locks, waiting to be put to use.

Biology, Inc.

Washington, DC, 1980

If there is some gold in the hills, and you happen to get a chunk, well, there is no point in leaving it in the ground if somebody is going to pay you for it. That's what the biological sciences are, a tremendous hotbed of intellectual and commercial ferment, with the possibility to discover one of the mysteries of life and also get rich quick.

—William J. Rutter, former chairman of biochemistry at the University
of California at San Francisco and cofounder in 1981 of the
biotechnology company Chiron, speaking in 1992[1]

H ayflick often describes the effects of his 1976 lawsuit against the U.S. government in grandiose terms. He credits it for "turning around the mind-set of almost the entire biological community" on the issue of whether academic scientists should be able to profit from their inventions.[2] When they saw the government go after him for laying claim to the cells he had derived from Mrs. X's fetus, he says, a vast swath of his colleagues became convinced that they should all have the right to make money on their inventions.

In fact, right through the 1970s many biologists remained adamantly opposed to sullying what they saw as a pure pursuit with the taint, and incentives, of commerce. One venture capitalist described Stanford's faculty in the late 1970s as "the bastion of anti-entrepreneurial activity," notwithstanding the university administration's efforts to patent Cohen and Boyer's gene-splicing technology.[3] In 1980, Stanford's new president, Donald Kennedy, a biologist by training, worried aloud that the university's new interest in patenting could jeopardize an academic tradition that treasured free and open inquiry.[4] Another venture capitalist, Brook Byers, recalled the reaction when in 1978 a professor of medicine at the University of California at San Diego named Ivor Royston founded a biotechnology company called Hybritech to produce specialized antibodies to treat cancer. "It was sort of like when Bob Dylan went electric in the sixties," said Byers. "We were seeing [Ivor] as a pioneer, but his colleagues were seeing him as a turncoat."[5]

Still, change was coming, and quickly: 1980 would be the year of a seismic shift in what was legally possible for biologists who wanted to become entrepreneurs and for the universities where they were based. The forces that converged to cause this landscape-shifting earthquake were far bigger than Hayflick and his lawsuit.

In the late 1970s the U.S. economy was on the rocks, ground down by high unemployment, alarming inflation, declining productivity, and growing international competition. Politicians were casting about for remedies. They became amenable to arguments that universities were parking lots of scientific inventions blocked in by the fusty U.S. government's title to discoveries made on its dime—discoveries that, when they were patented, the government licensed nonexclusively, depriving companies of the commercial incentives to develop them. The numbers spoke for themselves: prior to 1980 a paltry 5 percent of government-owned patents were ever used in the private sector.[6]

As Elizabeth Popp Berman explains in her excellent book, *Creating the Market University*, most university-based inventors were funded at least in part by the government, so the impact of government title was pervasive. Those who wanted to see change—an activist group of patent attorneys and patent managers both inside the government and at universities—had run into a concrete wall in the administration of President Jimmy Carter. Carter officials believed that handing title to government grantees amounted to a giveaway of inventions that taxpayers had already paid for and should be able to use at will.[7] In 1977 the new Health, Education and Welfare secretary, Joseph Califano, shut down even the waivers that had allowed inventions like Plotkin's rubella vaccine and Koprowski's rabies vaccine to be owned by the institutions where they were developed.[8]

The collision of the deteriorating economy and the Carter administration's unyielding opposition—the shutdown of the waiver process provoked agitated calls to their congressional representatives from universities—opened the minds of two key senators to the argument that U.S. innovation was being hurt by Uncle Sam's nearly ubiquitous title. Republican U.S. senator Robert Dole of Kansas and Democratic senator Birch Bayh of Indiana became convinced that allowing universities to claim title to inventions would speed commercialization, boosting U.S. competitiveness, creating jobs, and bringing badly needed economic growth.[9]

In September 1978 Bayh and Dole introduced legislation that granted universities, along with nonprofit organizations and small businesses, the right to claim title to U.S.-funded inventions, provided that they tried to patent them—and provided that the universities and nonprofits shared any resulting royalties with the individual inventor.[10] Big businesses were left out of the bill, to weaken the "government giveaway" argument.*[11]

The first congressional hearings on the Bayh-Dole bill were held in May of 1979.

In the meantime, in a majestic, colonnaded building one hundred yards from where the senators were pushing the new bill, the U.S. Supreme Court was preparing to hear arguments in a case that was of monumental significance for what we today call the biotechnology industry—an industry that scarcely existed in 1980.[12] In March of that year, the nine high-court justices heard the arguments of a microbiologist named Ananda Chakrabarty who was working for General Electric when, in 1972, he filed a patent application on a genetically engineered, oil-destroying bacterium (*Pseudomonas*) that he had created. Unlike its naturally occurring brethren, Chakrabarty's microbe was able to break down the crude oil from spills into harmless components that actually provided food for aquatic life—rapidly, efficiently, and in diverse environments. The U.S. Patent and Trademark Office in 1977 rejected his claim, on the grounds that living things were not patentable. In a five-to-four decision, the high court disagreed with the patent office, famously opining that Congress intended patent law to encompass "anything under the sun that is made by man."[13] So long as a living thing was truly man-made, in that it had been the product of human manipulations, it was patentable.

The decision in *Diamond v. Chakrabarty* was handed down in June 1980. It opened the floodgates that launched the biotechnology industry. Patents followed on everything from a gene that directs the production of human growth hormone to a mouse genetically engineered to be prone to cancer to the controversial human embryonic stem cells that were isolated at the turn of the twenty-first century. In just one measure of the 1980 decision's impact, consider that in 1979, before the high court made clear to potential investors that biotech products would definitely be patentable, there were about a dozen mostly nascent U.S. biotechnology companies with virtually no revenue. In 2015, there were 394 publicly held U.S. biotech companies, with $117 billion in revenue—not counting the successful biotechs that have been bought up by

*In 1983 President Ronald Reagan used an executive order to extend the Bayh-Dole Act to cover big businesses, too. Because any future president can reverse it, an executive order is not as ironclad as a bill passed by Congress and signed into law. In 1986 a new law (the Federal Technology Transfer Act) allowed big businesses with special contracts with government-operated labs to be awarded patents.

big pharmaceutical firms.[14] Other developments in the late 1970s also fed the biotech boom that began in earnest in 1980. They included a freeing up of venture capital stimulated by a capital gains tax cut and by a relaxing of Department of Labor rules that allowed big pension funds to begin investing in venture capital. They also included Congress's decision not to control the new gene-splicing technology with legislation.[15]

Encouraged in part by the Chakrabarty ruling, Genentech, the company that Herbert Boyer, the innovative UCSF scientist had launched in 1976, despite the opprobrium of his colleagues, went public. The four-year-old company had yet to produce a product—its genetically engineered insulin (the first biotech product) wouldn't come to market until 1982—but on its first day on the market, in October 1980, Genentech soared to a value of $532 million, or $1.5 billion in 2016 dollars. By the end of 1980, Boyer was worth $65 million.[16] Genentech would be the first among many biotech companies, and Boyer the most visible and successful of a new breed of entrepreneurial biologists who made the leap into business.

But even the high court's decision could not in itself make entrepreneurs of academics. Genentech and dozens of other companies that were soon founded by business-inclined biologists, backed by venture capitalists, would have been seriously handicapped had it not been for two other critical events that transpired in the closing weeks of 1980.

The first was the issuance by the Patent and Trademark Office on December 2, 1980, of the first of several patents on Cohen and Boyer's revolutionary gene-splicing technology—the tools that, among many other things, allowed scientists to target desirable genes with molecular "scissors" and splice them into bacteria that would then pump out the corresponding proteins in quantity. Boyer's Genentech and scores of other mostly brand-new firms rushed to license the technology from Stanford and the University of California, both of which had negotiated agreements with the U.S. government allowing them to own the patents well before the Carter administration came to power. By the end of 1981, seventy-two companies had each paid $20,000 to license the technology from Stanford and UC.*[17]

Ultimately, 468 companies would license the technology.[18]

*Normally, exclusive licenses were needed to lure commercial investors. However, the Cohen-Boyer patents were so broadly applicable and the technology they enabled so important and so much in demand that Neils Reimers, the Stanford patent guru who was managing the patents on behalf of both Stanford and UC, refused Genentech's demand that they be licensed exclusively to the company.

The second event transpired ten days later, on December 12, when a reluctant President Jimmy Carter, in his waning days in office and on the last possible day before the legislation would otherwise have died, signed into law the Bayh-Dole Act, which had been pushed through a lame-duck session of the U.S. Congress. At the stroke of his pen universities won title to any promising U.S.-funded inventions that they wanted to try to commercialize. Their only obligations were to direct a portion of the resulting royalties to the academic inventors and to put the rest of the money back into their institutions. No longer would discoveries and inventions, many of them being enabled almost daily by the new tools of molecular biology, remain parked in the academy. And no longer would academic scientists be looked at askance for collecting on their discoveries.

Universities rushed to establish technology-transfer offices. They staffed them with smart professionals deployed to hunt out and apply for patents on their scientists' most promising-looking inventions. Everyone wanted to get in on what looked to be a lucrative new revenue stream.

For some, it was. By the time they expired, the Cohen-Boyer patents had earned Stanford and the University of California more than $250 million.[19] Columbia University earned $790 million for its patents on a process for inserting foreign DNA into host cells that led to new medicines for illnesses including anemia, strokes, and multiple sclerosis.[20]

In the first thirty-four years after the Bayh-Dole law was enacted, the number of patents issued to U.S. universities each year grew more than sixteenfold to an all-time high in 2014 of nearly 6,400.[21] (This is the most recent year for which data is available.) Most have gone to big institutions.[22] And many of them—far out of proportion to the number of U.S. patents generally, and also far more than were found, proportionately, in universities' modest patent portfolios prior to 1980—are in the medical sciences. For instance, in 2012 biotech patents comprised 1 percent of all U.S. patents but 25 percent of the patents granted to U.S. universities. Universities' income from licensing, which had hardly amounted to chump change in 1980, exceeded $2.7 billion in 2014.

The 1980 changes marked the beginning of a shift to the acceptance within academia of biologists as businessmen, especially when venture capitalists started coming out of the woodwork to woo expert molecular biologists in the only place they could find them—the academy.

The opportunities didn't present themselves to everyone, by any means. In a 1985 survey of biotechnology faculty at major universities, just 8 percent reported owning equity in a company based on their own work.[23] And there was certainly ironic grumbling among some university scientists, on whom it was not lost that they had been asked, in scarcely the space of time needed to run one or two experiments, to do an about-face. They were no longer supposed to be motivated only by the love of science and the greater good; they were also to be advance agents with eyes ever peeled for the next commercial opportunity for their institutions—and themselves. Still, the adjustment was made, and today when academic biologists found, run, advise, or consult for companies, and become wealthy doing so, they are seen not as traitors to their calling but as paragons of success.

Hayflick watched the events of 1980 and beyond play out with a great sense of irony, first as he struggled to keep his family afloat and to find a better work situation than his insecure post in Oakland, and then as he moved in 1982 to a position as director of the Center for Gerontological Studies at the University of Florida in Gainesville.

His lawsuit against the government looked set to head to trial when, in September 1981, after numerous failed attempts, the two sides agreed to settle out of court. By this time Hayflick and his attorneys had been battling the NIH for nearly five years. The government's willingness to come to the settlement table may have been informed, in part, by Reagan's people taking over at the recently renamed Department of Health and Human Services, the NIH's parent department, where under the Carter administration the anti-Hayflick animus had extended to the highest levels. Those people were now gone. Bringing Hayflick to trial was also going to be tough for the government politically, against the backdrop of the brand-new Bayh-Dole law, with its emphasis on universities commercializing their scientists' inventions and directing some of the cash payoff to the scientists themselves.

That being said, there was also a central, if prosaic, reason that made the government willing to settle in the summer of 1981: Schriver, the NIH investigator who had pursued Hayflick so zealously, had retired in March 1980. His impending departure dismayed Vincent Terlep, the lead Department of Justice attorney handling the case, who wrote to the NIH's deputy director, Thomas Malone, that Schriver "has been invaluable in assisting me in the development of this case, and it is my opinion that his continued participation is essential to a successful outcome."[24] Terlep persuaded the NIH to hire the retired Schriver as a consultant. But Schriver was expensive, and paying him as the lawsuit dragged on began to look like too high a price to pay for an uncertain outcome.[25]

Hayflick, his company Cell Associates, and the NIH signed an out-of-court settlement on September 15, 1981. The agreement described itself as a "compromise." Under its terms both the question of whether the NIH had violated the Privacy Act in releasing the Schriver report and the question of who owned the WI-38 cells were "in reasonable dispute," the signatories conceded. The parties signed "without any party admitting or conceding liability."[26]

Under the terms of the settlement, Hayflick was granted title to six of the original eighth-passage ampules of WI-38 and their progeny—for use, the settlement specified, in his current grant to study cellular aging. The NIH was granted title to the rest of the eighth-passage ampules in its possession.

Hayflick renounced title to any WI-38 cells of any age in the possession of anyone else, including the government—with the exception of a handful of ampules of older WI-38 cells that the government had left behind after cleaning out his Stanford freezer in the summer of 1975. These ampules he was granted title to "without restrictions on use."

Thanks to sky-high interest rates, the money from Hayflick's sales of the cells had grown from $67,000 to about $90,000.[27] The government ceded all of it to Hayflick. He promptly signed it over to Bill Fenwick to pay for nearly six years of legal services.

A few months later a letter from a group of Hayflick supporters appeared in the pages of *Science*, calling attention to the settlement. It is significant for the vehemence with which it takes the NIH to task for its treatment of Hayflick—and for the number of its signatories: eighty-five scientists, including some lesser-knowns but also some leading lights of U.S. biology like Lewis Thomas, the chancellor and immediate past president of the Memorial Sloan-Kettering Cancer Center in New York City, and the esteemed virologist Joseph Melnick.[28]

The letter writers claimed that the settlement "exonerated" Hayflick and wrote that many of them wished he had not settled but had continued in court to press his case against the government. Nonetheless, the settlement was a "happy outcome of Dr. Hayflick's courageous, sometimes lonely, emotionally damaging, and professionally destructive ordeal."

The signatories reflected the strong feeling in part of the biological community that Hayflick had been extremely poorly dealt with by the NIH—and it's certainly clear that those driving the government's investigation harbored a nasty, personal, take-no-prisoners approach to Hayflick. (Even after he had retired, the lead NIH investigator, Schriver, would personally intervene to try to derail Hayflick's hiring at the University of Florida.) But like Hayflick in a letter he penned to *Science* several years earlier, defending himself and excoriating the government, the indignant biologists who wrote to the illustrious magazine did not acknowledge or address his sales of the WI-38 cells for tens of thousands of dollars.[29] Even the sea changes in the law governing patent ownership did not in 1981, and do not today, allow university biologists to walk away from their campuses with unpatented cells developed with U.S. support and begin selling them.

CHAPTER TWENTY-FIVE

Hayflick's Limit Explained

Stanford, California, 1973–Summer 1974
Berkeley, Boston, and Beyond, 1978–2009

It remains for us to discuss youth and age, and life and death. To come to a definite understanding about these matters would complete our course of study on animals.

—Aristotle, *On Longevity and Shortness of Life*, 350 BC

In the paper he wrote with Paul Moorhead, Hayflick had laid down a huge challenge to a half century of received wisdom that said cells in culture would keep dividing month after month and year after year, if only they were properly treated.[1] He had gone even further in a 1965 paper by suggesting that normal cells' lab-dish mortality might be related to the aging of whole organisms, including human beings.[2]

He met with plenty of push-back from the start. His findings were publicly challenged by Theodore Puck, the influential cell culturist whom the young Hayflick had stood up to question at that packed lecture in Atlantic City in 1961. "Sharp divisions were evident" between Puck and Hayflick about whether cells aged in culture, *Science* magazine reported from a 1965 meeting where Puck claimed that he had grown normal cells in culture through five hundred divisions without their dying.[3] Like other noted biologists, Puck suggested that technical failures in Hayflick's cell-culture methods were responsible for the cells' deaths.[4] Only time would definitively prove that Puck, whatever his eminence, was patently wrong. By 1974 the widely respected Nobel laureate Sir Frank Macfarlane Burnet allowed that "there is probably a majority opinion" that the lab-dish mortality that Hayflick had observed "measures an important biological quality of . . . cells which is significant for the interpretation of ageing."[5]

Burnet coined the term "the Hayflick limit" to describe the number of replications that normal, noncancerous cells can undergo before they cease dividing. It took lasting root not only in science but also in popular culture. Today you can buy Hayflick limit T-shirts; the limit plays a role in the 2004 horror film *Anacondas: The Hunt for the Blood Orchid*; and it has inspired the naming of at least one garage band.

Hayflick's findings opened a whole new field, the laboratory study of cellular aging. By the early 1970s the subject was enticing a growing number of biologists. In 1974 Congress created the National Institute on Aging as the newest among the cluster of research institutes that comprise the National Institutes of Health. (Congress launched the new institute in response to a relentless lobbying effort led by a wealthy Georgetown widow named Florence Mahoney. President Richard Nixon twice vetoed the bill establishing the new institute. He finally signed it when it was sent to his desk a third time, weeks before he was forced to resign in August 1974.)

Like others, Hayflick was drawn to the study of cellular aging and the raft of questions cracked open by his discovery. Was cellular aging in a lab dish simply the microequivalent of aging in human beings? What caused it? Was some essential factor that was needed for cell replication being depleted over time? Did cells age in lockstep, or were there individual differences between them? Were genes involved? How was it that the frozen fetal cells in his original experiments "remembered" their ages when they were thawed? Was there a different Hayflick limit for every species, and was it the number that one would expect, given the animal's natural life span? The questions were endless.

Hayflick had tackled one of the most obvious and tantalizing of them in a 1965 paper that appeared in *Experimental Cell Research*, the same journal that had published his groundbreaking 1961 report. Did cells taken from adult humans—as opposed to fetuses—divide fewer times before reaching the Hayflick limit? From a colleague at the Mayo Clinic in Rochester, Minnesota, Hayflick obtained lung cells from the cadavers of eight people, ranging in age from a twenty-six-year-old who was killed in a car crash to an eighty-seven-year-old who died of heart failure. He found that they replicated far fewer times before growing old and ceasing to divide than had the thirteen lung-cell strains he had derived over time from human fetuses. The fetal lung cells ceased dividing, on average, after forty-eight replications; the eight adult strains petered out after an average of twenty divisions.[6]

It made intuitive sense, and yet his data showed that our cells don't simply tick away toward the Hayflick limit on a predictable path as we age. The lung cells from the young auto crash victim doubled just twenty times in culture, while those from the eighty-seven-year-old who died of heart failure replicated themselves twenty-nine times before hitting the Hayflick wall.[7] These cells from the most elderly lungs in fact doubled more times than the lung cells from any of the other seven cadavers, whose donors were aged mostly in their fifties and sixties when they died.

"There appears to be no exact correlation between the age of the [adult] donor and the doubling potential of the derived strain," Hayflick wrote. Or if such a relationship did exist, "it cannot be detected by the present crude methods." What *was* clear, he wrote, was the general observation that adult lung cells doubled far fewer times than their fetal counterparts before they ground to a halt in their cultures. Papers by other scientists over the coming fifteen years would establish definitively that a cell's doubling potential did vary inversely with donor age, and that this was true for cell types as various as skin, liver, and the smooth muscle in the walls of arteries.[8]

Adding more complexity, a paper published in 1980 in *Science* showed that even *which* cells happened to have been randomly sliced from the lungs of those eight cadavers doubtless influenced Hayflick's findings. That report, published by biologists James R. Smith and Ronald G. Whitney of the W. Alton Jones Cell Science Center in Lake Placid, New York, blew away any notion that cells within a tissue are identical, preprogrammed entities all marching rigidly toward a predetermined finish line.[9] Smith and Whitney took a single lung cell from an aborted human fetus and, as it multiplied in culture, extracted one hundred or two hundred cells now and then. Then they tested each cell within that subgroup to see how many divisions it had left in it. They found a wide variation in the remaining doublings.

Digging deeper, they next took the two daughter cells that arose from the division of a *single* fetal lung cell and counted how many times the population of cells arising from each daughter was able to double itself. They did this over and over, beginning with many single cells. To their amazement, they found that the proliferative potential of the two daughter cells of each single cell was not identical and, in fact, varied by as many as eight population doublings.[10] When it came to theories about what caused the Hayflick limit, "clearly, all current hypotheses should be reexamined in light of our data," they concluded.

In the midst of the puzzlement, at least one thing had become clear, thanks to the efforts of Hayflick's red-bearded graduate student, Woody Wright, the young biologist who was also completing an MD at Stanford's medical school. In an elegant series of experiments under Hayflick's oversight, Wright used a chemical called cytochalasin B, which he applied to WI-38 cells in culture. At high doses the chemical ejected the nuclei from the WI-38 cells, resulting in WI-38 cells dubbed cytoplasts, full of cytoplasm but without nuclei.[11] Separately, Wright treated normal WI-38 cells with a different pair of chemicals that inactivated a broad range of components in their cytoplasm. Then he fused the nucleus-free cytoplasts—which still had functioning cytoplasm—with the WI-38 cells that still had nuclei.

This allowed for a series of mix-and-match experiments in which Wright created cells with young nuclei but old cytoplasm; young nuclei and young cytoplasm; old nuclei and young cytoplasm; and old nuclei and old cytoplasm. With each kind of hybrid he watched to see how many more divisions occurred before it hit the Hayflick limit and stopped dividing. If the control seat of cellular aging was in fact in the cytoplasm, then the hybrid WI-38 cells with "young" cytoplasm should live much longer than those with "old" cytoplasm, regardless of the age of the nucleus. But if the control seat of cellular aging was in the nucleus, then the cells with younger nuclei should live longest, regardless of the age of the cytoplasm. The latter is what Wright found.

His experiment was published, with Hayflick as coauthor, in 1975, as Wright completed his MD and PhD and launched a long career pursuing the mysteries of aging.*[12]

He had aptly demonstrated that whatever was controlling the number of cell doublings in normal diploid cells like WI-38 was located in the nucleus and not in the cytoplasm. Years later Hayflick would dub this whatever-it-was a "replicometer."[13] As distinct from a clock, which measures time, the replicometer measured replications. When a cell wasn't replicating—for instance, when it was in the freezer—the replicometer stopped chalking up replications.

That the replication-counting mechanism resided in the nucleus wasn't surprising. The nucleus was known to govern myriad important functions of cellular life. But locating the mechanism and understanding it were two vastly different projects. How it was teased out is one of the great stories of late-twentieth-century biology, because it shows how profound findings with huge implications for human health can emerge from the unlikeliest of organisms, the depths of basic biology, the powerful engine of scientific curiosity, and the cross-pollination that happens when scientists share ideas. It also connected

---

*Wright would go on to propose that the fundamental reason that normal cells stop dividing is to prevent cancer from forming. Cancer cells require many mutations to become malignant, and normal cells can't accumulate these if they stop dividing.

Hayflick's iconoclastic observation in the early 1960s with the discoveries that would lead to a Nobel Prize for others nearly fifty years later.

In 1938 the Nobel Prize–winning American geneticist Hermann Muller described what seemed like protective caps of DNA located at the ends of chromosomes, something like the bits of plastic banding that cap the ends of shoelaces to keep them from fraying. He named them "telomeres," from the Greek *telos* + *meros*, meaning "end part."[14] Three years later another great American geneticist, Barbara McClintock, who would also win a Nobel, described a crucial role for telomeres in maintaining the integrity of chromosomes.[15]

Both biologists worked with broken chromosomes, and both noted that chromosomes in their natural state had ends that were somehow protected from the abnormalities that occurred at unnatural chromosome breaks induced by radiation or rupture. Their tantalizing observations sat unexplained for most of the next forty years. There simply weren't the tools available to probe them further.

The revelation of the molecular structure of DNA by James Watson and Francis Crick in 1953 began the stepwise process that would produce those tools. It was soon followed by the discovery of a vital enzyme by Arthur Kornberg—one of the "Bergs" whose reputations drew Hayflick to Stanford. Kornberg described DNA polymerase—the enzyme that duplicates a cell's DNA in preparation for cell division—and teased out how this essential enzyme works, by attaching itself to a long strand of DNA and moving along its length, transcribing the letters of the DNA alphabet as it goes. Kornberg won a Nobel Prize in 1959 for his discovery.

One autumn evening in 1966, a young Russian biologist had a flash of insight that linked Kornberg's all-important enzyme to the Hayflick limit. Alexey Olovnikov was a postdoctoral student at the Gamelaya Institute of the Academy of Sciences of the USSR. He was on his way home from a lecture at Moscow University on the radical new findings of the American scientist Leonard Hayflick. "I was simply thunder-struck by the novelty and beauty of the Hayflick Limit," he later wrote.[16]

Mulling what he had heard, Olovnikov descended into the subway. As he stood on the platform and heard the roar of an approaching train, he thought in a flash that the tracks made an apt analogy to DNA and the train running along them to DNA polymerase, the enzyme that copies DNA in preparation for cell division by traveling along the length of the DNA in a chromosome, making a new copy as it passes. But just as there was a "dead zone" between the front of the train and the first passenger door—"dead" because the whole purpose of the train was to carry passengers—perhaps there was a dead zone at the very tip of the chromosome that DNA polymerase couldn't copy, although the enzyme's entire purpose was to copy DNA. If this was true, a cell's genome—its instruction book for a life—would be slowly shortened through the multiple cell divisions of a lifetime, with DNA at one end of its chromosomes getting lost bit by bit, copy by copy. A railway that kept losing track was untenable.

Olovnikov spent the next four years thinking about the Hayflick limit and developing his theory as he rode the Moscow subway to and from work, waiting for the approaching train in the same subway station where he had his epiphany. He finally published his theory in Russian in 1971.[17] He proposed that DNA polymerase could not copy the DNA at one end of each chromosome, so that chromosomes were inexorably shortened with each cell division. But he added that the DNA at the ends—the telomeres, that is—might be expendable, not spelling out any information that was vital for the life of the cell. If so, the DNA of the telomeres could be sacrificed, chopped down bit by bit, replication by replication, while the important DNA—the DNA that directed the life of the cell—survived intact.[18] Telomeres, in other words, played a buffering role. But they were finite buffers. Buffers that could be shortened and shortened until, when the last bit of telomeric DNA was reached, so was the Hayflick limit. Because the cell at this point could no longer replicate without eating into its life-directing DNA, it would stop dividing. Olovnikov also conjectured, presciently, that cancer cells might have a mechanism allowing them to escape this telomere shortening, making them immortal.[19]

In the Western-dominated world of science, Olovnikov's theory remained all but invisible, even after an English translation was published in 1973.[20] However, the theory that the copying enzyme, DNA polymerase, could not replicate that troublesome end bit of the chromosome was separately described in 1972 by James Watson, the American scientist who had twenty years earlier discovered the structure of DNA with Francis Crick.[21] While Watson did not make the connections to the Hayflick limit and cancer that Olovnikov did, his paper nonetheless put on the map the phenomenon that became widely known as the "end replication problem."

Olovnikov's ideas were, in any event, theoretical when he published them. They were not corroborated by any experimental evidence. The quest that would produce that evidence—and a Nobel Prize almost four decades later—was led by a trio of American scientists: an ambitious Californian graduate student with dyslexia; an exuberant, cerebral daughter of Australian doctors; and a competition-averse Canadian migrant to Boston.

Elizabeth Blackburn had been fascinated with the natural world from her earliest days growing up as the child of physicians in Tasmania, when she used to hold and even sing to jellyfish that washed up on the beach.[22] In the mid-1970s, as a postdoctoral student at Yale University, she began working with *Tetrahymena thermophila*, a single-celled, pond-dwelling organism that is covered with fine, hairlike projections called cilia. *Tetrahymena* are a favorite of research scientists because they are cheap, they grow rapidly, and they engage in many of the cellular processes common in more complicated organisms, like humans. Most important for Blackburn, *Tetrahymena* have a multitude of mini chromosomes. This provided plenty of material for her studies of telomeres, whose function had mystified scientists for decades, and which also captivated Blackburn.

She set out to determine the letter-by-letter sequence of *Tetrahymena* telomeres. While gene sequencing was a new tool at the time, Blackburn was ideally placed to use it—she had learned the technique from its pioneer, Nobel Prize winner Frederick Sanger, with whom she completed her PhD at Cambridge University. The paper she produced was coauthored by Joseph Gall, her mentor at Yale—although it was published in 1978, the year that Blackburn became an associate professor of molecular biology at the University of California at Berkeley. It showed that the ends of *Tetrahymena* chromosomes contained a short sequence of six letters of DNA code—CCCCAA—that was repeated over and over, from twenty to seventy times.[23] Her discovery was groundbreaking. No one had described the molecular structure of the chromosome tips in *any* species. Soon scientists were sequencing the telomeres of other simple organisms and finding that these too consisted of short DNA sequences repeated many times over at chromosome ends.

Across the country in his lab at the Sidney Farber Cancer Institute in Boston, a young Harvard Medical School professor named Jack Szostak had been struggling to understand how a very different single-celled organism, the humble baker's and brewer's yeast, handled broken bits of DNA. When Szostak inserted linear, lab-manufactured DNA strands, unprotected by telomeres, into the yeast, the DNA was either inserted into a chromosome or completely destroyed. Or, occasionally, the two ends of a strand would be joined together to form a circular bit of DNA. Szostak's pieces of DNA never survived as linear strands, for unlike naturally occurring chromosomes, their ends weren't protected by telomeres. Then Szostak heard Blackburn discuss her work at a conference in the summer of 1980. He approached her, and during an intense discussion the pair came up with what Szostak later recalled as a "wild idea."[24] They would graft *Tetrahymena* telomeres onto the end of Szostak's lab-created mini chromosomes and insert them in yeast. Just maybe the grafted-on telomeres would protect Szostak's chromosomes.

Blackburn and Szostak found their long-shot hunch confirmed. The mini chromosomes with their grafted-on telomeres were not degraded. Somehow that six-letter repeating sequence of DNA in the *Tetrahymena* telomeres was protecting the chromosome ends, even in a very different organism. (It's now known that in humans and other mammals the specific DNA sequence of telomeres attracts a group of proteins—appropriately named "shelterin"—that form a cap at the fragile chromosome end, preventing the cell from "thinking" that it is an open, broken end and trying to repair it, a process that can end in cell suicide if the repair work fails.)

Blackburn and Szostak published their findings in the eminent journal *Cell* in 1982.[25] Now another question presented itself. The end-replication problem dictated that the yeast telomeres should get shorter and shorter with each division, until they were gone and essential DNA began to be lost. In a single-celled organism, this wasn't compatible with survival. An organism of just one cell, like yeast, had to be able to replicate infinitely, or perish. Such organisms must have some mechanism for fighting back against the end-replication problem. Soon after, working with Blackburn's graduate student Janis Shampay, Blackburn and Szostak demonstrated that in yeast the telomere ends were being extended with additional tiny DNA snippets while they replicated.[26] In effect, the cells seemed to be "trying" to maintain their telomeres at a roughly constant length.

The scientists proposed that there might be an enzyme capable of lengthening chromosomes at their tips by adding those small, repetitive snippets of DNA. This enzyme would have to work quite differently from DNA polymerase, the powerful enzyme that replicated the whole of a cell's DNA during cell division—except for that troublesome end bit—but which required a DNA template from which to copy.

On Christmas Day 1984 Carol Greider, a twenty-three-year-old PhD student in Blackburn's Berkeley lab, peered at a newly developed X-ray film and saw something that seemed like it was too good to be true.[27] It was the first hard evidence that a telomere-lengthening enzyme existed. It showed up in a series of neat bands that climbed, ladderlike, up the X-ray film. The regularity of the bands made it clear that the putative enzyme was adding six-base snippets of DNA—telomeric repeats—to the telomeres in a test-tube mix that Greider had made, containing synthetic telomeres and extracts of *Tetrahymena* cells.[28] (After further experiments confirmed that her seemingly too-perfect findings were in fact real, Greider went home and danced to the whole of Bruce Springsteen's *Born in the U.S.A.*)

It would be 1990 before Blackburn and her colleagues at Berkeley definitively demonstrated the enzyme's activity in a living organism.[29] By then it had been given a name: telomerase. It was a complicated, unusual enzyme, and Greider and Blackburn had shown that it carried its own template of genetic code, allowing it to affix itself to the end of the chromosome—the bit that DNA polymerase can't manage to duplicate—and begin adding genetic sequence in the form of the telomeric repeats specific to that species.[30]

How does all of this circle back to Hayflick's fibroblasts aging in their bottles half a century ago? It would take a conversation between Greider and a scientist named Calvin Harley, at McMaster University in Hamilton, Ontario, to finally connect those dots. Greider often traveled from Berkeley to McMaster to visit her boyfriend, Bruce Futcher, who shared a lab with Harley, a mustachioed, mild-mannered Canadian who had been fascinated with aging since he was a boy.[31]

It was Harley who told Greider about an obscure theory that she had never heard of. It had been proposed back in 1971 by a young Russian who had an epiphany in the Moscow subway. Greider learned about Alexey Olovnikov's theory that cells hit the Hayflick limit because of the end-replication problem—the cells finally stopped dividing when they somehow discerned that more divisions would begin to eat into their critical, life-directing DNA. Olovnikov had not had the tools to test his theory almost two decades earlier. But in 1988 the human telomere sequence—TTAGGG—was discovered.[32] Greider, who had completed her PhD and taken a position at the Cold Spring Harbor Laboratory on Long Island, New York, heard of the discovery early, at a conference, and called Harley with the news. She, Harley, and Futcher now had the means at hand to find out if Olovnikov had been right. They could ask and answer the question: what happens to the length of human telomeres over time?

They grew fibroblast cells that had been derived from several types of human tissue: fetal lung tissue; newborn skin; and skin from adults aged twenty-four, seventy-one, and ninety-one years. As the cells replicated, the scientists regularly measured the length of their telomeres. No matter what their starting lengths—and the fetal and newborn cells had substantially longer telomeres to begin with—all of the cells' chromosome tips grew progressively shorter with time. On average their telomeres were clipped of two thousand DNA letters before they died, or roughly fifty letters each time the cells divided. The findings comported beautifully with the end-replication problem and left the cells with, on average, about two thousand letters in their telomeres when they finally stopped dividing—an amount, the scientists inferred, that could be merely a nonfunctional stub. Their findings "could be biologically significant," the three biologists wrote, with typical scientific restraint, in the 1990 *Nature* paper that reported on the experiment.[33]

Why didn't telomerase "save" the cells in this experiment, protecting their telomeres by adding to them to maintain their lengths? The answer had to be that telomerase was not active in all human cells. And in fact other scientists would soon demonstrate that telomerase is lacking in most human cell types.

With the publication of the Harley-Futcher-Greider paper, a simple, elegant concept captured the public imagination: telomeres in effect burn down, like a candle wick, until the end is reached, the light goes out, and the cell stops dividing and, finally, dies. The notion was a tantalizing and seductive one, not least for questers after the fountain of youth. For it wasn't unthinkable that what happened in a test tube might happen in the human body: that the ravages of growing old, from sagging skin and thinning hair to all manner of age-related diseases, resulted from the clipping down of our telomeres; and that telomerase might rescue humanity from all of that. That notion, Blackburn later told the writer Stephen Hall, "was a lovely idea—so simple that nobody but an idiot would believe it."[34]

This did not dissuade the enthusiastic. In 1990 a company called Geron was established in Menlo Park, California, to chase the promise of telomerase in battling aging. The scientists at Geron were keenly aware that Harley, Futcher, and Greider had demonstrated a *correlation* between telomere shortening and cellular aging but that, as the trio themselves pointed out, their findings did not establish that the shrinking telomeres *caused* cells to become elderly. A few years later the Geron scientists, working with Hayflick's former graduate student Woody Wright and his group at the University of Texas Southwestern Medical Center at Dallas, conducted an experiment designed to reveal whether telomere shortening *caused* cells to age. They added telomerase to two kinds of cells: fibroblasts from human foreskin and from specialized cells that line the retina in the human eye. They also kept control cultures of each of these cells, without telomerase added.

The control cells proceeded to divide to their Hayflick limit, take on the appearance of elderly cells, and stop dividing. Those with the telomerase did not. On the contrary, they flourished, maintaining long telomeres, dividing prolifically well beyond their Hayflick limit, and yet maintaining normal chromosomes; these were not cancer cells that had somehow bypassed the normal brakes on cell division. By the time the resulting paper went to press, the telomerase-rich cells had exceeded their normal number of replications by at least twenty doublings. The scientists had established beyond doubt a causal relationship between telomere shortening and the aging of cells in the lab. The new paper, published in *Science* in 1998, had an eye-grabbing title: "Extension of Life-span by Introduction of Telomerase into Normal Human Cells."[35]

A media frenzy followed. The *New York Times* splashed the news on its front page, then ran two more articles in the space of six days.[36] "I didn't think I'd live long enough to see this," Hayflick told the *San Francisco Chronicle* in another front-page article.[37] The price of a share in Geron rose from $7.80 on the last day of 1997 to $12.70 on January 16, 1998, the day the *Science* paper was published. Twenty months later, as the price still hovered around $11.00, *Fortune* observed that "the bold little company has achieved the highest buzz-to-equity ratio in biotech history."[38] But the story turned out to be commercially disappointing. In the summer of 2016 the twenty-six-year-old Geron had yet to win FDA approval for a single product and had abandoned its attack on aging for an attack on cancer. Its stock was trading at less than $3.00 per share.

The relationship between telomeres and aging has emerged as immensely complex. The running-down of telomeres in cells in the lab does not translate as the simple, unitary cause of the aging of entire organisms, including human beings. For one thing, unlike cells in culture that grow to their Hayflick limits and then stop dividing, most of the cells in our bodies divide rarely. These cells probably do not come close to running down their telomeres in the course of our life spans—something that Hayflick's work with those specimens from multiaged cadavers back in 1965 pointed to quite clearly. (Remember the lung cells from the eighty-seven-year-old cadaver that divided another twenty-nine times in culture?) What's more, we now know that the aging of our cells is also likely mediated by mechanisms that can bypass telomeres entirely, like chronic inflammation and something called oxidative stress, in which toxic by-products of cellular metabolism directly damage proteins and other cell components.

Nonetheless, a group of patients with rare, inherited genetic disorders have provided important evidence of a direct effect of telomere length on aging in living human beings. People with these disorders, which have been collectively dubbed "short telomere syndromes," have very short telomeres and suffer from a characteristic set of age-related degenerative diseases.[39] Often these diseases make themselves manifest in cells that, unlike most, divide frequently, like the stem cells in organs that repair and maintain them in the face of wear and tear, or the stem cells in bone marrow that give rise to billions of blood cells each day. For instance, one disease that appears in people with short telomere syndrome is aplastic anemia, a condition in which the blood cell–manufacturing cells in the bone marrow fail and the numbers of all kinds of circulating blood cells fall dangerously low.[40]

Is the converse then true? Do long telomeres bode well for health? Some tantalizing findings have suggested as much. One particularly striking study of scores of healthy centenarians showed that both they and their offspring maintained longer telomeres as they aged, compared with subjects of ordinary longevity. It also showed that longer telomeres were associated with better mental functioning, healthy levels of blood fat, and fewer age-related diseases.[41] However, this study did not establish cause and effect. It simply showed a correlation.

Importantly, telomerase has been implicated in cancer. The scientists at Geron—along with other groups working elsewhere—showed that in stark contrast to most normal cells, telomerase is active in cancer cells. Working with Wright and his colleague Jerry Shay at the University of Texas Southwestern Medical Center at Dallas, they measured telomerase activity in 101 biopsies from human cancers and in 50 biopsies from normal human tissues. Telomerase was active in 90 of the cancer samples and none of the normal ones.[42] After their paper was published in *Science* in 1994, research on the role of telomerase in cancer skyrocketed.[43] Recent findings include several papers showing that in families with inherited mutations that kick telomerase into high gear, the risks of malignant melanoma (a skin cancer) and of glioma (a brain cancer) are elevated.[44] Biologists have also found evidence that a cell's ability to shut down its own division—to short-circuit itself into a post–Hayflick limit, nondividing state—may have evolved as an essential defense *against* cancer.

Like the complex role of telomerase in cancer, the relationship between telomere shortening and aging is still being studied intensively today. We now know that there are a variety of influences, internal and external, on telomere length that make it clear that telomeres aren't like wind-up clocks that start ticking when we are in the womb and march in lockstep toward a predictable, predetermined finish line. Heredity plays a role: some of us have the good fortune to inherit longer telomeres than the average human being. So do environmental exposures.

Today biologists are probing basic questions like these: What exactly tips telomeres from being dangerously short into being nonfunctional? What precisely happens to them, at the molecular level, as they hit the Hayflick limit? And when this happens, what complicated symphony of cellular signaling actually causes the cell to stop dividing? Each new answer they find opens up an ever-deeper set of questions.

In December 2009 in Stockholm, King Carl XVI Gustaf of Sweden presented Blackburn, Greider, and Szostak with the Nobel Prize in physiology or medicine for their discovery of how chromosomes are protected by telomeres and of the enzyme telomerase.

Hayflick had described a phenomenon. The Nobel trio had explained it.

Boot-Camp Bugs and Vatican Entreaties

Largo, Florida
Naval Station Great Lakes, Illinois
Cleveland, Dayton, and Centerville, Ohio
1999–2012

I am fond of saying that rubella vaccine has prevented thousands more abortions than have ever been prevented by Catholic religionists.

—Stanley Plotkin, June 27, 2013[1]

In 1999 Debi Vinnedge, a forty-four-year-old grandmother living in Clearwater, Florida, founded a nonprofit group called Children of God for Life to advocate against human embryonic stem cell research. Scientists had announced the isolation of the versatile cells only one year earlier and they were very much in the news. Because they can grow into virtually any cell type, it was hoped that they could be used to grow cells and organs to replace diseased ones. The cells are obtained from days-old embryos left over at fertility clinics, destroying the embryos in the process. The cells' origins bothered Vinnedge greatly. She was a devout cradle Catholic and abhorred the destruction of human life at any point after conception.[2]

But soon after launching Children of God for Life, Vinnedge stumbled on an article in her local diocesan newspaper that caused her to turn her focus away from human embryonic stem cells. It asserted that several childhood vaccines used in the United States were produced using cells from two fetuses aborted decades earlier. She began reading everything she could find about the vaccines. The more she learned, the more distressed she became.

Vinnedge was a grandmother of two, and she soon figured out that both her children and her grandchildren had been vaccinated with Merck's MMR II vaccine, the only one available in the United States—and that the rubella component had been made using WI-38 cells. (Her children, born in 1975 and 1977, had received booster shots of Merck's MMR II vaccine in the 1980s.) As a young mother she had had no idea that fetal cells were used to make the vaccine, and this made her angry. She also learned that the company made the individual components of the MMR vaccine as separate vaccines, meaning that people who objected to the fetal origins of the rubella vaccine could boycott it but still vaccinate their children against measles and mumps.

This was something, but it was not a long-term solution in Vinnedge's view. It left children unprotected against rubella. She learned that in Japan the Kitasato Institute made what she calls an "ethical" rubella vaccine, growing the virus in rabbit kidney cells.[3] Similarly, another Japanese company made a hepatitis A vaccine, as Merck was now doing using MRC-5 cells, without using fetal cells.[4] She wrote to Merck expressing her concerns and asking it to drop the fetal cell–manufactured injections and replace them with animal cell–propagated alternatives. The company ignored her first two letters but responded to a third. In that response Isabelle Claxton, the company's executive director of public affairs, noted that both WI-38 and MRC-5 came from two legal abortions in the 1960s that were not undertaken with the intent of producing vaccines. No new abortions would be needed, now or in the future, for Merck to continue production, she added. "We use these cell lines because they are the best science has to offer in terms of producing a safe and effective vaccine against certain diseases."[5]

Not satisfied, Vinnedge, with support from the Catholic Medical Association and another antiabortion group, Human Life International, bought the requisite stock in Merck—$2,000 worth—and developed a shareholder resolution. It proposed that, because Merck was "materially benefiting from the destruction of human life," a special committee of the company's board be formed to link executive compensation to the company's "ethical and social performance."

In an appeal to the Securities and Exchange Commission, Merck tried but failed to prevent the resolution from appearing on the proxy statement on which shareholders would vote at the company's annual general meeting in 2003.[6] The SEC did agree to remove the language stating that the company was materially benefiting from the destruction of human life. The resolution that went before shareholders instead read that "Merck has violated its basic Statement on Values because of our company's use of cell culture lines from aborted humans." The resolution received 4.96 percent of the vote, above the 3 percent threshold that would allow Vinnedge to return with the resolution the following year.[7] (Such resolutions are not binding on companies; they are simply requests.)

Vinnedge attended the 2003 meeting in Rahway, New Jersey. At a break she approached the stage and presented Merck's then-CEO, Ray Gilmartin, with a book of some 375,000 signatures, names, and addresses she had gathered, asking the company to provide what the signatories described as moral alternatives to the company's rubella vaccine, and to the hepatitis A and chicken pox injections that it also produced using fetal cells.

The company did not do the bidding of the book's signatories or respond to the shareholder resolution. At the company's next annual general meeting, in 2004, the resolution failed to gain the requisite votes—5 percent on a second attempt—that would have allowed it to be put before stockholders again in 2005.

In June 2003, not long after the first Merck meeting she attended, Vinnedge took her campaign to an authority more likely to receive her warmly: the Vatican. There a group of cardinals and bishops is charged with promoting and defending Catholic doctrine; they are known as the Sacred Congregation for the Doctrine of the Faith, or, informally, the Holy Office. When Vinnedge wrote to them, their chief, or prefect, was Cardinal Joseph Ratzinger, the conservative German prelate who two years later would be inaugurated Pope Benedict XVI. Vinnedge asked the congregation to restate the church's teaching that Catholics shouldn't be forced to act contrary to their conscience.[8] She noted that some parents who opposed fetal cell–produced vaccines were challenged by state courts, health officials, and school administrators when they sought religious exemptions. "The permission to abstain from medical treatments and vaccines cultivated on aborted fetal cell lines needs to be addressed," she wrote. The congregation sent the question to another Vatican body, the Pontifical Academy for Life, launched by Pope John Paul II less than a decade earlier, for "the promotion and the defense of human life." Its roughly seventy members included medical scientists and its mandate was to study and pronounce on bioethics questions. (It also helped develop the church's response to sexual abuse allegations.) The academy began a study that would take two years to complete. In Florida, Vinnedge waited patiently for an answer.

In the early summer of 2000, as Vinnedge was becoming absorbed in her work launching Children of God for Life, Adam Wood, a previously healthy twenty-one-year-old recruit in boot camp at Naval Station Great Lakes in Illinois, appeared at the base infirmary complaining of a cold. He went back to the infirmary the next day, and the next. He was prescribed an antibiotic but couldn't seem to shake the illness. On the fourth morning he couldn't get out of bed. He lost vision, and then consciousness. He was taken to the hospital in an ambulance. A chest X-ray of the comatose young man showed that one of his lungs was infected, and a CT scan revealed the same of his sinuses, the bony spaces behind the forehead and cheeks. Two weeks after he first fell ill, Wood's parents, who had rushed to his bedside from a Colorado vacation, told staff to remove the respirator that was keeping their son alive.[9,10]

The pathologist who conducted an autopsy found that Wood's brain was severely inflamed—a condition called encephalitis. Researchers from the CDC and the military tested pieces of his lungs and brain and found the DNA of a respiratory virus that had very likely caused the encephalitis. It was adenovirus.

Death is a rare outcome of infection with adenovirus, a common virus that typically causes fevers, runny noses, sneezing, coughing, and pink eye where stressed people live in cramped quarters—think boot-camp trainees in military barracks. But it visited another formerly healthy recruit at the Great Lakes base later that same summer. He was nineteen-year-old Jess Duden, who died of respiratory collapse one month after developing symptoms of a garden-variety nose and throat infection that were initially treated with Tylenol and decongestants. Scientists found the DNA of adenovirus in his ravaged lungs, and concluded that his death was probably associated with the virus.[11]

Duden and Wood were, to an extent, victims of the success of the adenovirus vaccine. Six years earlier, in 1994, Wyeth, which made the vaccine for the military, had told the Pentagon that it was stopping production. The company had used WI-38 cells to make enough vaccine for every incoming recruit—some 7.5 million people—since 1971. It was administered within hours of their arrival at boot camp, as two oral tablets, coated to keep stomach acid from destroying the virus. (There is one pill for each of the two most common strains of adenovirus, types 4 and 7.) But by 1994 the firm had been pleading with the Pentagon for a decade for cash for a new facility; in 1984 FDA inspectors had warned that its dated existing plant in Marietta, Pennsylvania, needed upgrades. The military had answered the company's request for $3 million to $5 million with inaction. In 1995 Wyeth told the Pentagon that it had ceased production and that its existing vaccine stores would be depleted within several years. The supply ran out in 1999. Still nothing was done. Senior military leaders had become complacent.

Within a short time the rate of adenovirus infections in boot-camp barracks returned to what it had been in the 1950s, prior to mass vaccination, when physicians estimated that 10 percent of all new enlistees were infected and that the virus caused 90 percent of the pneumonia in military infirmaries.[12]

"When the vaccine was in use, a busy [basic-training] camp might see as many as 200 cases of respiratory infection a week, fewer than 10% of which were caused by adenovirus," the *Wall Street Journal* reported in 2001. "Without the vaccine, the same camp sees as many as 800 cases a week, with 90% caused by adenovirus."[13]

It would take the Pentagon another decade and about $100 million to restart the adenovirus vaccination program, after Israel-based Teva Pharmaceuticals finally agreed to manufacture the vaccine for the military.[14] (Barr Laboratories, part of Teva, packs the weakened, freeze-dried virus into pills at a plant in Forest, Virginia.) Since October 2011 every incoming recruit—more than a million of them through late 2016—has once again upon arriving at boot camp received the WI-38–propagated vaccine. Within two months of the program restarting, the rate of new adenovirus infections at the military's basic-training centers fell sevenfold and remained there.[15]

In August of 2000, as Jess Duden sickened and died, another life was beginning.

In Cleveland, Ohio, a thirty-one-year-old woman named Elizabeth Graham was singing the role of the Countess in *The Marriage of Figaro* at the Lyric Opera Cleveland, winning plaudits from the local press for the passion of her arias and looking stunning onstage, her red lips encircling the perfect *O* of her mouth.

At postperformance receptions Graham—everyone calls her Betsy—met many admirers, shook many hands, and kissed many cheeks. She felt fine except for something that seemed like allergies that mildly bothered her nose and throat for a few days. They were gone by the time her husband, Chip MacConnell, a physician assistant, traveled from the couple's home in Dayton to take in some performances. After the run the pair enjoyed a vacation in the Smoky Mountains of Tennessee. A month or so later Graham was pretty sure that she was pregnant. She was excited; she and Chip were ready to start a family.[16]

Betsy MacConnell (she uses the name Graham only professionally) saw her family physician in early October 2000. He ran routine pregnancy blood tests, including one for antirubella antibodies. The lab reported that she was immune. This computed; MacConnell had been vaccinated against rubella as a three-year-old in 1972 with one of the vaccines that pediatricians were using then: Merck's HPV-77 duck-embryo injection or the Cendehill vaccine.

On May 7, 2001, at Miami Valley Hospital in Dayton, after a marathon labor, Betsy gave birth to a girl with pink cheeks and a full head of dark hair. She was crying lustily. She weighed five pounds, twelve ounces—about the 5th per-centile for newborn girls. They named her Anna Gabrielle.

Within a day the MacConnells learned that Anna had the four-part heart defect known as tetralogy of Fallot. Over the next few weeks it emerged that she was also profoundly deaf and blinded by cataracts. Tests of Anna's blood not long after her birth showed antirubella antibodies of a kind that do not cross the placenta, meaning that Anna had made them herself, in the womb, in response to exposure to the virus. Later, virologists at the CDC isolated rubella virus from the cataract material that surgeons removed from Anna's eyes.

Anna underwent corrective heart surgery at eleven weeks of age. Several years later she would require another heart surgery to replace one of her heart valves. Postoperative complications after the first surgery led to a tracheostomy—a hole cut into her windpipe, on the front of her neck, through which she breathed for the first three years of her life. She had cataract surgeries at three and five months of age and later multiple eye surgeries to extract scar tissue and to reattach a detached retina. She was also put under anesthesia several times so that doctors could take pressure measurements for glaucoma, raised pressure inside the eye. Glaucoma affected both of Anna's eyes, as it had those of Stephen Wenzler, the boy born with congenital rubella in Toms River, New Jersey, in 1964.

Anna benefited from one medical advance that was not available to Wen-zler. At the age of two she received a cochlear implant that gave her a degree of hearing. Her father remembers how his heart leaped the first time that she reached up and turned her "ear" off when he scolded her—it meant that she was hearing him.

Anna's cataract surgery provided her with what is called navigational vision—in Chip MacConnell's definition, she could see a wall just before she ran into it. As she grew—and was joined by a healthy baby sister, Danielle, in 2004—she attended her local elementary school in Centerville, Ohio, where the family had since moved. There she became something of a rock star; the whole school learned to sing the school song in sign language.

Anna took great pleasure in life. She gleefully slid along a backyard zip line and loved to lie under the kitchen window, watching the play of the sun on translucent toys she held up to the light. She played soccer with her service dog, a goldendoodle named Cadi, at her side. She fought with her sister from time to time; there was pinching and pulling of hair. She identified people she knew by feeling their thumbs. She had a radiant smile and an extraordinary bullshit detector.

In the autumn of 2011, when she was ten years old, Anna began flagging on the soccer field and having a hard time staying awake at school. When walking, she leaned on Cadi heavily in a way she hadn't done before. It became clear that her heart was failing. She was turned down for a heart transplant, in part because of the uncertainty inherent in putting a new heart in someone with congenital rubella; it had never been done, and there was a question as to whether lurking rubella in Anna's system would infect and damage a new heart.

In September 2012, a few months after her eleventh birthday, Anna died. Her last hours, in her hospital bed, were peaceful. Cadi was curled in a ball in the corner of the room. Her family was around her, and the ceaselessly beeping machines had been turned off. Anna repeatedly rotated her hands so that her palms faced out, and then in, and then out, and then in. She was signing the word "finished."

Betsy MacConnell's positive blood test for antirubella antibodies in October 2000 seems likely to have resulted from an infection when she first became pregnant two months earlier, not from her 1972 vaccination. Either she had not produced antibodies in response to that long-ago injection or they had waned over time or they were simply ineffective—precisely the outcome that the Yale pediatrician Dorothy Horstmann had worried about all those decades earlier as she harangued Merck's Maurice Hilleman to switch to Plotkin's rubella vaccine.

Debi Vinnedge at Children of God for Life received a reply from the Vatican in June 2005. The Pontifical Academy for Life had produced a study documenting the use of cells from aborted fetuses in vaccine making and addressing whether it was sinful for Catholics to have their children immunized with such vaccines.[17] The powerful Congregation for the Doctrine of the Faith had signed off on the study, and it had been translated from the original Italian for transmission to Vinnedge.

The report was carefully argued, using arcane concepts from moral philosophy. But it began by noting that, among the vaccines made using human fetal cells, the rubella vaccine was "perhaps the most important due to its vast distribution and its use on an almost universal level"—and because of the particular risk at which rubella put fetuses.

The report's bottom line was this: so long as no alternative vaccines existed, it was "lawful" for Catholic parents to have their children immunized with vaccines made in fetal cells, "in order to avoid a serious risk not only for one's own children but also . . . for the health conditions of the population as a whole—especially for pregnant women."

The Academy admonished the companies for producing vaccines that forced parents to choose between acting against their conscience and putting the health of their children, and others, at risk. And it urged Catholics to fight for alternatives and to do their all "to make life difficult for the pharmaceutical industries which act unscrupulously and unethically." But it did not call for Catholics to refuse all vaccinations with fetal-cell–produced vaccines.

In 2009 Vinnedge and her constituents were dealt a blow when Merck said it would no longer make the measles, mumps, and rubella vaccines as separate injections. The company said that making the vaccines, which accounted for just 2 percent of its measles, mumps, and rubella vaccine output, was not in the best interests of public health.[18] It pointed to recommendations from the CDC's Advisory Committee on Immunization Practices and from the American Academy of Pediatrics, which preferred the combined MMR vaccine on the grounds that it meant fewer shots and better vaccine coverage.[19] Asking parents to make two or three doctor visits for two or three different vaccines presents a greater risk of missed injections, the reasoning goes, and during the protracted process, children are at risk for the diseases that they haven't yet been vaccinated against. (The CDC advisers also endorsed a four-in-one vaccine called Pro-Quad, in which Merck added its chicken pox vaccine to MMR.)[20]

Merck had initially indicated to a parent affiliated with Children of God for Life that it would continue production of monovalent vaccines.[21] When it stopped, Vinnedge and her group felt betrayed. In 2015, when there was an outbreak of measles that began at Disneyland, Children of God for Life issued a press release under the headline "Blame Merck—Not the Parents!" The company's "rash" decision, the group wrote, "left families who cannot use the aborted fetal MMR in good conscience, unprotected."[22]

The Afterlife of a Cell

Philadelphia, 2014
Bethesda, 2016

Developing vaccines is probably one of the most productive things you can do, simply because if you succeed in getting one made, you watch a disease disappear.

—Alan Shaw, former executive director of vaccine research at Merck[1]

In his laboratory at the Wistar Institute on a spring day not long ago, Rugang Zhang was waxing enthusiastic about some cells luxuriating in what looks like an overgrown fridge but is in fact a tissue incubator. The cells are an important, familiar tool for Zhang, a Chinese-born scientist who studies how cells age and how ordinary cells transform into cancerous ones. "Cancer cells have so many alterations they are hard to track, but WI-38 is essentially normal. Which is why it's a very important biological system, even in our days," Zhang explains.

Look at a photo of Zhang's WI-38 cells on this late April day, and you might be underwhelmed: it shows a motley assortment of purple-stained, spindly bodies; they are typical fibroblasts, pumping out the ingredients of the extracellular matrix—the noncellular scaffolding that holds tissues together. Turn to a second photo, and those same skinny cells are hardly recognizable. Their once-streamlined bodies are bloated and bright turquoise, their shapes various, their neat uniformity gone. They look unruly. And they are. These are elderly cells, their powers waning, their interiors increasingly decrepit and disorganized.

It hasn't always been so easy to identify aging cells. The turquoise is a dye that Zhang added because it stains an enzyme that accumulates when cells have hit the Hayflick limit and stopped dividing. How do scientists know that this enzyme reliably marks a cell's old age? They learned this in part by studying WI-38.[2] This is one of the many ways that WI-38 cells have pushed aging research forward, because they so reliably age in the lab, allowing scientists to study how that process unfolds. The importance of this work across the spectrum of human health is enormous; it can also be heartbreakingly compelling. Robert Goldman, the chairman of molecular and cell biology at Northwestern University's Feinberg School of Medicine in Chicago, Illinois, is using WI-38 cells to try to understand why cells age prematurely in patients with progeria, a rare, inevitably fatal disease that ages children so rapidly that they look like they're eighty years old before they're ten.[3]

Sometimes WI-38 cells are used in basic biological discoveries that may seem esoteric but end up being relevant in ways that nonscientists can easily appreciate. A few years ago researchers at the Weizmann Institute of Science in Rehovot, Israel, used WI-38 cells to discover a key function of an enzyme called QSOX1 that is thought to have a role in helping prostate, breast, and pancreatic cancers to invade normal tissues. They found that the enzyme, which is secreted by WI-38 cells, catalyzes the formation of laminin, an important component of the thin, fibrous basement membrane that anchors many kinds of cells to the connective tissue underneath them. Cancer cells depend on laminin to spread and invade tissues. When the Israeli scientists used an antibody to block QSOX1, this shut down cancer-cell migration: lung-cancer cells that are normally mobile couldn't move across a layer of WI-38 cells. This is the kind of far-reaching, basic discovery that appears in the esteemed journal *Science*, which is where the scientists published their paper.[4]

It's hard to quantify the impact of fundamental findings like the Rehovot researchers'. Biology builds on itself, discovery after discovery, and apportioning credit in just the right measure to individual revelations is not terribly useful and often impossible. What's more, findings that may at the time seem to be of only limited interest—think of Elizabeth Blackburn's first, obscure description of the telomeres of that single-celled freshwater organism, *Tetrahymena*—can launch whole new fields of endeavor and lead to major discoveries and ever-deepening understandings. So can findings that are dismissed as "rash" and wrongheaded, like Hayflick's observation that the normal cells from many different fetal organs stopped dividing after several months in lab bottles and his bold suggestion that this phenomenon might be related to human aging. It took decades for the importance of his observations and thinking to be fully appreciated. And because they were lab-based findings, far removed from patient bedsides, their impact can't be neatly measured in terms of medicines made or lives saved.

It's quite different with the vaccines that Hayflick enabled, not only by launching the WI-38 cells and then turning his lab into a one-stop shop for vaccine scientists and companies in search of them, but also by relentlessly fighting for their acceptance against the United States' obdurate, ultraconservative, self-protective vaccine regulators. Few people, if any, thanked him for his efforts at the time, and by the time those efforts did come to wide public attention, it was in the worst possible way—under the damning cloud of a government investigation announced on page one of the *New York Times*. Certainly Hayflick brought his downfall on himself through his stubborn defiance and his inability to give an inch when he is sure he is right, which is often. But none of that takes away from the enormous impact on public health that is due to this one man's endeavors and persistence with his beloved WI-38 cells—and to the many vaccines that resulted.

Merck's chicken pox vaccine won FDA approval in the mid-1990s, making the company the only chicken pox vaccine maker in the world. The company makes the vaccine at a new plant in North Carolina, using MRC-5 cells—the cells launched by the British in imitation of WI-38, using Hayflick's methods. The weakened vaccine virus itself traces its roots to its passage long ago in WI-38 cells, in the Osaka University lab of the pediatrician Michiaki Takahashi, from whose organization Merck later licensed it. The same is true of Merck's vaccine against shingles, which was approved in 2006 for Americans over sixty and that is simply a higher-dose version of the chicken pox vaccine. Both illnesses are caused by the same virus, varicella zoster. People over sixty need a higher dose because their immune systems don't respond as vigorously to vaccination as do young people's.

In the first decade after 1995, when American infants began receiving a single chicken pox injection at between twelve and eighteen months of age, the incidence of disease caused by varicella zoster virus—that is, the incidence of chicken pox and shingles—and the associated number of U.S. hospitalizations and deaths declined by about 90 percent.[5] In 2006 the CDC recommended that children receive a second chicken pox shot between the ages of four and six, just before school entry. The incidence of varicella-caused disease then fell another 81 percent through 2013.[6] No one less than twenty years old has died of chicken pox in the United States since 2010.[7]

Merck finally won FDA approval for its hepatitis A vaccine in 1996, twenty years after Hilleman ordered his vaccine scientist Philip Provost to scrap his work developing the vaccine in WI-38 cells and begin again using MRC-5. The company shares the U.S. market with GlaxoSmithKline; both companies use MRC-5 cells to manufacture their vaccines. Beginning in 1996, the injection was given to infants in geographical areas with high rates of the disease, which is transmitted when people ingest food or water contaminated with the feces of an infected person. The vaccine was made part of the routine schedule for all infants in 2006.[8] Between 2000 and 2013 the number of cases of hepatitis A reported to the CDC declined by 86.7 percent, from 13,397 in 2000 to 1,781 in 2013.[9]

The number of human deaths from rabies in the United States has declined steadily since the 1970s, thanks in part to animal control and immunization efforts that have more or less eliminated domestic dogs and cats as reservoirs of the disease—and thanks, too, to the improved human cell–propagated vaccine that arrived in 1980 courtesy of Hilary Koprowski, his colleagues, and Hayflick's WI-38 cells. After Wyeth was forced to withdraw its vaccine in the mid-1980s, the Institut Mérieux in Lyon, France—now part of the huge vaccine-making company Sanofi Pasteur—enjoyed years dominating the U.S. market with Imovax, the rabies vaccine that is made using MRC-5 cells but that owes its existence to Koprowski and Wiktor's tweaking of the virus in WI-38 cells half a century ago. Today, Imovax is facing stiff competition from newer vaccines that are equally effective but much cheaper to manufacture. Like Imovax, the newer vaccines are a vast improvement on the animal-nerve-tissue vaccines of the early twentieth century, which, sadly, are still used in a handful of developing countries because they are still cheaper to make.

In the United States between thirty thousand and sixty thousand people are vaccinated each year after possible exposure to rabies, with either the Sanofi human diploid cell vaccine or with a competing vaccine by Novartis, made in chick embryo cells.[10] Thousands more—veterinary students, animal handlers, rabies researchers—receive preventive vaccinations.

Wild animals, particularly raccoons, bats, and skunks, are still reservoirs of the disease in the United States; the CDC receives around six thousand reports of rabid animals annually, 92 percent of them in wildlife.[11] Each year several hundred domestic dogs and cats will contract rabies from wild animals because their owners have not had them vaccinated.[12]

In most years between one and three people die of rabies in the United States—it is a disease so rare that doctors may fail to recognize it. For people, bats are a particularly dangerous reservoir, both because they can harbor the disease without being evidently rabid and because their teeth are so tiny that people who are bitten—often while they are asleep—may not realize it.

In September 2015, a seventy-seven-year-old Wyoming woman appeared at her local hospital after five days of growing weakness and loss of balance. Her speech was slurred and she couldn't swallow. Eight days later, her family told doctors that one month earlier she had waked with a bat on her neck. By the time they mentioned the bat, she was paralyzed, comatose, and on a breathing machine. Three days after that, she was dead. Experts at the CDC tested the woman's saliva and skin from the nape of her neck and found rabies.

When the woman woke with the bat on her neck, her husband had examined her for a bite wound and found none, so she had not sought medical help. Her husband had reported the incident to local invasive species authorities, but they had not raised the risk of rabies.[13]

In 2014 a Missouri man who lived in a thickly wooded area and who had spotted a bat in his trailer appeared at a hospital emergency room with a sudden onset of severe neck pain and tingling in his left arm. He became fearful and began hallucinating; he communicated an aversion to water. Within three days he was on a ventilator, and nine days after that he was dead. It had taken his medical team six days to consider that he might have rabies, and confirmatory lab results weren't returned for an additional six days. By then it was far too late for a vaccine to save him.[14]

In Wisconsin in 2004 a fifteen-year-old girl named Jeanna Giese became the first person known to survive rabies without being vaccinated.[15] She developed neurological symptoms thirty-seven days after being bitten by a bat. Doctors put her in a chemically induced coma and administered antiviral drugs. Today she is a married college graduate.

In 2007 researchers from the Pasteur Institute in Tehran went back to the villages of northwestern Iran. There they managed to find twenty-six of the forty-five people who had been bitten by rabid animals and vaccinated thirty-one years earlier with the WI-38–propagated Koprowski vaccine. All of them still had antirabies antibodies, many at levels deemed protective by the World Health Organization.[16]

The U.S. rubella epidemic that was expected in the early 1970s never materialized. By 1979, ten years after the first rubella vaccine was licensed in the United States, between 75 million and 80 million children had been vaccinated in a U.S. population that numbered around 225 million. In 1969, not an epidemic year, there had been 55,549 cases of rubella reported to the CDC.[17] By 1979 the number had fallen to an all-time low of 11,795.[18, 19] (Both the 1969 and 1979 numbers were probably gross underestimates because so many rubella infections don't cause obvious disease.)[20]

Between 1969 and 1980 reports of babies born in the United States with congenital rubella declined by 36 percent to 50 infants.[21] At the turn of the twenty-first century, by which time Plotkin's RA 27/3 rubella vaccine had been immunizing American toddlers for twenty years, there were 176 reported cases of rubella in children and adults and 9 reports of babies born with congenital rubella in the United States.[22] By that time more than 9 in 10 cases of congenital rubella in the United States occurred in infants born to foreign-born mothers from countries without rubella vaccination programs, or with programs only recently put in place.[23]

In 2005 the Centers for Disease Control and Prevention announced that endemic rubella—meaning homegrown rubella not imported with unvaccinated immigrants—had been eliminated from the United States.

In April 2015 the Pan American Health Organization reported that endemic rubella had been eliminated in the Western Hemisphere. By that time Merck had shipped more than 660 million doses of Plotkin's WI-38–based rubella vaccine.[24]

Globally rubella remains a serious problem. In 2014, 28 percent of countries did not vaccinate against rubella.[25] Japan, which did not begin vaccinating young boys against rubella until 1995—in the beginning, several countries vaccinated only girls because only girls go on to get pregnant—experienced a serious epidemic from 2012 to 2014, with most of the cases in adult men.[26,27] Forty-five infants were born with congenital rubella.[28]

Sporadic imported cases of congenital rubella continue to appear in the United States. For instance, in 2012 three babies were diagnosed with the syndrome in the United States. They were born to women from African countries that do not vaccinate against the disease; each mother had been in Africa early in her pregnancy.[29]

Experts estimate that a hundred thousand babies are born with congenital rubella each year, most of them in developing countries.[30]

On a cold, crisp, sunny day in November of 2014 I visited the Merck campus at West Point, near Philadelphia—a low-lying, four-hundred-acre complex with its own trash-hauling and fire and emergency service where, since 1979, the company has made the rubella vaccine that has now been given to some 140 million U.S. preschoolers. Here, in the supersterile building 29, the company was in the midst of what is termed in the industry a "campaign" in which, over several months, it makes all of the rubella vaccine that it will produce for the coming year. (Measles and mumps vaccines, which are combined with the rubella vaccine to make the MMR injection, are made in the same facility at different times of the year to avoid cross-contamination.)

The company had been making the vaccine since September, in shifts that run daily from 8:00 a.m. to 12:30 a.m. Behind a series of gowning rooms and air locks that ensure that outside air never makes its way in, hooded technicians in white jumpsuits and steel-toed shoes with green shoelaces that mark them for sterile rooms were overseeing scores of cylindrical half-gallon plastic bottles that were rotating slowly, their sides made hazy by the WI-38 cells growing on them, the medium inside them awash with rubella virus. The cells incubate at 86 degrees Fahrenheit, the temperature that Stanley Plotkin discovered long ago would weaken the RA 27/3 vaccine virus just enough, but not too much. After several days the virus-laden medium that will become the vaccine is harvested, pooled, and filtered several times to remove the larger cellular debris of the WI-38 cells. Each batch of vaccine fills most of a sixty-six-gallon, stainless-steel tank; each will be safety-tested on a long list of parameters on which the FDA must sign off. It will then be freeze-dried, packaged with measles and mumps vaccines, labeled, and sent out, along with the sterile water to reconstitute it, to untold numbers of doctors and health workers in forty-two countries on five continents.

I met in a conference room with several key people involved with rubella vaccine production. They included Michael Lynn, a biotechnician and father of four young children. Lynn had a shaved head and bulging biceps and declared of his long hours working behind goggles, hood, and mask: "You're helping children. To me, there's nothing really better than that." They also included an engineer and molecular biologist, Vic Johnston, who looked to be in his forties and who is an expert on the rubella vaccine. It was Johnston who explained to me that it was 2008 when Merck last retrieved a single sterile ampule of WI-38 cells from the scores of vaccine-ready ampules held at the American Type Culture Collection. Before that the company hadn't needed one since 1995.

Johnston said that when on these occasions the tiny, decades-old ampule arrives at Merck from the ATCC, it contains about three million cells. The company expands these cells to create a working bank of cells—dozens of ampules—with population doubling levels in the low twenties; these it freezes. They will last eight or ten or thirteen years. Each year, during its annual campaign, the company will draw on the cells from this bank. It will begin with about 120 million such cells; by the end of the campaign, some 37 billion infected WI-38 cells will have been conscripted to produce the vaccine virus. By this point the cell populations will have reached a doubling level in the low thirties. While these cells aren't killed by the virus, they will become "sort of exhausted" after producing vast quantities of it, Johnston said. So there, cling-ing to the plastic sides of the glass roller bottles, they will end a long journey—a mostly deeply frozen trip that took their progenitors from deep in Mrs. X's womb to a Swedish hospital to the Karolinska Institute to the Wistar Institute to Stanford University to the American Type Culture Collection and finally to West Point, Pennsylvania.

"Do you ever lose sleep," I asked Johnston, "worrying about running out of WI-38?"

He hesitated for a beat. Then he said that if the company could begin each rubella campaign with somewhat older WI-38 cells than it starts with now—a change that would require FDA approval—"that would essentially make the supply infinite."*

Nanobits of the cells used to make them are present in all viral vaccines, whether or not they are made with fetal cells. These bits of protein and DNA are so small that they pass through the filters that catch larger cell remnants. In the case of the WI-38 cells used to make rubella vaccine, the minuscule DNA snippets that are present in one dose of vaccine weigh in the neighborhood of 180 nanograms, which is about 0.6 millionths of an ounce. (The blood circu-lating in the vessels of a twenty-two-pound toddler weighs about five billion times as much.) These WI-38 remnants are considered so safe, because of the cells' normalcy—the fact that they never turn cancerous—and because of the decades-long safety records of the vaccines made in them, that neither the Food and Drug Administration nor the World Health Organization sets an upper limit on the levels of WI-38 DNA allowed to be present in the rubella vaccine.†[31,32]

I am moved by the intimate interaction between the WI-38 cells and the hundreds of millions of people who have benefited from the rubella and adeno-virus vaccines made using them. When they are vaccinated, they are literally,

*The World Health Organization says that WI-38 cells can be used for vaccine making until about ten generations before they stop dividing.

†For the same reasons, the WHO and the FDA do not limit residual DNA in vaccines made in MRC-5 cells, derived from a British fetus in 1966.

physically connected to Mrs. X's fetus. Taking this idea further, to the level of infinitesimal bits of WI-38, one could argue that this physical relationship extends to every adult and child vaccinated against shingles and chicken pox with the Merck vaccine that once passed through WI-38 cells, and to everyone who has been saved from a horrifying death by Sanofi Pasteur's rabies vaccine, created by Koprowski and his colleagues at the Wistar half a century ago, using WI-38. Admittedly, at so many removes the notion of a physical link between these vaccinees and the WI-38 fetus is almost completely symbolic—it's a single drop in the Pacific Ocean, perhaps. But "philosophically" it is true, says Alan Shaw, who developed the shingles and chicken pox vaccines for Merck.[33]

Mrs. X has never been compensated for the use of her fetus. But the institutions and companies that market the WI-38 cells as research tools and turn them into vaccines have made a lot of money. In the spring of 2016, scientists ordering a tiny ampule of WI-38 cells from a cell bank paid as much as $467. In the mid-1980s, when the WI-38–based rabies and rubella vaccines were still under patent, the Wistar Institute collected more than $3 million in royalties annually; some 15 percent of these were shared by the inventors: Plotkin, Koprowski, Fernandes, and Wiktor, before he passed away in 1986.[34] The money in vaccine making today is orders of magnitude bigger. Teva, the company that uses WI-38 cells to make adenovirus vaccine for the Pentagon, earned about $30 million from sales of that vaccine in 2012.[35] (The company declined to provide a more recent figure.) Merck's sales of its rubella- and chicken pox–containing vaccines in 2015 grew 10 percent to $1.5 billion.

Should Mrs. X be paid, even at this late date, for the use of her fetus? Some ethicists argue that paying people like Mrs. X is not warranted, because she wasn't put to any extra inconvenience—she was going to have the abortion anyway—and because she, unlike Hayflick, Plotkin, Koprowski, and others, didn't do any of the work that turned the cells of the fetus into lasting tools for improving human health.[36] Others, like Alta Charo, a lawyer and bioethicist at the University of Wisconsin, point out that an abortion so long ago and so far away leaves Mrs. X with virtually no legal route to claim compensation, even if she was motivated to pursue it. That doesn't mean, Charo adds, that compensating Mrs. X might not be the moral thing to do.[37]

Going forward, there is still a live debate about whether tissue donors should be compensated when their contributions are not a by-product of routine medical care but are specifically collected, with their consent, for research—and are later turned into lucrative research tools or therapies. Some experts argue that sorting out who is owed what would be a legal nightmare that would bring science to a standstill.[38] Others say that the prospect of profiting would be an incentive to people to donate and would speed research.[39]

What about research tools or medicines developed from the hundreds of millions of tissue samples that are already stored in biobanks all over the United States—samples, from pancreas to placenta, that are left over from surgeries and medical procedures?[40] They will never financially benefit their donors; researchers can use these samples—as long as they have been stripped of information that could lead a scientist back to the person they came from—freely, without the consent or knowledge of the donor. In the spring of 2016, the U.S. government was finalizing rules that would change that. They would generally require patients to give a one-time informed consent for the use of their surgical leftovers for future research.[41] As this book went to press, the proposed change was generating a wave of protest from medical researchers who argued, backed by a new report from the National Academy of Sciences, that the new consent requirement would create an expensive bureaucratic morass without benefiting the people it is supposed to protect.[42] Ironically in the context of this book, the rule change, if it does go through, would not apply to fetal tissue, which is governed by a different part of U.S. regulations. This part says that fetal tissue can be used without consent as long as it does not have identifying information attached to it that would allow it to be traced back to a living human being—and as long as using it without consent would not break state or local laws.*[43] In fact, in most places, doing so would indeed break state laws: At least thirty-seven states and the District of Columbia have enacted laws that require a woman's informed consent for the use of tissue from her fetus in research.[44]

The unregulated, exploitative experiments on orphans, prisoners, newborns, and intellectually disabled children that populate these pages are no longer permitted in the United States. In 1966, after the publication of Henry Beecher's *New England Journal of Medicine* report documenting two decades of appalling abuses of human subjects, the U.S. surgeon general told government-funded

---

*There is one exception. Under a 1993 law fetal tissue from an abortion can't be transplanted into a human being for therapeutic purposes—think, for instance, of Parkinson's disease, for which fetal tissue transplants have been tried as a treatment—without the informed consent of the woman who had the abortion, even if the tissue has been stripped of identifiers.

medical researchers that they would have to obtain the informed consent of research participants, and that an independent "committee of associates" at their hospital or university would have to preapprove human studies. Those committees were instructed to scrutinize the risks and benefits to the people participating and just how the informed consent was going to be obtained. Their charge was to protect the participants' rights and welfare.[45]

Then, in 1972, journalist Jean Heller of the Associated Press revealed the particulars of the Tuskegee Syphilis Study.[46] Her story appeared in national newspapers, reporting that since 1932 U.S. government researchers had been studying the effects of untreated syphilis in 399 poor, illiterate African American men. Even after penicillin became available in the 1940s and was widely and successfully used to treat the slowly progressive, devastating venereal disease, the researchers lied to the men, saying they were getting treatment while withholding the drug.[47]

In 1974, responding in part to an avalanche of protest that followed the Tuskegee revelations, Congress enshrined in law the 1966 requirements that until then had simply been a policy.[48] The Department of Health, Education and Welfare issued new regulations interpreting the law. They required that laypeople—lawyers, bioethicists, religious figures, community members—be included on the committees that preapprove human studies before they go ahead. (A 1968 survey had revealed that 73 percent of medical research institutions had constituted their committees entirely of scientists and physicians.)[49] The regulations also laid out very specifically what constituted informed consent.[50] Those 1974 regulations have since been added to with extra protections for pregnant women, newborns, fetuses in the womb, prisoners, and children— including institutionalized children.*

Today independent ethics committees typically meet for several hours to scrutinize proposed human trials—like the committee that met in 2015 to assess a study of a new Ebola vaccine being put forward by scientists at the National Institutes of Health. (The vaccine contains not the Ebola virus but a snippet of its DNA that cannot cause the disease.) The committee grilled the trial's leaders on points of safety and told them to make the informed-consent

---

*Because of forceful protests a quarter century ago from intellectually disabled people and their advocates—a right-to-participate-in-research campaign, in effect—there are no special protections in the U.S. regulations for this group. The issue is not simple. The advocates were concerned, among other things, that if researchers had to jump through too many hoops to conduct studies, clinical research on mental illnesses would be neglected.

form less long and complicated. An eighth grader should be able to read and understand it, one of the committee members insisted.

After the form was revised, the trial participants—140 healthy volunteers who would be among the first human beings injected with the vaccine—were given days to scour both the study plan and the informed-consent form. They asked the researchers as many questions as they wanted to and were given tutorials on the trial. They were told repeatedly, by various members of the research team, that they could withdraw from the trial at any time for any reason. They were closely watched on site for hours after their injections and commanded to call a 24-7 hotline to the trial investigators if they were experiencing any serious symptoms. They returned for a next-day follow-up and then at intervals of weeks and finally months for blood draws so that their anti-Ebola antibody levels could be measured. They were compensated modestly for their time and inconvenience.

One of the volunteers, Grant, a twenty-six-year-old apartment manager, says that he began volunteering for vaccine trials as a way to earn some extra cash but has come to consider it something of a civic duty. "It's not this weird, cold, archaic process anymore," he says when prompted to compare the trial with experiments on prisoners and orphans fifty and sixty years ago. "Why not donate my body and my time?"

For some, the use of cells from Mrs. X's fetus will continue to be an abomination, no matter how long ago the abortion was and no matter that it would have happened anyway. Antiabortion activists recently took their fight beyond the use of fetal cells in vaccine-making, to attack any use of fetal tissue in research. In the summer of 2015 David Daleiden, an antiabortion activist who founded an organization that he calls the Center for Medical Progress, released covertly filmed, heavily edited videos in which he and a colleague posed as biotechnology company executives. They showed senior physicians from the women's health provider Planned Parenthood discussing how they performed abortions and provided fetal tissue for medical research.

Fetal tissue research is legal in the United States. The NIH spent $80 million in 2015 on scores of projects that use fetal tissue. The research is increasing scientists' understanding of HIV/AIDS, hepatitis, blinding diseases, and what happens when things go wrong during fetal development.[51] Separately, companies have used fetal cells to create medicines like the arthritis drug, Enbrel, which was developed using the WI-26 cells that Hayflick derived before he launched WI-38; Pulmozyme, which helps children with cystic fibrosis clear the thick mucus that clogs their lungs; and Nuwiq, an improved treatment for hemophilia, a life-threatening bleeding disorder that occurs in boys and men.

The law stipulates that abortion clinics can be reimbursed only for costs when they provide fetal tissue for research. Daleiden's videos purported to show Planned Parenthood trafficking illegally, for profit, in fetal body parts. More than a dozen investigations were launched by outraged conservative lawmakers on Capitol Hill and in the states. None of the investigations have found evidence that Planned Parenthood profited by providing fetal tissue for research—although Daleiden was indicted by a Texas grand jury on a felony charge of tampering with a government record, for creating a false driver's license to mask his real identity. Nonetheless, in the spring of 2016, a specially-convened panel of the House of Representatives, the House Committee on Infant Lives, subpoenaed scientists across the country who work with fetal tissue, asking for voluminous amounts of information.[52] The episode has cast a chill on fetal tissue research in the United States. One institution posted a guard outside a lab where the research was being conducted.[53] Fetal tissue has become harder for scientists to procure.[54]

In other countries, the research proceeds apace. In one recent paper, Chinese scientists reported the launch of a new cell line, derived specifically to give the Chinese people their own normal, WI-38-like cells for vaccine-making. The new human fetal cell strain is called walvax-2, after the biotechnology company in Yunnan where the cells were derived. They came from the lungs of a female fetus aborted at three months of pregnancy because the mother, a healthy twenty-seven-year-old, had scar tissue from a previous Caesarian section affecting her uterus. The scientists reported that they dissected nine electively aborted fetuses, with the consent of the women who had the abortions, before choosing the lungs that they did.[55] When Debi Vinnedge at Children of God for Life learned of the paper, she published an article on a website called LifeNews.com with the headline "Scientists in China Create New Vaccines Using Body Parts From Nine Aborted Babies."[56]

Later on the same cold November day that I visited Merck, I attended a special event at the American Philosophical Society's Benjamin Franklin Hall, a grand, Italianate building with a white marble facade in old, historic Philadelphia. Hayflick and his colleague Moorhead, who coauthored the 1961 paper asserting that normal cells aged in their lab dishes, were there to receive an award. The John Scott Award was endowed by a Scottish druggist who was reportedly an admirer of Benjamin Franklin and thus chose the city of Philadelphia to administer the prize. It recognizes "ingenious men or women who make useful inventions" for the "comfort, welfare and happiness" of mankind. Since 1822 inventors of a door lock, a wheelbarrow, and a tooth extractor have been honored. But there have also been scientific and engineering superstars among the recipients: Thomas Edison; Marie Curie; the Wright brothers; Frederick Banting, the codiscoverer of insulin; and Saul Perlmutter, who won the Nobel Prize in physics in 2011. The award was the latest in a long list of honors that have been bestowed on Hayflick, who was eighty-eight years old as this book went to press.

The crowd that night was exceedingly gray haired, and the event was something of a reunion for those who were still standing among Hayflick's former colleagues. Plotkin was not there; he was in Germany, presenting a paper on the development of vaccines against cytomegalovirus, today the most common cause in the United States of viral-induced injury to fetuses in the womb. Robert Roosa, the Wistar administrator who had been called back from Toronto when Koprowski discovered the missing WI-38 cells, was there, in a wheelchair, clearly happy to be among old friends; he would pass away a few months later. Also present was Anthony Girardi, the scientist then at Merck who tested Hayflick's first fetal cell strains on hamsters to see if cancers sprouted in their cheek pouches. Eero Saksela, the visiting Finnish scientist who worked with Moorhead in 1962 and 1963 to study whether WI-38 cells turned cancerous late in their lives, had traveled from Helsinki to be there. And Moorhead himself attended, albeit in poor health. It was doubtless the last time that many of those assembled would see one another.

When Hayflick's turn at the podium came, he outlined his many accomplishments, from the discovery of the Hayflick limit, to the identification of the microbe that causes walking pneumonia, to the launch of WI-38 and its widespread use to make vaccines. He exaggerated, telling his elderly audience that if they had ever received a viral vaccine of any kind, "it's almost a certainty that it was produced in WI-38." He claimed that the cells had been used to make vaccines that have immunized more than two billion people. (Merck's rubella vaccine is by far the largest product made with WI-38 cells, and the company says that as of March, 2016, it has shipped some 677 million doses.)[57] And he ended his talk by saying that he felt his greatest achievement was "having reversed the universal belief and the laws that biologists have no intellectual property rights."

Hayflick didn't need to exaggerate. His accomplishments were remarkable enough unadorned.

As Hayflick spoke that night, it had been seven years since he was in possession of any WI-38 cells. The six ampules of original eighth-passage cells that he received under the 1981 settlement with the NIH he first kept in his lab at Oakland Children's Hospital. When he moved to the University of Florida in Gainesville, they went with him in his car, in a portable liquid nitrogen freezer. In 1988, when he moved back to California, the cells drove cross-country with him again. Back in the Bay Area they lived in the portable freezer at his home in San Carlos and then at another home in Hillsborough. At last, in 1991, he deposited the cells in his garage overlooking the Pacific on the beautiful bluffs at the Sea Ranch.[58]

For the next sixteen years, every month or so, Hayflick drove 130 miles round-trip to Santa Rosa, the nearest city of any size, to buy liquid nitrogen to top up the freezer. In 2006 he at last had had enough. He donated the cells to the Coriell Institute for Medical Research in Camden, New Jersey, where they reside to this day.

"It was about time," he told *Nature* in 2013, "that my 'children'—now adults—should leave home."[59]

Epilogue: Where They Are Now

Studying history, my friend, is no joke and no irresponsible game.
—Hermann Hesse, *The Glass Bead Game*[1]

Margareta Böttiger, the young physician and scientist who in 1962 was dispatched by her boss, Sven Gard, to find the medical history of Mrs. X, played a key role in Sweden's successful polio eradication effort, and continued follow-up studies on vaccinated people for forty years. In 1976 she became the Swedish government's top epidemiologist. In that position, she was deeply involved in responding to the HIV/AIDS epidemic in its earliest days and worked to improve vaccine coverage against several other infectious diseases. Now eighty-nine years old, she lives part of the year in a historic seaside mansion near Stockholm that was once a favorite escape for close friends of the Böttiger family: Sweden's Crown Princess Margaret and Crown Prince Gustav Adolf—later king Gustav VI Adolf.

Bernice Eddy, the NIH microbiologist who was demoted for discovering a cancer-causing "substance" in the monkey kidney cells used to make polio vaccine, worked at the agency's Division of Biologics Standards until she retired in 1973 at age seventy. She continued to publish papers on the tumor viruses until her retirement. She passed away in 1989.

The question of whether Eddy's "substance," the silent monkey virus SV40, has caused cancers in Americans who were vaccinated against adenovirus and polio between 1955 and 1963, when vaccine makers were obligated to switch to using a species of monkey that does not naturally harbor SV40, has become, and remains, a contentious one.

The most authoritative statement to date on the matter comes from the Institute of Medicine, a branch of the National Academy of Sciences. The Institute concluded in 2002 that although studies that followed vaccine recipients over the decades provide no evidence of increased cancer risk, these studies were "sufficiently flawed" that the question of whether or not contaminated polio vaccine caused cancer couldn't be answered. It recommended continued analysis and research on the question.[2]

In 2013 a fact sheet on the Web site of the Centers for Disease Control and Prevention stated that "more than 98 million Americans received one or more doses of polio vaccine from 1955 to 1963 when a proportion of vaccine was contaminated with SV40; it has been estimated that 10–30 million Americans could have received an SV40 contaminated dose of vaccine."

The CDC fact sheet also stated, "The majority of scientific evidence suggests that SV40-contaminated vaccine did not cause cancer; however, some research results are conflicting and more studies are needed."[3]

In July 2015 a CDC official said that the fact sheet had been removed for updating. The agency has not posted a new one.

Eva Ernholm, who performed the WI-38 abortion, practiced gynecology in Sweden until 1992. She also taught sex education in high schools and vocally cautioned young women against approaching abortion casually. When she moved away from her first private practice in the town of Kristinehamn, patients there continued to seek her out, driving the four hours to Ernholm's private practice in Falkenberg. She died in 2011 at age eighty-six.

After leaving the University of Florida in 1988, Leonard Hayflick became an unpaid adjunct professor of anatomy at the University of California at San Francisco. He no longer had a lab but spent a portion of his time consulting for nearby Genentech, the first big biotechnology company, advising labs there on cell-culture techniques. "I was involved essentially in research by proxy, but it still gave me the joy of doing research," he said in a 2013 interview.[4] He was elected a fellow of the American Association for the Advancement of Science in 1986 and served as editor in chief of the journal *Experimental Gerontology* for thirteen years, until 1998. He served as a scientific adviser to several companies, including Geron, the firm that was founded in 1990 to exploit telomeres and telomerase. He won many awards. His 1994 book *How and Why We Age* has been translated into nine languages. Today he is eighty-eight years old and lives in Sea Ranch, California. His wife, Ruth, passed away in March 2016.

Eva Herrström, who dissected fetal organs for shipping to the Wistar Institute, was the chief technician in Sven Gard's laboratory at the Karolinska Institute until he retired in 1972. She then worked for Margareta Böttiger at Sweden's National Bacteriological Laboratory for another twenty years. Today Herrström is ninety-one, lives in Stockholm, and enjoys spending time at a family-owned seaside house in the Stockholm archipelago.

The monkey house on the grounds of the National Bacteriological Laboratory, where Herrström worked and where the fetal dissections were done, is still standing. In 2013 the second-floor space that once housed Sven Gard's lab was lined with offices occupied by quality-control employees of Crucell, a Dutch vaccine-making company that is an arm of Janssen Pharmaceutical Companies. Crucell's flagship technology, which it is using to develop candidate vaccines against the Ebola and Marburg viruses, was developed using retinal cells from a fetus that was electively aborted in the Netherlands in 1985.

Hilary Koprowski in the 1980s made a $15 million personal fortune by launching a company, Centocor, using antibody technology developed—but not patented—by other scientists. Koprowski obtained it for free from one of the inventors, then filed for two far-reaching patents.[5]

Koprowski led the Wistar Institute for thirty-four years, until 1991, when the board of managers ousted him in the face of dire financial problems. They were caused by the expiration in the 1980s of the patents on the rubella and rabies vaccines, and the attendant loss of more than $3 million in annual income.[6] Koprowski sued for age discrimination; the lawsuit was settled out of court in 1993.[7] He moved to Thomas Jefferson University in Philadelphia, where he was director of the Center for Neurovirology. The last of his 822 scientific papers was published in 2013, the year that he died at age ninety-six. It assessed whether bat ticks could transmit rabies and found that they were unlikely vectors.[8]

Chip and Betsy MacConnell, the parents of Anna MacConnell, who died from the effects of congenital rubella in 2012, established a charity in her memory called Angels for Anna (www.angelsforanna.com). It provides "parent packs" of pajamas and toiletries for parents facing the unexpected admission of a child to the hospital. Their long-term goal is to build a comprehensive pediatric rehabilitation facility in the Dayton area.

Stanley Plotkin, who invented the rubella vaccine that is used in almost every country that vaccinates against the disease, became the director of infectious diseases at the Children's Hospital of Philadelphia and later medical and scientific director for the vaccine maker now called Sanofi Pasteur. At eighty-four he is today an independent consultant for vaccine manufacturers and nonprofits. He has worked for years to develop a vaccine against cytomegalovirus (CMV), today the most common cause of virus-induced injury to fetuses in the United States. Thanks in part to his efforts, the major vaccine makers now have CMV vaccines in development. Plotkin lives with his wife, Susan, outside Philadelphia. He got his pilot's license at age seventy-four and began piano lessons at age eighty-one.

The St. Vincent's Home for Children, where Plotkin conducted the first trials of his rubella vaccine in the mid-1960s, was closed by the Archdiocese of Philadelphia in 1981. Today it serves as a truancy court for the School District of Philadelphia. Across the lane the archdiocese's home and maternity hospital for unwed mothers, where Hayflick's polio vaccine was tested on newborns in 1962, houses a day-care center, the Early Years Development Center.

The Hamburg State School and Hospital, which housed some 950 people when Plotkin tested his rubella vaccine there, is still a state-run institution. Today called the Hamburg Center, it houses and provides services for 122 people with intellectual disabilities and physical challenges.

Philadelphia General Hospital, where both Koprowski's and Hayflick's polio vaccines were tested on premature babies, was closed in 1977. Today, the grounds where it stood are occupied by the medical campuses of the University of Pennsylvania and the Children's Hospital of Philadelphia.

Clinton State Farms, the New Jersey women's prison where both polio vaccines were tested on newborns, is still a state penitentiary. It has been renamed the Edna Mahan Correctional Facility for Women, in honor of the warden who ran it for forty years.

Jim Poupard, the school-hating nineteen-year-old who worked for the Wistar in 1962 vaccinating babies with the Hayflick polio vaccine, was lured away to a better-paying job with a drug company after just six months. He went on to earn a PhD at the University of Pennsylvania and became the supervisor of clinical microbiology at the Hospital of the University of Pennsylvania. Later he was the head of global strategic microbiology at GlaxoSmithKline Pharmaceuticals. Retired since 2003, he is chairman of the American Society for Microbiology's Center for the History of Microbiology.

Roderick Murray retired in 1973 and died of a heart attack in 1980. He was seventy years old.

James Schriver died in 1999.

The Ebola vaccine that was tested at the NIH in 2015 proved to be safe in all 140 volunteers. As this book went to press, the investigators were waiting to learn whether it generates effective levels of antibodies. If it does, the trial will continue, with the recruitment of another larger group of volunteers and then another group of hundreds of people at many sites. The process, in the best case, will take years.

Stephen Wenzler, who was born blind and deaf from congenital rubella in Toms River, New Jersey, in 1964, graduated from the Perkins School for the Blind in 1985. He briefly attended Gallaudet University in Washington, DC, but left after being harassed by deaf students; he was one of the few students there who was also blind. He enrolled at Camden County College in New Jersey, where he earned a certificate in computer operations. He excelled with electronics and computers and in the two years before he graduated saved the college an estimated $20,000 on computer repairs. But he ended up in a series of jobs mopping floors and cleaning bathrooms at Roy Rogers, McDonald's, and Denny's because his social skills were poor and employers were reluctant to hire him.

In 2004 the Helen Keller National Center for Deaf-Blind Youths and Adults featured Wenzler's photo on a poster that read "Help fight the 1964 rubella epidemic . . . by hiring one of the thousands of Americans it left deaf and blind. Like Stephen Wenzler."

In November 2006, on the evening of his forty-second birthday, Wenzler was walking on the shoulder of the road in front of his apartment in suburban New Jersey. A speeding car swerved and hit him. He died instantly.

Kathy Earp, who worked for the program for the deaf and hard of hearing at Camden County College and knew Wenzler well, spoke at the funeral. "Stephen maintained a sense of humor, patience, and dignity in spite of all life's challenges that far exceeded that of the average person," she said. "His life taught me about true courage."

Mrs. X was living in Sweden in 2013.

After sending two unanswered letters, I tried to reach her during a research trip to Sweden that year. In a brief conversation with my Swedish research assistant and translator, Mrs. X said that she had no interest in being interviewed.

She also said: "They were doing this without my knowledge. That cannot be allowed today."

Soon after this conversation, I wrote to Mrs. X, and my assistant translated and printed the letter. I mailed it. In it I assured her that I would never make her name public or contact her again.

ACKNOWLEDGMENTS

Many, many people helped to make this book a reality. I am thankful to every one of them. I have undoubtedly forgotten some people in what follows. I ask their indulgence.

These pages would never have seen the light of day were it not for New America. Fellowships there funded by Eric and Wendy Schmidt and by Bernard Schwartz supported me for two invaluable years of full-time work. I am deeply grateful. The community of writers and scholars at New America was a great source of inspiration and a healthy complement to the lonely writer's slog. I would especially like to thank Becky Shafer and Kirsten Berg for their first-rate research assistance and their support of the fellows program; Fuzz Hogan and Andrés Martinez, who, with Becky, believed in this book from the beginning; my fellow fellows, whose enthusiasm and companionship were a boost and a delight; and the following associates and interns who gamely and capably helped with my research: David Allen, Jacob Glenn, Madhu Ramankutty, Courtney Schuster, and Andrew Small.

My agent, Gail Ross, and my editor at Viking, Wendy Wolf, are two formidable women whom I am lucky to have in my corner. Gail and her partner Howard Yoon turned the enthusiastic wanderings of my mind into a streamlined, compelling book proposal. Their assistant Jennifer Manguera saved me during several crises with crashing and obstreperous versions of the manuscript. Wendy Wolf's brilliant and precise scalpel turned my rough-edged efforts into a cohesive, meaningful whole, Viking's staff, including Francesca Belanger, Georgia Bodnar, Victoria Klose, Min Lee, Eric Wechter, and Hilary Roberts, ably dealt with the myriad tasks that finally got the words onto the page.

Leonard Hayflick gave unstintingly of his time, his sharp memories, his talent for lucid explication, and his entrancing ability to recount tales at fifty years' distance as if they had happened yesterday. In addition to submitting to three dozen interviews, he answered more e-mails than any human being should have to. Without Leonard Hayflick there would have been no story; I am deeply indebted to him. I am also grateful to his late wife, Ruth Hayflick, and his son, Joel Hayflick, who obligingly searched their memories at length and in depth. Joel was also a willing guide to the former Hayflick neighborhood and the onetime site of Leonard Hayflick's lab on the Stanford campus. Joel also generously shared his genealogical research on the Hayflick family.

Burton and Shulamith Caine, Norman Cohen, and Moselio Schaecter kindly offered their memories of Leonard Hayflick's youth and young adulthood.

Stanley Plotkin spent many hours being interviewed and literally retracing the steps he took in the 1960s at and around the Wistar Institute and the Children's Hospital of Philadelphia, even trooping through the rain on one cold March afternoon in an ill-fated effort to find St. Vincent's Home for Children. (I found the still-standing building several months later, on a warm June evening.) Stanley answered innumerable follow-up questions and patiently explained myriad points of detail. Crucially, he opened to me his hitherto-unplumbed papers. They were essential to the telling of his important story. Stanley and his wife, Susan Plotkin, graciously allowed me to turn their sunroom into a makeshift archive in this effort. I am deeply grateful.

Steve and Mary Wenzler opened their hearts, their home, and their memories of their beloved son Stephen. Mary particularly gave of her time and recollections, despite the pain renewed by my many questions about Stephen's life. Chip and Betsy MacConnell did the same in recounting the brief life of their precious daughter Anna and in providing many medical records, photos, and memories. To both of these couples I am deeply grateful. Danielle MacConnell also helpfully shared her memories of her sister. Janet Gilsdorf went several extra miles to help track down Anna's medical records. Pamela Ryan, Nancy O'Donnell, and Kathy Earp kindly gave their time helping me to understand Stephen Wenzler as a boy and a man.

Fran Scalise, Dee Krewson, and one anonymous woman generously shared their memories of living through months of pregnancy and through childbirth at St. Vincent's Hospital for Women and Children. Sisters Rosa Hoflacher and Mary Thérèse Hasson of the Missionary Sisters of the Precious Blood kindly recalled their time at the St. Vincent's Home for Children; the Mother House in Reading, Pennsylvania, generously provided a bed. In Hamburg, Pennsylvania, Pat Pitkin offered guidance and Janet Barr and Francis Muller went out of their way to be helpful.

Leonard Warren graciously shared his lengthy, deeply researched history of the Wistar Institute. Edward Hooper made available a key World Health Organization document and otherwise-unobtainable records of polio vaccine testing of newborns at Clinton Farms. Paul Offit allowed me to use his irreplaceable interviews with Maurice Hilleman. He also shared his vaccine expertise and his support as a fellow book writer. Daniel Wilson kindly provided the transcript of his 1990 interview with Dorothy Horstmann. Frank Simione at the American Type Culture Collection willingly tracked down key historical documents and information; at the Coriell Institute for Medical Research,

Christine Beiswanger was equally helpful, as was Howard Garrison at the Federation of American Societies for Experimental Biology. At the University of California at Davis, Scott Simon offered his statistics expertise. David Segal, Amy Swift, and Narisu Narisu went out of their way to probe the genetic underpinnings of WI-38.

I am deeply grateful to the following scientists for their expert help and the gift of their time reviewing sections of the manuscript: Margaret Burgess, Francis Collins, Ahmad Fayaz, Carol Greider, Julie Ledgerwood, Leonard Norkin, Gary Stein, Stanley Plotkin, Werner Slenczka, Jack Szostak, Harold Varmus, Woodring Wright, and Bill Wunner. Stanley Gartler generously reviewed the entire manuscript and offered timely, incisive comments. Wudan Yan was a first-rate fact-checker. Any errors in the book are mine alone.

John F. O'Neill helped me understand the serious challenges of pediatric eye surgery in the early 1960s and the devastating effects of rubella on fetal and newborn eyes. Maggi Buterbaugh kindly spoke with me about her experience as the parent of a child affected by congenital rubella. Rugang Zhang explained the importance of WI-38 cells in research today; his colleagues at the Wistar Institute, Hildegund Ertl and Meenhard Herlyn, were gracious and informative hosts. Virginia Berwick, Richard Carp, Barbara Cohen, Vittorio Defendi, William Elkins, Anthony Girardi, Fred Jacks, Anne Kamrin, Betsy Meredith, Paul Moorhead, Jim Poupard, the late Robert Roosa, Eero Saksela, and Kaighn Smith shared their memories of science and life at the Wistar Institute and the University of Pennsylvania in the 1950s and 1960s. Jim Poupard completed his recollections with a full walking tour of the grounds where Philadelphia General Hospital once stood. All of these people were generous with their time, and I am grateful to every one of them.

Nancy Pleibel kindly shared her time, her technical expertise, and her memories of the Hayflick lab both at the Wistar and at Stanford. Robert Stevenson helped me with his expertise about the history of cell culture, the NIH in the 1960s, and the NIH contract that funded the production of WI-38. Paul Parkman helpfully recounted developing the HPV-77 rubella vaccine and working with Roderick Murray in the Division of Biologic Standards; John Finlayson also remembered the 1960s-era DBS. Richard Dugas and Nicholas Wade shared their memories of James Schriver and his investigation of Leonard Hayflick.

Suresh Jadhav at the Serum Institute of India was consistently helpful. Alan Shaw, who for fifteen years steered the making of live virus vaccines at Merck, patiently answered many nuts-and-bolts questions about vaccine making. Philip Provost recalled developing Merck's hepatitis A vaccine. Pamela Eisele in media relations at Merck was unfailingly helpful; Colleen Lange, Michael Lynn, Vic Johnston, and other experts involved with making the rubella vaccine at Merck's West Point campus shared hours of their time. I am grateful to all of them, as I am to Gwynne Oosterbaan at GlaxoSmithKline, who gamely tracked down the seemingly long-lost facts from the history of the Cendehill vaccine, and to Sean Clements at Sanofi Pasteur.

I am also grateful to Debi Vinnedge for sharing her story, her wide-ranging knowledge, her correspondence with the Vatican, and her perspective. She was unfailingly cordial, prompt, and incredibly thorough.

In Sweden Lisa Tallroth provided outstanding research assistance and translation. She and her husband, Erik Tallroth, kindly hosted me in their home, cheerfully and despite their duties with work and three small children. Erling Norrby was an attentive and fascinating guide to Swedish science; his persistence allowed me to gain entry to the former monkey house. There Crucell Sweden's Agneta Strömberg was a helpful and gracious host.

Erling's insights helped me to understand Sven Gard and the scene in his lab in the early 1960s, as did the thoughtful recollections of Margareta Böttiger. Eva Herrström's memories were hugely helpful, as was her generous sharing of her diaries and her photos from the Gard lab. Elisabeth, Håkan, and Lars Ernholm kindly provided hospitality, transportation, and fascinating recollections of their aunt, Eva Ernholm, as did Osborne Carlson. Elisabeth helpfully dug out Eva's photos, work records, and news clippings. Solveig Jülich, Lena Lennerhed, Niels Lynöe, and Jane Reichel shared their knowledge of Swedish abortion law, politics, and policy in the midtwentieth century. Georg and Eva Klein recalled their colleague and friend Hilary Koprowski. Their colleagues at the Karolinska Institute, Hans-Gustaf Ljunggren and Sabina Bossi, made me welcome. Petter Byström offered additional help with translation.

I'm indebted to many archivists and librarians. I would like in particular to thank the dedicated experts at the University Archives and Records Center at the University of Pennsylvania and at the National Library of Medicine; David Baugh at the Philadelphia City Archive; Shawn Weldon at the Philadelphia Archdiocesan Historical Research Center; Richard Mandel, the archivist for the Executive Secretariat at the National Institutes of Health; Barbara Faye Harkin in the Office of NIH History; Michael Horvath at the Karolinska Institute; Daniel Hartwig at Stanford University; Greg Lester, Nina Long, and April Miller, all formerly of the Wistar Institute, and Jennifer Evans Stacey, who works there now. Caroline Brogan at the *Lancet* kindly dug out many decades-old papers. Christopher Koprowski willingly and cheerfully recalled his father, Hilary Koprowski, and family history.

He gamely tried to find his father's scientific correspondence, the whereabouts of which remain unknown. Sue Jones also did her best to help in this endeavor.

My brilliant brother-in-law David Weih rendered crucial calculations when my rusty math skills failed me. Leslie Smith provided invaluable help formatting the manuscript. The excellent graphic designer Lynne Smyers generously donated her time and talent preparing the book's photos for submission. Sarah Kellogg kindly spent tens of hours helping to compile and organize the photos. Sarah, Leslie Smith, and my other dear friends Ruth Barzel and Katie Smith Milway read the manuscript, as did my hugely supportive sister, Andrea Wadman. I am deeply grateful to each of them for their thoughtful, incisive feedback and for their unfailing faith in me and my project. Helen Pearson, then my editor at *Nature*, immediately understood my compelling interest in the story of the WI-38 cells and ably guided the magazine feature that eventually grew into this book. She has since become a loyal book-writing colleague and wise friend. Alex Witze and Chris Leonard also offered strength and wisdom from the writerly trenches, as did Sarah Kellogg. Other far-flung friends and family too numerous to name have asked enthusiastically about the book over these many months; their interest and support too have kept me going.

My mother, Barbara Greenfield Wadman, and my late father, Hamilton Wadman, believed in me, loved me, and taught me a great deal about medicine and life. Their belief in the power of medicine to do good and their compassion for the poor, the unlucky, and the forgotten set me on the path to writing this book. No words can express my gratitude for their love and support, and for that of my sister Andrea Wadman.

My sons Bobby and Christopher Wells put up with many dinner-table conversations about cells and vaccines and offered finely honed teenage wisdom on several matters in these pages. I thank them for tolerating and even cheering on my commitment to this book, and for forgiving me when my motherly duties fell by the wayside. Tim Wells, my husband, fellow voyager, and friend, gave me the quiet support of a man of few words—and the inspiring example of the most serious, disciplined, and committed writer I know.

NOTES

In the following notes, certain documents obtained under the Freedom of Information Act are referenced using the followng titles: Factual Chronology, Hayflick Rebuttal to Schriver Report, Riseberg Memo, Schriver Report, and Schriver Rebuttal to Hayflick Rebuttal. Full citations for these documents are in the Selected Bibliography.

Epigraph
Helen Keller, *The Open Door* (Garden City, New York: Doubleday & Company, Inc., 1957), 31.

Prologue
1. Quoted in William A. Clark and Dorothy H. Geary, "The Story of the American Type Culture Collection: Its History and Development (1899–1973)," *Advances in Applied Microbiology* 17 (1974): 295.
2. Pathological Laboratory Autopsy Report, Philadelphia General Hospital, autopsy no. 74681, February 21, 1966, Post Mortem Records, vol. 465 (January 18–April 8, 1966), Collection 80-101.24, City Archives, City of Philadelphia Department of Records.
3. Leonard B. Seeff et al., "A Serologic Follow-up of the 1942 Epidemic of Post-vaccination Hepatitis in the United States Army," *New England Journal of Medicine* 316 (1987): 965–70.
4. Neal Nathanson and Alexander Langmuir, "The Cutter Incident: Poliomyelitis Following Formaldehyde-Inactivated Poliovirus Vaccination in the United States During the Spring of 1955 II: The Relationship of Poliomyelitis to Cutter Vaccine," *American Journal of Hygiene* 78 (1963): 39; Paul A. Offit, "The Cutter Incident, 50 Years Later," *New England Journal of Medicine* 352 (2005): 1411.
5. David M. Oshinsky, *Polio: An American Story* (New York: Oxford University Press, Inc., 2005), 238.
6. The description of the launch of WI-38 is taken from personal interviews with Leonard Hayflick listed in the bibliography including, especially, a telephone interview on July 1, 2013.
7. The number of U.S. recruits vaccinated from October 2011 through November 2015 (826,317) was provided by the Defense Press Office in the Office of the Assistant Secretary of Defense; for the number vaccinated between 1971 and 1999, some 8.5 million, see: CNA Analysis and Solutions, *Population Representation in the Military Services: Fiscal Year 2014 Summary Report*, Appendix D, Table D–4 (Washington, DC: Office of the Undersecretary of Defense for Personnel and Readiness, 2014), https://www.cna.org/pop-rep/2014/appendixd/d_04.html; several hundred thousand U.S. recruits also received adenovirus vaccine made in WI-38 cells during testing of the vaccine in the mid- and late-1960s: see: "Conference on Cell Cultures for Virus Vaccine Production Nov. 6–8 1967, National Institutes of Health, Bethesda, Maryland," *NCI Monographs*, no. 29 (1968), 499.
8. Nicholas Wade, "Hayflick's Tragedy: The Rise and Fall of a Human Cell Line," *Science* 194, no. 4235 (1976): 125.
9. For an illuminating analysis of this era, see David J. Rothman, *Strangers at the Bedside: A History of How Law and Bioethics Transformed Medical Decision Making* (New York: Aldine de Gruyter, 1991, 2003), 30–50.

10. See for example Marcia Angell, "Medical Research: The Dangers to the Human Subjects," *New York Review of Books* 62, no. 18 (November 19, 2015): 50.

11. Henry K. Beecher, "Ethics and Clinical Research," *New England Journal of Medicine* 274, no. 24 (1966): 1354–60.

12. William H. Stewart, "Surgeon General's Directives on Human Experimentation," PPO #129 (Bethesda, MD: US Public Health Service, revised July 1, 1966), https://history.nih .gov/research/downloads/Surgeongeneraldirective1966.pdf.

13. See, for instance, Angell, "Medical Research," 48–51; and Marcia Angell, "Medical Research on Humans: Making It Ethical," *New York Review of Books* 62, no. 19 (December 3, 2015): 30–32.

14. World Health Organization, "Situation Report: Zika Virus Microcephaly Guillain-Barré Syndrome, July 7, 2016," 7.

Chapter One: Beginnings

1. Nicole Krauss, *The History of Love* (New York: W. W. Norton, 2006), 11.

2. Maxwell Whiteman, "Philadelphia's Jewish Neighborhoods," in *The Peoples of Philadelphia: A History of Ethnic Groups and Lower-Class Life, 1790–1940*, Allen F. Davis and Mark H. Haller, eds. (Philadelphia: Temple University Press, 1973), 250; Richard A. Varbero, "Philadelphia's South Italians in the 1920s," in *Peoples of Philadelphia*, Davis and Haller, eds., 260–61.

3. "13th US Federal Census (1910) ED 71, Sh. 3B, line 4: Res. 411 Lombard St., Phila. PA," cited at http://freepages.genealogy.rootsweb.ancestry.com/~jhayflick/ps08/ps08_281.htm; Mark H. Haller, "Recurring Themes," in *Peoples of Philadelphia*, Davis and Haller, eds., 283–84.

4. John F. Sutherland, "Housing the Poor," in *Peoples of Philadelphia*, Davis and Haller, eds., 184–85.

5. Dennis J. Clark, "The Philadelphia Irish: Persistant Presence," in *Peoples of Philadelphia*, Davis and Haller, eds.; Whiteman "Jewish Neighborhoods," in ibid, 242.

6. John M. Barry, *The Great Influenza: The Story of the Deadliest Pandemic in History* (New York: Viking Penguin, 2004), 326.

7. Roger D. Simon, "Great Depression," *The Encyclopedia of Greater Philadelphia*, http:// philadelphiaencyclopedia.org/archive/great-depression/ (reprinted and adapted from Roger D. Simon, "Philadelphia During the Great Depression, 1929–1941," Historical Society of Pennsylvania, "Closed for Business: The Story of Bankers Trust During the Great Depression" http://www.hsp.org/bankers-trust).

8. Philadelphia County Relief Board, *Office Manual of the Philadelphia County Relief Board*, August 1934, section V, page 1, Temple University Urban Archive, Jewish Family Service, series 4, box 7, volume 16.

9. Simon, "Great Depression."

10. Leonard Hayflick, telephone interview with the author, March 23, 2014.

11. Vincent J. Cristofalo, "Profile in Gerontology: Leonard Hayflick, PhD," *Contemporary Gerontology* 9, no. 3 (2003): 86.

12. Leonard Hayflick, interview with the author, October 3, 2012.

13. Leonard Hayflick, telephone interview with the author, March 31, 2014.

14. Ibid.

Chapter Two: Discovery

1. D. Ivanovsky, 1892. "Concerning the Mosaic Disease of the Tobacco Plant," *St. Petsb. Acad. Imp. Sci. Bul.* 35 (1892): 67–70. Cited in Alice Lustig and Arnold J. Levine, "One Hundred Years of Virology," *Journal of Virology* 66, no. 8 (August 1992): 4629–31.

2. M. W. Beijerinck, "Concerning a Contagium Vivum Fluidum as a Cause of the Spot-Disease of Tobacco Leaves," *Verh. Akad. Wet. Amsterdam* 2, no. 6 (1898): 3–21. This is the third reference in Lustig and Levine, "One Hundred Years of Virology."

3. F. Loeffler and P. Frosch, *Zentralbl. Bakteriol. Parasitenkd. Infektionskr. Hyg. Abt. 1 Orig.* 28 (1898): 371. Cited as reference 13 in Lustig and Levine, "One Hundred Years of Virology."

4. Edward Rybicki, *A Short History of the Discovery of Viruses* (Buglet Press e-book, 2015), 6. This book is available for purchase at https://itunes.apple.com/us/book/short-history -discovery-viruses/id1001627125?mt=13. Part one is also freely available here: https://ry bicki.wordpress.com/2012/02/06/a-short-history-of-the-discovery-of-viruses-part-1/. And part 2 is here: https://rybicki.wordpress.com/2012/02/07/a-short-history-of-the -discovery-of-viruses-part-2/

5. F. Loeffler and P. Frosch, "Summarischer Bericht über die Ergebnisse der Untersuchungen der Kommission zur Erforschung der Maul- und Klauenseuche bei dem Institut fu$ r In-fektionskrankheiten in Berlin," *Centralblatt für Bakteriologie, Parasitenkunde und Infektionskrankheiten, Abt. I* 22 (1897): 257–59. Loeffler and Frosch's achievement is also described in Rudolf Rott and Stuart Siddell, "One Hundred Years of Animal Virology," *Journal of General Virology* 79 (1998): 2871–72, http://jgv.microbiologyresearch.org/con tent/journal/jgv/10.1099/0022-1317-79-11-2871?crawler=true&mimetype=applica tion/pdf.

6. Thomas M. Rivers, "Filterable Viruses: A Critical Review," *Journal of Bacteriology* xiv, no. 4 (1927): 228.

7. Thomas M. Rivers, ed., *Filterable Viruses* (Baltimore: Williams and Wilkins, 1928), 1–418.

8. Albert B. Sabin and Peter K. Olitsky, "Cultivation of Poliomyelitis Virus *in Vitro* in Human Embryonic Nervous Tissue," *Proceedings of the Society for Experimental Biology and Medicine* 34 (1936): 357–59.

9. Rachel Benson Gold, "Lessons from Before Roe: Will Past Be Prologue?" *Guttmacher Report on Public Policy* 6 (2003): 8–11, www.guttmacher.org/about/gpr/2003/03/lessons -roe-will-past-be-prologue; Rosemary Nossif, *Before Roe: Abortion Policy in the States* (Philadelphia: Temple University Press, 2001), 33–35.

10. Carole R. McCann, "Abortion," in *The Oxford Companion to United States History*, ed. Paul S. Boyer (New York: Oxford University Press, 2001), 3; Henry J. Sangmeister, "A Survey of Abortion Deaths in Philadelphia from 1931 to 1940 Inclusive," *American Journal of Obstetrics and Gynecology* 46 (1943): 757–58.

11. John F. Enders, Thomas H. Weller, and Frederick C. Robbins, "Cultivation of the Lansing Strain of Poliomyelitis Virus in Cultures of Various Human Embryonic Tissues," *Science* 109 (1949): 85–87.

12. Saul Benison, *Tom Rivers: Reflections on a Life in Medicine and Science: An Oral History Memoir Prepared by Saul Benison* (Cambridge, MA, and London: MIT Press, 1967), 446.

13. Leonard Hayflick, interview with the author, October 3, 2012.

14. Leonard Hayflick, "Early Days at Merck," *Web of Stories*, July, 2011, www.webofstories .com/play/leonard.hayflick/13.

15. Ruth Hayflick, telephone interview with the author, September 29, 2014.

16. Wistar Institute, "Our History," www.wistar.org/the-institute/our-history (accessed June 7, 2016).

17. Leonard Warren, "A History of the Wistar Institute of Anatomy and Biology" (unpub-lished manuscript, March 25, 2014), 25, 40, 60; Erik Larson, *The Devil in the White City: Murder, Magic, and Madness at the Fair That Changed America* (New York: Random House, 2003).

18. Warren, "A History of the Wistar Institute" 2, 47, 61, 78, 79; Paul Offit, *Vaccinated: One Man's Quest to Defeat the World's Deadliest Diseases* (New York: HarperCollins, 2007), 79–80.

19. Warren, "History of the Wistar Institute," 106–12.

20. Ibid., 111.

21. Ibid., 101.

22. Leonard Hayflick, interview.
23. Leonard Hayflick, "The Growth of Human and Poultry Pleuropneumonia-like Organisms in Tissue Cultures and in Ovo and the Characterization of an Infectious Agent Causing Tendovaginitis with Arthritis in Chickens" (PhD diss., University of Pennsylvania, 1956).
24. Leonard Hayflick, interview with the author, March 3, 2013.

Chapter Three: The Wistar Reborn
1. Maurice Hilleman, interview with Paul Offit, Nov. 30, 2004, audio file courtesy of Paul Offit.
2. Barbara Cohen, interview with the author, November 20, 2014.
3. John Rowan Wilson, *Margin of Safety* (Garden City, NY: Doubleday, 1963), 140.
4. Ursula Roth, eighty-fifth-birthday tribute to Hilary Koprowski, 2001. Courtesy of Sue Jones.
5. The Belgian virologist Lise Thiry related this tale in a written tribute for a book of memories compiled to celebrate Hilary Koprowski's eighty-fifth birthday in 2001. Thiry's recollections were provided courtesy of Sue Jones, Koprowski's former assistant.
6. Minutes of Wistar Institute Board of Managers Meeting, February 24, 1967, page 4, UPT 50 R252, box 68, file folder 12 (Wistar Institute 1966–67), Isidor Schwaner Ravdin Papers, University Archives and Records Center, University of Pennsylvania; Trustee Minutes, vol. 26, page 313, collection no. UPA 1.1, University Archives and Records Center, University of Pennsylvania; Vittorio Defendi, interview with the author, March 18, 2014.
7. Peter Doherty, telephone interview with the author, January 20, 2015.
8. Michael Katz, "A Devotion to 'Real Science,'" in "Symposium in Honor of Hilary Koprowski's Achievements," *Journal of Infectious Diseases* 212, Supplement 1 (2015): S6.
9. Stacey Burling, "Hilary Koprowski, Polio Vaccine Pioneer, Dead at 96," *Philadelphia Inquirer*, April 14, 2013, http://articles.philly.com/2013-04-15/news/38559037_1_polio-vaccine-hilary-koprowski-rabies-vaccine.
10. Roger Vaughan, *Listen to the Music: The Life of Hilary Koprowski* (New York: Springer-Verlag, 2000), 5.
11. Bob Gallo, "Lessons from a World Trip," in "Symposium in Honor of Hilary Koprowski's Achievements," S6.
12. Vaughan, *Listen to the Music*, 111.
13. I. S. Ravdin to Jonathan Rhoads, January 27, 1960, UPT 50 R252, box 65, file folder 56 (Wistar Institute, 1959–1961), Isidor Schwaner Ravdin Papers, University Archives and Records Center, University of Pennsylvania.
14. "Draft of Minutes of Meeting of the Executive Committee of the Board of Managers of the Wistar Institute," September 21 1960, and I. S. Ravdin, "Memo," December 15, 1960, both in UPC 1.4, box 6, file folder "Wistar Institute Board of Managers," Vice President for Medical Affairs Correspondence and Records, University Archives and Records Center, University of Pennsylvania; I. S. Ravdin to Francis Boyer, July 1, 1960, UPT 50 R252, box 65, file folder 56 (Wistar Institute, 1959–1961), Isidor Schwaner Ravdin Papers, University Archives and Records Center, University of Pennsylvania; Committee on the Wistar Institute to President Gaylord P. Harnwell, November 17, 1959, p. 3, UPA4, box 114, file folder 24 (Goddard Report: Committee on the Wistar Institute 1955–1960), Office of the President, Gaylord Probasco Harnwell Administration, University Archives and Records Center, University of Pennsylvania; "Tentative Minutes, Wistar Executive Committee, 11/18/59," pp. 2–3, UPA4, box 114, file folder 17 (Wistar Institute IV: 1955–1960), Office of the President, Gaylord Probasco Harnwell Administration, University Archives and Records Center, University of Pennsylvania; correspondence among Joseph Stokes Jr., Hilary Koprowski, Jonathan E. Rhoads, Albert Linton, and I. S. Ravdin, January 20, 1960, through February 22, 1960, call no. B.St65p, file folder "Koprowski, Hilary #2," Joseph Stokes Papers Series I, American Philosophical Society Archives.
15. "Resolution by the Senior Members of the Scientific Staff of the Wistar Institute of Anatomy and Biology, to Be Forwarded to the President and Board of Managers," December

30, 1959, UPA4, box 114, file folder 17 (Wistar Institute IV: 1955–1960), Office of the President, Gaylord Probasco Harnwell Administration, University Archives and Records Center, University of Pennsylvania.

16. Much of the biographical information in this section is rendered in lively detail in Vaughan, *Listen to the Music*, 17–40; this section also draws on this engaging memoir by Hilary Koprowski's wife: Irena Koprowska, *A Woman Wanders Through Life and Science* (Albany: State University of New York Press, 1997), 38–122; Christopher Koprowski, e-mail to the author, October 15, 2013.

17. Vaughan, *Listen to the Music*, 42, 67.

18. Ibid., 44.

19. David M. Oshinsky, *Polio: An American Story* (New York: Oxford University Press, 2005), 138–42.

20. Vaughan, *Listen to the Music*, 1–2.

21. Hilary Koprowski, Thomas W. Norton, and Walsh McDermott, "Isolation of Poliomyelitis Virus from Human Serum by Direct Inoculation into a Laboratory Mouse," *Public Health Reports* 62, no. 41 (1947): 1467–76.

22. Hilary Koprowski, George A. Jervis, and Thomas W. Norton, "Immune Responses in Human Volunteers upon Oral Administration of a Rodent-Adapted Strain of Poliomyelitis Virus," *American Journal of Hygiene* 55 (1952): 109.

23. Oshinsky, *Polio*, 135.

24. Hilary Koprowski, "A Condensed Version of an Unpublished Lecture, 'Frontiers of Virology: Development of Vaccines Against Polio Virus,' at the Medical School of the Hershey Medical Center, June 18, 1980," in Koprowska, *A Woman Wanders*, 298.

25. James W. Trent Jr., *Inventing the Feeble Mind: A History of Mental Retardation in the United States* (Berkeley, Los Angeles, and London: University of California Press, 1994), Kindle e-book, location 3312 of 5061.

26. Records of the Office of Scientific Research and Development, Committee on Medical Research, Contractor Records, C 450, R L2, Bimonthly Progress Report, PI Alf S. Alving (University of Chicago), August 1, 1944, cited in David J. Rothman, *Strangers at the Bedside: A History of How Law and Bioethics Transformed Medical Decision Making* (New York: Aldine de Gruyter, 1991, 2003), 36.

27. Ibid., 38; Werner Henle et al., "Experiments on Vaccination of Human Beings Against Epidemic Influenza," *Journal of Immunology* 53 (1946): 75–93.

28. Thomas Francis Jr. et al., "Protective Effect of Vaccination Against Induced Influenza A," *Journal of Clinical Investigation* 24, no. 4 (1945): 536–46; Jonas E. Salk et al., "Protective Effect of Vaccination Against Induced Influenza B," *Journal of Clinical Investigation* 24, no. 4 (1945): 547–53.

29. Koprowski, "Condensed Version of an Unpublished Lecture," 298.

30. Koprowski, Jervis, and Norton, "Immune Responses in Human Volunteers," 109.

31. Oshinsky, *Polio*, 136.

32. Koprowski, "Condensed Version of an Unpublished Lecture," 299.

33. Vaughan, *Listen to the Music*, 15–16.

34. Koprowski, Jervis and Norton, "Immune Responses in Human Volunteers," 125.

35. Norman Topping (with Gordon Cohn), *Recollections* (Los Angeles: University of California Press, 1990), 139–41.

36. Leonard Warren, "A History of the Wistar Institute of Anatomy and Biology" (unpublished manuscript, March 25, 2014), 121–22.

37. Vaughan, *Listen to the Music*, 77–78; Warren, "History of the Wistar Institute," 122.

38. William DuBarry, president of the Wistar Institute board of managers, to Hilary Koprowski, January 24, 1957, UPA4 (Office of the President Records), box 114, folder 19 (Wistar Institute: Administrative Appointments, 1955–1960), University Archives and Records Center, University of Pennsylvania.

39. Vaughan, *Listen to the Music*, 77.

40. Ibid., 81.

41. Ibid., 63–64.

42. Ibid., 79.

43. Warren, "History of the Wistar Institute," 127.

44. Wistar Institute of Anatomy and Biology, *Biennial Report, 1958–1959*, 9–11, Wistar Institute Archives, Philadelphia, PA.

45. Ibid., 24.

46. Elsa M. Zitcer, Jørgen Fogh, and Thelma H. Dunnebacke, "Human Amnion Cells for Large-Scale Production of Poliovirus," *Science* 122, no. 3157 (1955): 30; Jørgen Fogh and R. O. Lund, "Continuous Cultivation of Epithelial Cell Strain (FL) from Human Amniotic Membrane," *Proceedings of the Society for Experimental Biology and Medicine* 94, no. 3 (1957): 532–37; Thelma H. Dunnebacke and Elsa M. Zitcer, "Transformation of Cells from the Normal Human Amnion into Established Strains," *Cancer Research* 17 (1957): 1047–53.

47. Leonard Hayflick, "The Establishment of a Line (WISH) of Human Amnion Cells in Continuous Cultivation," *Experimental Cell Research* 23, no. 1 (1961): 14–20.

48. Stanley M. Gartler, *National Cancer Institute Monograph* 26, no. 167 (1967); Stanley M. Gartler, "Apparent HeLa Cell Contamination of Human Heteroploid Cell Lines," *Nature* 217 (1968): 750–51.

49. Wistar Institute of Anatomy and Biology, *Biennial Report, 1958–1959*, 24; Leonard Hayflick, telephone interview with the author, June 9, 2014.

Chapter Four: Abnormal Chromosomes and Abortions

1. Robert E. Hall, "Abortion in American Hospitals," *American Journal of Public Health* 57, no. 11 (1967): 1933.

2. Wistar Institute of Anatomy and Biology, *Biennial Report: 1958–1959*, 25, Wistar Institute Archives, Philadelphia, PA.

3. D. A. Rigoni-Stern, "Fatti Statistici Relativi alle Malattie Cancerose Che Servirono di Base Alle Poche Cose Dette dal Dott.," *G Servire Progr Path Tera* 2 (1842): 507–17, Daniel DiMaio, "Nuns, Warts, Viruses and Cancer," *Yale Journal of Biology and Medicine* 88, no. 2 (2015): 127.

4. V. Ellerman and O. Bang, "Experimentelle Leukämie bei Hühnern," *Zentralbl Bakteriol Parasitenkd Infectionskr Hyg Abt I* 46 (1908): 595–609. For an English-language description of Ellerman and Bang's paper, see Robin A. Weiss and Peter K. Vogt, "One Hundred Years of Rous Sarcoma Virus," *Journal of Experimental Medicine* 208, no. 12 (2011): 2352, www.ncbi.nlm.nih.gov/pmc/articles/PMC3256973/#bib12.

5. Peyton Rous, "A Sarcoma of the Fowl Transmissible by an Agent Separable from the Tumor Cells," *Journal of Experimental Medicine* 13, no. 4 (1911): 397–411, www.ncbi.nlm.nih.gov/pmc/articles/PMC2124874/pdf/397.pdf.

6. George Klein, "The Strange Road to the Tumor-Specific Transplantation Antigens," *Cancer Immunity* 1 (2001): 6, http://cancerimmunolres.aacrjournals.org/content/canimmarch/1/1/6.full.

7. "Medicine: Cornering the Killer," *Time*, July 27, 1959, http://content.time.com/time/magazine/article/0,9171,864777,00.html.

8. Hilary Koprowski, "The Viral Concept of the Etiology of Cancer," undated article from unidentified journal, c. 1960, p. 57, UPT 50 R252, box 22, file folder 24, "Professional Correspondence, Kop-Kos," Isidor Schwaner Ravdin Papers, University Archives and Records Center, University of Pennsylvania; Wistar Institute of Anatomy and Biology, *Biennial Report: 1958–1959*, 25, Wistar Institute Archives, Philadelphia, PA.

9. Joe Hin Tjio and Albert Levan, "The Chromosome Number of Man," *Hereditas* 42 (1956): 1–6, http://onlinelibrary.wiley.com/doi/10.1111/j.1601-5223.1956.tb03010.x/epdf.

10. T. C. Tsu, Charles S. Pomerat, and Paul S. Moorhead, "Mammalian Chromosomes in

Vitro VIII: Heteroploid Transformation in the Human Cell Strain Mayes," *Journal of the National Cancer Institute* 19 (1957): 867–73; R. S. Chang, "Continuous Subcultivation of Epithelial-like Cells from Normal Human Tissues," *Proceedings of the Society for Experimental Biology and Medicine* 87, no. 2 (1954): 440–43; J. Leighton et al., "Transformation of Normal Human Fibroblasts into Histologically Malignant Tissue in Vitro," *Science* 123, no. 3195 (1956): 502–3; L. Hayflick and P. S. Moorhead, "Cell Lines from Non-neoplastic Tissue," in P. L. Altman and D. S. Dittmer, eds., *Growth* (Bethesda, MD: Federation of American Societies for Experimental Biology and Medicine, 1962), 156–60.

11. D. von Hansemann, "Ueber asymmetrische Zellteilung in Epithelkrebsen und deren biologische Bedeutung," *Virchows Arch Path Anat Physiol Klin Med* 119 (1890): 299–326.

12. Henry Harris, trans. and ann., "'Concerning the Origin of Malignant Tumours' by Theodor Boveri," *Journal of Cell Science* 121 (2008): S1–S84. This is a translation of Boveri's original paper.

13. Peter C. Nowell and David A. Hungerford, "A Minute Chromosome in Human Chronic Granulocytic Leukemia," *Science* 132 (1960): 1497.

14. J. H. Tjio and Theodore T. Puck, "Genetics of Somatic Mammalian Cells II: Chromosomal Constitution of Cells in Tissue Culture," *Journal of Experimental Medicine* 108 (1958): 261.

15. Rachel Benson Gold, "Lessons from Before Roe: Will Past Be Prologue?" *Guttmacher Report on Public Policy* 6, no. 1 (2003): 9.

16. *Purdon's Pennsylvania Statutes and Consolidated Statutes*, Title 18, Crimes and Offenses, Chapter 2: The Penal Code of 1939, Article VII: Offenses Against the Person, Section 718: Abortion.

17. Ibid., Section 719: Abortion Causing Death.

18. *Commonwealth v. Zimmerman*, 214 Pa. Super. 61, 251 A.2d 819 (1969); *Commonwealth v. King*, No. 37, March Term 1968, Pa. Court of Common Pleas (Allegheny County, Criminal Division); Janet M. LaRue, "Is Kate Michelman Telling the Truth About Her Own Abortion Story," *Human Events*, January 24, 2006, http://humanevents.com/2006/01/24/is-kate-michelman-telling-the-truth-about-her-own-abortion-story/.

19. Leslie J. Reagan, *When Abortion Was a Crime: Women, Medicine and Law in the United States, 1867–1973* (Berkeley and Los Angeles: University of California Press, 1997): 13, 61–70.

20. Ibid., 15, 132, 135.

21. Henry J. Sangmeister, "A Survey of Abortion Deaths in Philadelphia from 1931 to 1940 Inclusive," *American Journal of Obstetrics and Gynecology* 46 (1943): 756–57.

22. Reagan, *When Abortion Was a Crime*, 15.

23. R. K. Denworth (Drinker, Biddle and Reath) to I. S. Ravdin, September 28, 1962, page 1, series UPC 1.4, box 14, file folder "Dept. of Ob/Gyn," University Archives and Records Center, University of Pennsylvania.

24. Name withheld, telephone interview with the author, June 30, 2014.

25. Department of Obstetrics and Gynecology, "Rules Governing Requests for Therapeutic Abortion and/or Sterilization," January 21, 1963, series UPC 1.4, box 14, file folder "Dept. of Ob/Gyn," University Archives and Records Center, University of Pennsylvania.

26. Reagan, *When Abortion Was a Crime*, 204–7.

27. Hall, "Abortion in American Hospitals," 1933.

28. Betsy Meredith, telephone interview with the author, July 16, 2014.

29. Photograph of I. S. Ravdin and Dwight D. Eisenhower, UPT 50 R252, box 182, file folder 13, Isidor Schwaner Ravdin Papers, University Archives and Records Center, University of Pennsylvania.

30. John M. Mitchell to I. S. Ravdin, October 29, 1952, series UPC 1.4, box 1, file folder "1952–Marriage Council of Philadelphia, Contract with University of Pennsylvania," University Archives and Records Center, University of Pennsylvania.

31. "The Roman Catholic Church and the Hospital of the University of Pennsylvania," un-dated, series UPC 1.4, box 2, file folder "Department of Ob-Gyn," University Archives and Records Center, University of Pennsylvania.

32. I. S. Ravdin to Franklin Payne, July 15, 1960, series UPC 1.4, box 2, file folder "Depart-ment of Ob-Gyn," University Archives and Records Center, University of Pennsylvania.

33. Photograph of I. S. Ravdin and Pope Pius XII, UPT 50 R252, box 182, file folder 56, Isidor Schwaner Ravdin Papers, University Archives and Records Center, University of Pennsylvania.

34. I. S. Ravdin to Mrs. Roland T. Addis, December 24, 1964, series UPC 1.4, box 15, file folder "VPMA," University Archives and Records Center, University of Pennsylvania.

35. I. S. Ravdin's growing distrust of Hilary Koprowski is on display in his correspondence from the period. "Dr. Koprowski may be a distinguished person, but I have previously ques-tioned his ethics and I am questioning them again," Ravdin wrote to University of Pennsyl-vania president Gaylord P. Harnwell on April 8, 1960, after Koprowski reneged on funding he had told the Wistar's board of managers that he would use to support a scientist that he had recruited to the Wistar from England. (See I. S. Ravdin to Gaylord Harnwell, April 8, 1960, UPA 4, box 114, file folder 17, "Wistar Institute IV 1955–1960," Office of the President Records, Gaylord Probasco Harnwell Administration, University Archives and Records Center, University of Pennsylvania.)

36. I. S. Ravdin to Margaret Reed Lewis, January 11, 1947, UPT 50 R252, box 68, file folder 15, Isidor Schwaner Ravdin Papers, University Archives and Records Center, University of Pennsylvania.

37. I. S. Ravdin to William F. McLimans, March 21, 1956, UPT 50 R252, box 68, file folder 15, Isidor Schwaner Ravdin Papers, University Archives and Records Center, University of Pennsylvania.

38. Hilary Koprowski, "The Viral Concept of the Etiology of Cancer" (publication title unknown), page 57, UPT 50 R252, box 22, file folder 24, Isidor Schwaner Ravdin Papers, University Ar-chives and Records Center, University of Pennsylvania; Betsy Meredith, telephone interview.

39. Leonard Hayflick, e-mail to the author, July 18, 2014.

40. Hugo Lagercrantz and Jean-Pierre Changeux, "The Emergence of Human Consciousness: From Fetal to Neonatal Life," *Pediatric Research* 65 (2009): 255–60, www.nature .com/pr/journal/v65/n3/full/pr200950a.html.

41. Leonard Hayflick, interview with the author, October 3, 2012.

42. Leonard Hayflick and Paul S. Moorhead, "The Serial Cultivation of Human Diploid Cell Strains," *Experimental Cell Research* 25, no. 3 (1961): 587.

43. Ibid., 588.

44. Ibid.

45. Wistar Institute of Anatomy and Biology, *Biennial Report: 1958–1959*, cover and "Contents" page.

46. Ibid., 26.

Chapter Five: Dying Cells and Dogma

1. Jules Verne, *Journey to the Centre of the Earth* (1864; New York: Bantam/Dell, Bantam Classic Reissue, May 2006), 143.

2. Alexis Carrel, "On the Permanent Life of Cells Outside the Organism," *Journal of Experimental Medicine* 15 (1912): 516–30, www.ncbi.nlm.nih.gov/pmc/articles/ PMC2124948/pdf/516.pdf.

3. "Dr. Carrel's Miracles in Surgery Win the Nobel Prize," *New York Times*, October 13, 1912, pp. 78–79. http://timesmachine.nytimes.com/timesmachine/1912/10/13/100594519 .html?pageNumber=78.

4. "Isolated Tissue Holds Life 12 Years in Test," *New York Tribune*, January 6, 1924.

5. "Medicine: Carrel's Man," *Time*, September 16, 1935; Alexis Carrel, *Man, the Unknown* (New York: Harper & Brothers, 1935).

6. Wilton R. Earle et al., "Production of Malignancy in Vitro IV: The Mouse Fibroblast Cultures and Changes Seen in the Living Cells," *Journal of the National Cancer Institute* 4 (1943): 165–69.

7. Leonard Hayflick and Paul S. Moorhead, "The Serial Cultivation of Human Diploid Cell Strains," *Experimental Cell Research* 25, no. 3 (1961): 591.

8. Leonard Hayflick, interview with the author, November 19, 2014.

9. John F. Kennedy to American Association of Newspaper Editors, Washington, DC, April 21, 1960, Papers of John F. Kennedy. Prepresidential Papers. Senate Files. Series 12.1. Speech Files, 1953–1960. Box 908, Folder "American Society of Newspaper Editors, Washington, DC, http://www.jfklibrary.org/Research/Research-Aids/JFK-Speeches/American-Society-of-Newspaper-Editors_19600421.aspx.

10. Paul Moorhead, interview with the author, November 14, 2012.

11. Hayflick and Moorhead, "Serial Cultivation," 601.

12. Ibid.; Leonard Hayflick, "Citation Classics," *Current Contents* 26 (1978): 144.

13. Leonard Hayflick, interview with the author, October 4, 2012.

14. Leonard Hayflick, *How and Why We Age* (New York: Ballantine Books, 1994), 127–29.

15. Associated Press, "Pneumonia Study Points to Vaccine: U.S. Scientists Isolate Agent That Causes a Prevalent Form of the Disease," *New York Times,* January 23, 1962, p. 1.

16. Hayflick and Moorhead, "Serial Cultivation," 619.

17. Ibid., 616–19.

18. Ibid., 605.

19. Ibid., 600–601.

20. Ibid., 589.

21. Ibid., 607–8.

22. Chester M. Southam, Alice E. Moore, and Cornelius P. Rhoads, "Homotransplantation of Human Cell Lines," *Science* 125 (1957): 158.

23. Elinor Langer, "Human Experimentation: Cancer Studies at Sloan-Kettering Stir Public Debate on Medical Ethics," *Science* 143 (1964): 552; Robert D. Mulford, "Experimentation on Human Beings," *Stanford Law Review* 20 (1967): 110.

24. Mulford, "Experimentation on Human Beings," 100; Rebecca Skloot, *The Immortal Life of Henrietta Lacks* (New York: Crown, 2010), 135.

25. Southam, Moore, and Rhoads, "Homotransplantation of Human Cell Lines," 158.

26. Hayflick and Moorhead, "Serial Cultivation," 608–9.

27. Steve Hale, "Noted Cancer Researcher Urges 'Imaginative' Tests on Human," *Deseret News and Salt Lake Telegram*, March 23, 1964, p. B1.

28. William Elkins, interview with the author, June 25, 2014.

29. Leonard Hayflick et al., "Preparation of Poliovirus Vaccines in a Human Fetal Diploid Cell Strain," *American Journal of Hygiene* 75 (1962): 250.

30. Ibid.

31. Hayflick and Moorhead, "Serial Cultivation," 609.

32. World Health Organization Scientific Working Group on the Human Diploid Cell, "Report to the Director-General," Geneva, July 16–18, 1962, 14.

33. Leonard Hayflick, "The Limited *in Vitro* Lifetime of Human Diploid Cell Strains," *Experimental Cell Research* 37 (1965): 628–29.

34. Ibid., 631.

35. Peyton Rous, "A Sarcoma of the Fowl Transmissible by an Agent Separable from the Tumor Cells," *Journal of Experimental Medicine* 13, no. 4 (1911): 397–411.

36. Paul Moorhead, interview.

37. Leonard Hayflick, interview with the author, November 19, 2014.

38. J. H. Tjio and Theodore T. Puck, "Genetics of Somatic Mammalian Cells II: Chromosomal Constitution of Cells in Tissue Culture," *Journal of Experimental Medicine* 108 (1958): 260–62.

39. Leonard Hayflick, interview with the author, October 4, 2012.
40. Peyton Rous to Hilary Koprowski, April 24, 1961. Courtesy of Leonard Hayflick.

Chapter Six: The Swedish Source

1. David Maraniss, *Barack Obama: The Story* (New York: Simon & Schuster, 2012), xxiii.
2. Leonard Hayflick and Paul S. Moorhead, "The Serial Cultivation of Human Diploid Cell Strains," *Experimental Cell Research* 25, no. 3 (1961): 585–621.
3. A. M. Rosenthal, "Red Tape Tangles India's Monkeys: Policy Shift Strands 5,000 Animals at Airport on Way Abroad for Medical Use," *New York Times*, March 2, 1958, p. 26.
4. Leonard Hayflick, interview with the author, October 3, 2012.
5. Erling Norrby, *Nobel Prizes and Life Sciences* (Singapore: World Scientific, 2010), 70, and 123.
6. Per Axelsson, "The Cutter Incident and the Development of a Swedish Polio Vaccine, 1952–1957," *Dynamis* 32, no. 2 (2012): 312.
7. Erik Lycke, telephone interview with the author, October 24, 2013; Erling Norrby, interview with the author, September 20, 2013.
8. World Health Organization, "Report to the Director-General," July 24, 1962, MHO/PA/140.62 (World Health Organization Scientific Group on the Human Diploid Cell, Geneva, July 16–18, 1962); Erik Lycke, telephone interview.
9. Erik Lycke, telephone interview.
10. Hayflick and Moorhead, "Serial Cultivation," 588.
11. Eva Herrström, interview with the author, September 26, 2013.
12. Eva Herrström, diary, April 24, 1961. Courtesy of Eva Herrström.
13. Ibid., May 2, 1961.
14. Leonard Hayflick et al., "Preparation of Poliovirus Vaccines in a Human Fetal Diploid Cell Strain," *American Journal of Hygiene* 75 (1962): 245.
15. Wistar Institute of Anatomy and Biology, *Biennial Research Report 1962–1963*, 22, Wistar Institute Archives, Philadelphia, PA.
16. Department of Health Education, and Welfare, Public Health Service, NIH, Contract No: PH43-62-157, Feb., 6, 1962, 1. Courtesy of Leonard Hayflick.
17. Ibid., 1; "Wistar Institute Comparative Balance Sheets as of 4/30/66," p. 4, UPT 50 R252, box 68, file folder 14, "Wistar Institute, 1966," Isidor Schwaner Ravdin Papers, University Archives and Records Center, University of Pennsylvania; "Wistar Institute Comparative Balance Sheets as of 10/31/67," pages 2 and 4, UPT 50 R252, box 68, file folder 12, "Wistar Institute, 1966–1967," Isidor Schwaner Ravdin Papers, University Archives and Records Center, University of Pennsylvania.
18. "The Wistar Institute Board of Managers Meeting, June 19, 1962," page 1, series UPC 4.1 VPMA, box 14, file folder "Wistar Institute," University Archives and Records Center, University of Pennsylvania.
19. National Institutes of Health, Contract no: PH43-62-157, Section 30: "Termination for the Convenience of Government," part (g), p. HEW-315-6.
20. This and other personal information about Eva Ernholm is from interviews with Elisabeth, Håkan, and Lars Ernholm and with Osborne Carlson on September 21 and 22, 2013, and from documents provided by Elisabeth Ernholm.
21. "Eva Ernholm: Bättre Upplysning—Färre Aborter Hävdar Ung Kristinehamnsgynekolog" ("Eva Ernholm: Better Education—Fewer Abortions, Claims Young Kristinehamn Gynecologist"), *Nya Wermlandstidningens* (Karlstad, Sweden), October 5, 1963. Original article in Swedish; translation by Lisa Tallroth.
22. Per Gunnar Cassel, "Induced Legal Abortion in Sweden During 1939–1974: Change in Practice and Legal Reform," Working Paper 2009:1 (Stockholm: Stockholm University Demography Unit Department of Sociology, 2009), 4–5.
23. "Swedish Medical Association: Rules for Physicians (Codex Ethicus Medicorum Svecorum)," *Swedish Medical Journal* 49, no. 1 (1951): 1–3.

24. Lena Lennerhed, *Historier om ett brott* (Stockholm: Bokförlaget Atlas, 2008), cited in English in Cassel, "Induced Legal Abortion in Sweden," 5.

25. Hayflick and Moorhead, "Serial Cultivation," 604.

26. This calculation is based on Merck's annual use of WI-38 cells for making rubella vaccine. The company begins making the vaccine with about 120 million cells with population-doubling levels in the low 20s.

27. Hayflick and Moorhead, "Serial Cultivation," 604.

28. "Phoenix Abortion Ruling Delayed," *New York Times*, July 28, 1962.

29. Eero Saksela and Paul S. Moorhead, "Aneuploidy in the Degenerative Phase of Serial Cultivation of Human Cell Strains," *Proceedings of the National Academy of Sciences* 50 (1963): 390.

30. J. P. Jacobs, C. M. Jones, and J. P. Baille, "Characteristics of a Human Diploid Cell Designated MRC-5," *Nature* 227 (1970): 169.

31. Riseberg Memo; Factual Chronology, 5–6.

32. Factual Chronology, 5–6; Hilary Koprowski to Dr. W. C. Cockburn, World Health Organization, October 6, 1962, 1, investigations 9, folder 4, Directors' Files, Office of the Director, National Institutes of Health.

33. Factual Chronology, 5–6.

34. D. G. Evans to Ronald Lamont-Havers, August 20, 1975, attachment, 1–2, investigations 9, folder 1, Directors' Files, Office of the Director, National Institutes of Health.

35. Schriver Report, 3.

36. Leonard Hayflick letter to the author, October 27, 2015.

Chapter Seven: Polio Vaccine "Passengers"

1. Edward Shorter, "The Health Century Oral History Collection," interviewee: Dr. Bernice Eddy, December 4, 1986, page 1, transcript at the National Library of Medicine, National Institutes of Health.

2. Ibid., 12–15.

3. J. A. Morris et al., Memo to Robert Q. Marston (director of the National Institutes of Health), September 27, 1971. This memo is reprinted in Senate Subcommittee on Executive Reorganization and Government Research of the Committee on Government Operations, *Consumer Safety Act of 1972: Hearings on Titles I and II of S. 3419*, 92nd Cong., 2nd sess., April 20 and 21 and May 3 and 4, 1972, 519, 779.

4. Shorter, "Health Century," Bernice Eddy interview, 3–4.

5. Neal Nathanson and Alexander Langmuir, "The Cutter Incident: Poliomyelitis Following Formaldehyde-Inactivated Poliovirus Vaccination in the United States During the Spring of 1955 II: The Relationship of Poliomyelitis to Cutter Vaccine," *American Journal of Hygiene* 78 (1963): 39.

6. Paul A. Offit, "The Cutter Incident, 50 Years Later," *New England Journal of Medicine* 352 (2005): 1411.

7. David M. Oshinsky, *Polio: An American Story* (New York: Oxford University Press, 2005), 237–238.

8. Ibid., 519.

9. Bernice E. Eddy and Sarah E. Stewart, "Characteristics of the SE Polyoma Virus," *American Journal of Public Health* 49, no. 11 (1959): 1492.

10. Hilary Koprowski, "Live Poliomyelitis Virus Vaccines: Present Status and Problems for the Future," *Journal of the American Medical Association* 178, no. 12 (1961): 1153–54.

11. Robert N. Hull, James R. Minner, and Carmine C. Mascoli, "New Viral Agents Recovered from Tissue Cultures of Monkey Kidney Cells III: Recovery of Additional Agents Both from Cultures of Monkey Tissues and Directly from Tissues and Excreta," *American Journal of Hygiene* 68 (1958): 40.

12. Bernice E. Eddy et al., "Tumors Induced in Hamsters by Injection of Rhesus Monkey

Kidney Cell Extracts," *Proceedings of the Society for Experimental Biology and Medicine* 107, no. 1 (1961): 191–97.

13. Ibid., 193.

14. Leo Morris et al., "Surveillance of Poliomyelitis in the United States, 1962–65," *Public Health Reports* 82, no. 5 (1967): 418.

15. Debbie Bookchin and Jim Schumacher, *The Virus and the Vaccine: Contaminated Vaccine, Deadly Cancers and Government Neglect* (New York: St. Martin's, 2004), 71–73. This book renders the SV40 story in far greater detail and is recommended for interested readers.

16. B. H. Sweet and M. R. Hilleman, "The Vacuolating Virus SV40," *Proceedings of the Society for Experimental Biology and Medicine* 105 (1960): 420–27, www.ncbi.nlm.nih.gov/pubmed/13774265. This paper is accessible at pages 561–68 of Senate Subcommittee on Executive Reorganization and Government Research of the Committee on Government Operations, *Consumer Safety Act of 1972: Hearings*.

17. Bernice E. Eddy, memo to Joseph Smadel, "The Presence of an Oncogenic Substance or Virus in Monkey Kidney Cell Cultures," July 6, 1960, Senate Subcommittee on Executive Reorganization and Government Research of the Committee on Government Operations, *Consumer Safety Act of 1972: Hearings*, 551.

18. Joseph E. Smadel, memo to Bernice Eddy, "Requirements for Outside Lectures," October 24, 1960, Senate Subcommittee on Executive Reorganization and Government Research of the Committee on Government Operations, *Consumer Safety Act of 1972: Hearings*, 549–50.

19. Shorter, "Health Century," Bernice Eddy interview, 9.

20. Sweet and Hilleman, "Vacuolating Virus SV40," 425–26.

21. Ibid.

22. H. Koprowski, "Tin Anniversary of the Development of Live Virus Vaccine," *Journal of the American Medical Association* 174 (1960): 975.

23. Hilary Koprowski and Stanley A. Plotkin, "Notes on Acceptance Criteria and Requirements for Live Poliovirus Vaccines" (World Health Organization Study Group on Requirements for Poliomyelitis Vaccine [Live, Attenuated Poliovirus], Geneva, November 7–12, 1960), 8–9. WHO/BS/IR/85/1 November 1960.

24. Ibid., 10.

25. Leonard Hayflick et al., "Preparation of Poliovirus Vaccines in a Human Fetal Diploid Cell Strain," *American Journal of Hygiene* 75 (1962): 240–58.

26. Alan P. Goffe, James Hale, and P. S. Gardner, "Letter to the Editor," *Lancet* 277, no. 7177 (1961): 612.

27. D. I. Magrath, Kate Russell, and J. O. Tobin, "Preliminary Communications: Vacuolating Agent," *British Medical Journal* 2, no. 5247 (1961): 287, http://www.ncbi.nlm.nih.gov/pmc/articles/PMC1969653/.

28. Hilary Koprowski letter to Representative Kenneth Roberts, Chairman, House Health and Safety Subcommittee, House Interstate and Foreign Commerce Committee, April 14, 1961, in *Polio Vaccines: Hearings Before a Subcommittee of the Committee on Interstate and Foreign Commerce of the House of Representatives, Eighty-Seventh Congress, First Session, on Developments with Respect to the Manufacture of Live Virus Polio Vaccine and Results of Utilization of Killed Virus Polio Vaccine, March 16–17, 1961*, 311–12.

29. S. A. Plotkin, H. Koprowski, and J. Stokes Jr., "Clinical Trials in Infants of Orally Administered Attenuated Poliomyelitis Viruses," *Pediatrics* 23, no. 6 (1959): 1060.

30. Keerti Shah and Neal Nathanson, "Human Exposure to SV40: Review and Comment," *American Journal of Epidemiology* 103, no. 1 (1976): 5.

31. Eleanor Roosevelt, "My Day," June 25, 1956. Reprinted by The Eleanor Roosevelt Papers Project, papers.columbian.gwu.edu.

32. Roger Vaughan, *Listen to the Music: The Life of Hilary Koprowski* (New York: Springer-Verlag, 2000), 52; Edward Hooper, *The River: A Journey to the Source of HIV and AIDS* (Boston, New York, and London: Little, Brown, 1999), 406.

33. Clinton Farms birth records for 1960, courtesy of Edward Hooper.

34. Mary Q. Hawkes, *Excellent Effect: The Edna Mahan Story* (Arlington, VA: American Correctional Association, 1994), 109–10.

35. Plotkin, Koprowski, and Stokes, "Clinical Trials in Infants," 1061.

36. Ibid., 1041.

37. Hooper, *The River*, 424.

38. Hayflick et al., "Preparation of Poliovirus Vaccines," 251, 257.

39. Ibid., 253.

40. Leonard Hayflick, interview with the author, November 19, 2014.

41. Koprowski, "Live Poliomyelitis Virus Vaccines," 1154–55.

42. Associated Press, "2 Companies Halt Salk-Shot Output, Seek to Eliminate a Monkey Virus, Believed Harmless, Found in Some Vaccine," *New York Times*, July 26, 1961.

43. Bookchin and Schumacher, *Virus and the Vaccine*, 103–4.

44. "The Great Polio Vaccine Cancer Cover-up: Polio Shots May Kill You," *National Enquirer*, August 20, 1961, 1, 14–15, 25.

45. Hayflick et al., "Preparation of Poliovirus Vaccines," 253.

46. Ibid., 240–58.

47. June 19, 1961, draft of Hayflick et al., "Preparation of Poliovirus Vaccines," page 24, file folder "CHAT-WIHL," Stanley Plotkin private papers, Doylestown, PA.

48. Hayflick et al., "Preparation of Poliovirus Vaccines," 256.

49. Shorter, "Health Century," Bernice Eddy interview, 10.

50. Roderick Murray (director, Division of Biologics Standards), memo to Dr. Bernice Eddy through Dr. J. E. Smadel, February 16, 1961, in Senate Subcommittee on Executive Reorganization and Government Research of the Committee on Government Operations, *Consumer Safety Act of 1972: Hearings*, 592.

51. Smadel to Eddy, "Requirements for Outside Lectures," 549.

52. Shorter, "Health Century," Bernice Eddy interview, 7.

53. Ruth Kirschstein, oral history, part 2, interview by Victoria Harden and Caroline Hannaway, October 29, 1998, page 31, Office of NIH History, National Institutes of Health, Bethesda, Maryland.

54. Edward Shorter, "The Health Century Oral History Collection," interviewee: Maurice Hilleman, February 6, 1987, page 8, transcript available at the National Library of Medicine, National Institutes of Health.

55. Bernice Eddy et al., "Identification of the Oncogenic Substance in Rhesus Monkey Kidney Cell Cultures as Simian Virus 40," *Virology* 17 (1962): 65–75.

56. Hilary Koprowski et al., "Transformation of Cultures of Human Tissue Infected with Simian Virus SV40," *Journal of Cellular and Comparative Physiology* 59, no. 3 (1962): 281–92, http://onlinelibrary.wiley.com/doi/10.1002/jcp.1030590308/abstract.

57. Harvey M. Shein and John F. Enders, "Transformation Induced by Simian Virus 40 in Human Renal Cell Cultures I: Morphology and Growth Characteristics," *Proceedings of the National Academy of Sciences* 48 (1962): 1164–72; Harvey M. Shein, John F. Enders, and Jeana D. Levinthal, "Transformation Induced by Simian Virus 40 in Human Renal Cell Cultures II: Cell-Virus Relationships," *Proceedings of the National Academy of Sciences* 48 (1962): 1350–57.

58. Shah and Nathanson, "Human Exposure to SV40," 3.

Chapter Eight: Trials

1. John Cardinal O'Hara (archbishop of Philadelphia) to Sister M. Jacob, June 26, 1959, Accession R1990.004, Chancery Files, Philadelphia Archdiocesan Historical Research Center.

2. James A. Poupard, interviews with the author, March 20 and November 19, 2014.

3. Lisa Levenstein, *A Movement Without Marches: African American Women and the Politics*

of Poverty in Postwar Philadelphia (Chapel Hill: University of North Carolina Press, 2009), 157–80.

4. Ibid., 167; Donna Gentile O'Donnell, *Provider of Last Resort: The Story of the Closure of the Philadelphia General Hospital* (Philadelphia: Camino Books, 2005): 96.

5. Levenstein, *A Movement Without Marches*, 166, 169, 175–76.

6. Ibid., 165; "Ten Year Report: Philadelphia General Hospital," October 1961, pages 1–10, collection 80-101.2 ("Philadelphia General Hospital, Ten Year Report, 1952–1962"), Philadelphia City Archive.

7. Ibid., 1.

8. Ibid., 4.

9. Joseph S. Pagano et al., "The Response of Premature Infants to Infection with Attenuated Poliovirus," *Journal of Pediatrics* 29, no. 5 (1962): 794–807; Joseph S. Pagano, Stanley A. Plotkin, and Donald Cornely, "The Response of Premature Infants to Infection with Type 3 Attenuated Poliovirus," *Journal of Pediatrics* 65, no. 2 (1964): 165–75.

10. Stanley Plotkin, e-mail to the author, July 20, 2016.

11. Pagano, Plotkin, and Cornely, "Response of Premature Infants," 174.

12. Ibid., 807.

13. Ibid., 806.

14. Stanley Plotkin, interview with the author, May 25, 2015.

15. Pagano, Plotkin, and Cornely, "Response of Premature Infants," 794.

16. Leo Morris et al., "Surveillance of Poliomyelitis in the United States, 1962–65," *Public Health Reports* 82, no. 5 (1967): 419.

17. James A. Poupard, interview with the author, March 20, 2014.

18. Katherine Auchy, St. Vincent's Hospital for Women and Children: Report of Inspection and Evaluation, Department of Public Welfare of the Commonwealth of Pennsylvania, March 1963, p. 2, Cardinal Krol Papers, Philadelphia Archdiocesan Historical Research Center; Ruth McClain, St. Vincent's Hospital for Women and Children: Report of Inspection and Evaluation, Department of Public Welfare of the Commonwealth of Pennsylvania, April 1966, p. 3, Cardinal Krol Papers, Philadelphia Archdiocesan Historical Research Center.

19. Auchy, St. Vincent's Hospital for Women and Children: Report of Inspection and Evaluation.

20. The material on life for the girls and women at St. Vincent's Hospital for Women and Children was drawn from: Dee Krewson, telephone interview with the author, May 5, 2014; name withheld, telephone interview with the author, April 12, 2014; Fran Scalise, telephone interviews with the author, April 14 and 15, 2014.

21. McClain, Report of Inspection and Evaluation, Department of Public Welfare of the Commonwealth of Pennsylvania, April 1966, p. 5.

22. Sister Mary Jacob (administrator) to Joseph Stokes, June 18, 1959, accession R1990.004, Chancery Files, Philadelphia Archdiocesan Historical Research Center.

23. Sister Mary Jacob to John Cardinal O'Hara, July 1, 1959, accession R1990.004, Chancery Files, Philadelphia Archdiocesan Historical Research Center.

24. "Milestones, Sep. 5, 1960," *Time*, September 5, 1960, http://content.time.com/time/sub scriber/article/0,33009,826583,00.html.

25. O'Hara to Jacob, June 26, 1959.

26. Hilary Koprowski to John Cardinal O'Hara, March 7, 1960, accession R1990.004, Chancery Files, Philadelphia Archdiocesan Historical Research Center.

27. Sister Mary Jacob to John Francis Cardinal O'Hara, March 17, 1960, accession R1990.004, Chancery Files, Philadelphia Archdiocesan Historical Research Center.

28. John Cardinal O'Hara to Sister Mary Jacob, March 22, 1960, accession R1990.004, Chancery Files, Philadelphia Archdiocesan Historical Research Center.

29. James Poupard, interview with the author, March 20, 2014.

30. T. W. Norton, Richard Carp, and Stanley A. Plotkin, "Summary of Feeding Results with

Attenuated Polioviruses Grown in Human Diploid Cell Strains," Virus Diseases/WP/6, July 5, 1962, 1–2 (World Health Organization Scientific Group on the Human Diploid Cell, Geneva, July 16–18, 1962).

31. "Ruth L. Kirschstein oral history interview, part 2," Victoria Harden and Caroline Hannaway, October 29, 1998, p. 5, Office of NIH History, Oral History Archive, Bethesda, MD, https://history.nih.gov/archives/oral_histories.html; Nicholas Wade, "Division of Biologics Standards: The Boat That Never Rocked," *Science* 175 (1972): 1226.

32. Leonard B. Seeff et al., "A Serologic Follow-up of the 1942 Epidemic of Post-vaccination Hepatitis in the United States Army," *New England Journal of Medicine* 316 (1987): 966.

33. John Farley, "*To Cast Out Disease: A History of the International Health Division of the Rockefeller Foundation (1913–1951)* (New York: Oxford University Press USA, 2004)173, 176.

34. Roderick Murray et al., "Hepatitis Carrier State II: Confirmation of Carrier State by Transmission Experiments in Volunteers," *Journal of the American Medical Association* 154, no. 13 (1954): 1072–74.

35. David J. Rothman, *Strangers at the Bedside: A History of How Law and Bioethics Transformed Medical Decision Making* (New York: Aldine de Gruyter, 1991, 2003), 30–69.

36. William L. Laurence, "Drugs to Combat Malaria Are Tested in Prisons for Army," *New York Times*, March 5, 1945, 1, 30.

37. Rothman, *Strangers at the Bedside*, 33–34.

38. Ibid., 38.

39. Ibid., 51–60.

40. Murray et al., "Hepatitis Carrier State II," 1072,

41. "Kirschstein oral history interview, part 2," October 29, 1998, p. 4.

42. John Finlayson, telephone interview with the author, June 24, 2016.

43. Kirschstein oral history interview, part 2," p. 4.

44. "Continuously Cultured Tissue Cells and Viral Vaccines: Potential Advantages May Be Realized and Potential Hazards Obviated by Careful Planning and Monitoring: Report of a Committee on Tissue Culture Viruses and Vaccines," *Science* 139 (1963): 15–20.

45. Ibid., 17.

46. L. Hayflick et al., "Choice of a Cell System for Vaccine Production," *Science* 140 (1963): 766–68.

47. World Health Organization, "Report to the Director-General," MOH/PA/140.62, July 24, 1962, p. 5 (World Health Organization Scientific Group on the Human Diploid Cell, Geneva, July 16–18 1962).

48. Ibid., 18, 24.

49. Hayflick et al., "Choice of a Cell System," 768.

50. World Health Organization, "Report to the Director-General," 19, 24.

51. Ibid., 24.

52. Joseph S. Pagano et al., "The Response and the Lack of Spread in Swedish School Children Given an Attenuated Poliovirus Vaccine Prepared in a Human Diploid Cell Strain," *American Journal of Hygiene* 79 (1964): 83.

53. F. Buser et al., "Immunization with Live Attenuated Polio Virus Prepared in Human Diploid Cell Strains, with Special Reference to the WM-3 Strain," in *Proceedings, Symposium on the Characterization and Uses of Human Diploid Cell Strains* (Opatija: International Association of Microbiological Societies, 1963), 386.

54. Drago Ikić et al., "Postvaccinal Reactions After Application of Poliovaccine Live, Oral Prepared in Human Diploid Cell Strains Wi-38," *Proceedings, Symposium on the Characterization and Uses of Human Diploid Cell Strains* (Opatija: International Association of Microbiological Societies, 1963), 405, 406, 413.

55. Hilary Koprowski to Vre C. Mackowiak, December 11, 1963, folder "Outgoing Correspondence," Stanley Plotkin private papers, Doylestown, PA.

Chapter Nine: An Emerging Enemy

1. Norman McAlister Gregg, "Congenital Cataract Following German Measles in the Mother," *Transactions of the Ophthalmological Society of Australia* 3 (1941): 40.
2. P. M. Dunn, "Perinatal Lessons from the Past: Sir Norman Gregg, ChM, MC, of Sydney (1892–1966) and Rubella Embryopathy," *Archives of Disease in Childhood, Fetal and Neonatal Edition* 92, no. 6 (2007): F513–14, www.ncbi.nlm.nih.gov/pmc/articles/PMC2675410/.
3. Margaret Burgess, e-mail to the author, February 23, 2015; Margaret Burgess, "Gregg's Rubella Legacy 1941–1991," *Medical Journal of Australia* 155 (1991): 355.
4. Erwin Heinz Ackerknecht, *A Short History of Medicine*, rev. ed. (Baltimore and London: The Johns Hopkins University Press, 1982), 129. This was originally published in English by the Ronald Press Company, New York, 1955.
5. William George Maton, "Some Account of a Rash Liable to Be Mistaken for Scarlatina," *Medical Transactions Published by the College of Physicians of London* 5 (1815): 149–65.
6. Henry Veale, "History of an Epidemic of Rötheln, with Observations on Its Pathology," *Edinburgh Medical Journal* 12 (1866): 404–14.
7. Dorothy Horstmann, "Maxwell Finland Lecture: Viral Vaccines and Their Ways," *Reviews of Infectious Diseases* 1, no. 3 (1979): 510.
8. Alfred D. Heggie and Frederick C. Robbins, "Natural Rubella Acquired After Birth: Clinical Features and Complications," *American Journal of Diseases of Children* 118 (1969): 15.
9. Gregg, "Congenital Cataract," 42.
10. Burgess, "Gregg's Rubella Legacy," 355.
11. Gregg, "Congenital Cataract," 42.
12. Ibid., 35–46.
13. C. Swan et al., "Congenital Defects in Infants Following Infectious Diseases During Pregnancy: With Special Attention to the Relationship Between German Measles and Cataract, Deaf-Mutism, Heart Disease and Microcephaly, and to the Period of Pregnancy in Which Occurrence of Rubella Is Followed by Congenital Abnormalities," *Medical Journal of Australia* 2 (1943): 201–10.
14. "Rubella and Congenital Malformations," *Lancet* 1 (1944): 316.
15. "Congenital Defects Following Maternal Rubella," *Journal of the American Medical Association* 130 (1946): 574–75.
16. Morris Greenberg, Ottavio Pellitteri, and Jerome Barton, "Frequency of Defects in Infants Whose Mothers Had Rubella During Pregnancy," *Journal of the American Medical Association* 165, no. 6 (1957): 675–76.
17. Stella Chess, "Autism in Children with Congenital Rubella," *Journal of Autism and Childhood Schizophrenia* 1, no. 1 (1971): 33–47; Brynn E. Berger, Ann Marie Navar-Boggan, and Saad B. Omer, "Congenital Rubella Syndrome and Autism Spectrum Disorder Prevented by Rubella Vaccination: United States, 2001–2010," *BMC Public Health* 11 (2011): 340, www.ncbi.nlm.nih.gov/pmc/articles/PMC3123590/#B7.
18. John Fry, J. B. Dillane, and Lionel Fry, "Rubella, 1962," *British Medical Journal* 2, no. 5308 (1962): 833–34.
19. Elizabeth Miller, "Rubella in the United Kingdom," *Epidemiology and Infection* 107 (1991): 34; C. S. Peckham, "Congenital Rubella in the United Kingdom Before 1970: The Prevaccine Era," *Reviews of Infectious Diseases* 7 (supp. 1) (1985): S11–21.
20. "News in Brief," *Times* (London), Thursday, March 15, 1962, 6.
21. "German Measles at Eton," *Times* (London), Tuesday, March 27, 1962, 16.
22. Eric Todd, "Shackleton's Old Tricks Serve Him Well Against Yorkshire: Wilson Consoles the Partisans," *Guardian*, June 28, 1962, 10.
23. A Mother, "Difficult Duty," *Guardian*, August 9, 1963, 6.
24. Thomas H. Weller and Franklin A. Neva, "Propagation in Tissue Culture of Cytopathic

Agents from Patients with Rubella-Like Illness," *Proceedings of the Society for Experimental Biology and Medicine* 111 (October 1962): 216–25.

25. Paul D. Parkman, Edward L. Buescher, and Malcolm S. Artenstein, "Recovery of Rubella Virus from Army Recruits," *Proceedings of the Society for Experimental Biology and Medicine* 111 (1962): 225–30.

26. Lee Plotkin, *Anecdotes of My Life*, a self-published memoir (Lee Plotkin, Stanley Plotkin, and Brenda Magalaner, 1986), 1986, 38–45, 57–61, 66–71.

27. The biographical material on Stanley Plotkin comes from interviews I conducted with Plotkin from 2013 to 2015, dates of which are detailed in the selected bibliography; from *Anecdotes of My Life*, a 1986 memoir by Stanley Plotkin's mother, Lee Plotkin; and from Stanley A. Plotkin, "The Late Sequelae of Arrowsmith," *Pediatric Infectious Disease Journal* 21 (2002): 807–8.

28. Ibid., 808.

29. "Memorandum to Editors Concerning Press, Radio and Television Conference," June 15, 1959, UPA 4, box 114, file folder "Lederle Laboratories National Drug Company (1955–1960)," Office of the President Records, Gaylord P. Harnwell Admin. 1955–1960, University Archives and Records Center, University of Pennsylvania.

30. Stanley A. Plotkin, John A. Dudgeon, and A. Melvin Ramsay, "Laboratory Studies on Rubella and the Rubella Syndrome," *British Medical Journal* 2, no. 5368 (1963): 1299.

31. Ibid.

32. J. A. Dudgeon, N. R. Butler, and Stanley A. Plotkin, "Further Serological Studies on the Rubella Syndrome," *British Medical Journal* 2, no. 5402 (1964): 159–60.

33. L. S. Oshiro, N. J. Schmidt, and E. H. Lennette, "Electron Microscopic Studies of Rubella Virus," *Journal of General Virology* 5 (1969): 205.

34. Robert S. Duszak, "Congenital Rubella Syndrome—Major Review," *Optometry* 80 (2009): 38–39; Van Hung Pham et al., "Rubella Epidemic in Vietnam: Characteristic of Rubella Virus Genes from Pregnant Women and Their Fetuses/Newborns with Congenital Rubella Syndrome," *Journal of Clinical Virology* 57 (2013): 152.

35. William S. Webster, "Teratogen Update: Congenital Rubella," *Teratology* 58 (1998): 16, http://teratology.org/updates/58pg13.pdf; J. E. Banatvala and D.W.G. Brown, "Seminar: Rubella," *Lancet* 363 (2004): 1129.

36. Joseph A. Bellanti et al., "Congenital Rubella: Clinicopathologic, Virologic, and Immunologic Studies," *American Journal of Diseases of Children* 110 (1965): 465, 470; Thong Van Nguyen, Van Hung Pham, and Kenji Abe, "Pathogenesis of Congenital Rubella Infection in Human Fetuses: Viral Infection in the Ciliary Body Could Play an Important Role in Cataractogenesis," *EBioMedicine* 2 (2015): 59–60.

37. Webster, "Teratogen Update," 17–20.

38. Stanley A. Plotkin and Antti Vaheri, "Human Fibroblasts Infected with Rubella Produce a Growth Inhibitor," *Science* 156 (1967): 659–61.

39. W. E. Rawls, J. Desmyter, and J. L. Melnick, "Virus Carrier Cells and Virus Free Cells in Fetal Rubella," *Proceedings of the Society for Experimental Biology and Medicine* 129 (1968): 477–83; Webster, "Teratogen Update," 21.

40. W. Dimech et al., "Evaluation of Three Immunoassays Used for Detection of Anti-Rubella Virus Immunoglobulin M Antibodies," *Clinical and Diagnostic Laboratory Immunology* 12, no. 9 (September 2005): 1104–8, www.ncbi.nlm.nih.gov/pmc/articles/PMC1235794/.

41. Gisella Enders et al., "Outcome of Confirmed Periconceptual Maternal Rubella," *Lancet* 331, no. 8600 (1988): 1445–47 (originally published as vol. 1, no. 8600).

42. M. M. Desmond, "Congenital Rubella Encephalitis, Course and Early Sequelae," *Journal of Pediatrics* 71, no. 3 (1967): 311–31.

43. Jill M. Forrest, Margaret A. Menser, and J. A. Burgess, "High Frequency of Diabetes Mellitus in Young Adults with Congenital Rubella," *Lancet* 297, no. 7720 (1971): 332–34.

44. John F. O'Neill, "The Ocular Manifestations of Congenital Infection: A Study of the Early Effect and Long-Term Outcome of Maternally Transmitted Rubella and Toxoplasmosis," *Transactions of the American Ophthalmological Society* 96 (1998): 858–68.

45. Stephen A. Winchester et al., "Persistent Intraocular Rubella Infection in a Patient with Fuchs' Uveitis and Congenital Rubella Syndrome," *Journal of Clinical Microbiology* 51, no. 5 (2013): 1622–24, www.ncbi.nlm.nih.gov/pmc/articles/PMC3647901/.

46. Margaret A. Menser, Lorimer Dods, and J. D. Harley, "A Twenty-five-Year Follow-up of Congenital Rubella," *Lancet* 290, no. 7530 (1967): 1347–50.

47. Miller, "Rubella in the United Kingdom," 32; John J. Witte et al., "Epidemiology of Rubella," *American Journal of Diseases of Children* 118 (1969): 107.

48. Stanley Plotkin to Alistair Dudgeon, November 4, 1963, folder, "Correspondence-out, October 1963–December 1964," Stanley Plotkin private papers, Doylestown, PA.

Chapter Ten: Plague of the Pregnant

1. William S. Webster, "Teratogen Update: Congenital Rubella," *Teratology* 58 (1998): 13, http://teratology.org/updates/58pg13.pdf.

2. Stanley A. Plotkin memo to Hilary Koprowski, "Reference: Christmas Party," December 11, 1963, folder "Correspondence-out, October 1963–December 1964," Stanley Plotkin private papers, Doylestown, PA.

3. Vincent Cristofalo, "Profile in Gerontology: Leonard Hayflick, PhD," *Contemporary Gerontology* 9, no. 3(2003): 83.

4. Eugene B. Buynak et al., "Live Attenuated Rubella Virus Vaccines Prepared in Duck Embryo Cell Culture, I: Development and Clinical Testing, *Journal of the American Medical Association* 204, no. 3 (1968): 195.

5. David J. Rothman, *Strangers at the Bedside: A History of How Law and Bioethics Transformed Medical Decision Making* (New York: Aldine de Gruyter, 1991, 2003), 40, 51–59.

6. "Wistar Institute Comparative Balance Sheets as of 4/30/66," p. 4, UPT 50 R252, box 68, file folder 14, "Wistar Institute, 1966," Isidor Schwaner Ravdin Papers, University Archives and Records Center, University of Pennsylvania; "Wistar Institute Comparative Balance Sheets as of 10/31/67," p. 4, UPT 50 R252, box 68, file folder 12, "Wistar Institute, 1966–1967," Isidor Schwaner Ravdin Papers, University Archives and Records Center, University of Pennsylvania; Robert Dechert, "Memorandum of RD's Discussion with Dr. Thomas Norton and Dr. Stanley Plotkin About the Work of the Latter," p. 2, January 12, 1968, folder "DBS," Stanley Plotkin private papers.

7. Alan D. Lourie to Thomas Norton, "Discoveries or Inventions Developed Under Public Health Service Research Grants and Awards," March 13, 1968, p. 3, folder "SKF Correspondence 1968," Stanley Plotkin Private papers, Doylestown, PA.

8. Stanley Plotkin to Alistair Dudgeon, November 4, 1963, folder "Correspondence-Out," Stanley Plotkin private papers.

9. Wolfgang Saxon, "Harry Martin Meyer Jr., 72; Helped Create Rubella Vaccine," *New York Times*, August 25, 2001, www.nytimes.com/2001/08/25/us/harry-martin-meyer-jr-72-helped-create-rubella-vaccine.html.

10. Paul D. Parkman et al., "Attenuated Rubella Virus I.: Development and Laboratory Characterization," *New England Journal of Medicine* 275 (1966): 569–74; Harry M. Meyer Jr, Paul D. Parkman, and Theodore C. Panos, "Attenuated Rubella Virus II: Production of an Experimental Live-Virus Vaccine and Clinical Trial," *New England Journal of Medicine* 275, no. 11 (1966): 575.

11. Stanley A. Plotkin, Andre Boué and Joelle G. Boué, "The In Vitro Growth of Rubella Virus in Human Embryo Cells," *American Journal of Epidemiology* 81, no. 1 (1965): 71–85; Stanley A. Plotkin and Antti Vaheri, "Human Fibroblasts Infected with Rubella Produce a Growth Inhibitor," *Science* 156 (1967): 659–61.

12. Minutes of the Meeting of the Board of Managers, the Wistar Institute, June 22, 1964.

These minutes are found on page 195 of a bound volume of Board of Managers' minutes housed at the Wistar Institute, Philadelphia, PA.

13. Communicable Disease Center, "Rubella," *Morbidity and Mortality Weekly Report* 13, no. 12 (1964): 93–101.

14. "Surveillance Summary: Rubella—United States," *Morbidity and Mortality Weekly Report* 19, no. 34 (1970): 335; J. J. Witte et al., "Epidemiology of Rubella," *American Journal of Diseases of Children* 118, no. 1 (July 1969): 107–11.

15. Stanley Plotkin, index cards cataloging infants born with congenital rubella syndrome, Stanley Plotkin private papers.

16. Stanley Plotkin, "List of Patients," file folder "Rubella Patients," Stanley Plotkin private papers.

17. J. M. Lindquist et al., "Congenital Rubella Syndrome as a Systemic Infection: Studies of Affected Infants Born in Philadelphia, USA," *British Medical Journal* 2 (1965): 1402.

18. Stanley A. Plotkin, "Virologic Assistance in the Management of German Measles in Pregnancy," *Journal of the American Medical Association* 190 (1964): 268.

19. Name withheld, letter to Stanley Plotkin, April 2, 1964, folder "Correspondence-in," Stanley Plotkin private papers.

20. Plotkin, "Virologic Assistance," 266–67.

21. Ibid., 267.

22. R. Beaver, letter to the editor, and J. D. Pryce, letter to the editor (both under the heading "Rubella and Termination of Pregnancy"), *British Medical Journal* 2, no. 5416 (October 24, 1964): 1075–76.

23. Pryce, letter to the editor, 1076.

24. Stanley A. Plotkin, "Rubella and Termination of Pregnancy" (unpublished letter to the editor of the *British Medical Journal*), November 20, 1964, folder "Correspondence-out," Stanley Plotkin private papers.

25. Stanley Plotkin, interview with the author, December 18, 2012.

26. Minutes, "Meeting of the Committee on Congenital Malformations, American Academy of Pediatrics," May 8, 1965, Philadelphia, file folder "Correspondence-in," Stanley Plotkin private papers.

27. Robert E. Hall, "Abortion in American Hospitals," *American Journal of Public Health* 57, no. 11 (1967): 1934.

28. Robert E. Hall, "Therapeutic Abortion, Sterilization, and Contraception," *American Journal of Obstetrics and Gynecology* 91, no. 4 (1965): 523.

29. Leslie J. Reagan, *Dangerous Pregnancies: Mothers, Disabilities, and Abortion in Modern America* (Berkeley and Los Angeles: University of California Press, 2010), 73–74.

30. Ibid., 74–75.

31. Ibid., 73.

32. Stanley Plotkin to Dr. Henry Fetterman, February 5, 1965, folder "Correspondence-out," Stanley Plotkin private papers.

33. Stanley Plotkin to Dr. Leonard M. Popowich, October 13, 1964, folder "Correspondence-out," Stanley Plotkin private papers.

34. Alfred D. Heggie and Frederick C. Robbins, "Natural Rubella Acquired After Birth: Clinical Features and Complications," *American Journal of Diseases of Children* 118 (1969): 15.

35. National Communicable Disease Center, "Estimated Morbidity Associated with the 1964–1965 U.S. Rubella Epidemic," in *Rubella Surveillance Report No. 1* (Washington, DC: U.S. Department of Health, Education and Welfare, Public Health Service, June 1969), 12.

36. Ibid., "Preface," 1.

37. Stanley Plotkin to Franklin Payne (head, Department of Pediatrics, Hospital of the University of Pennsylvania), January 23, 1964, folder "Correspondence-out," Stanley Plotkin private papers.

38. Stanley Plotkin to Dr. Lester Eisenberg (Department of Obstetrics, Cherry Hill Hospital), November 24, 1964, folder "Correspondence-out," Stanley Plotkin private papers.

39. Ibid.
40. Stanley A. Plotkin, David Cornfeld, and Theodore H. Ingalls, "Studies of Immunization with Living Rubella Virus: Trials in Children of a Strain Cultured from an Aborted Fetus," *American Journal of Diseases of Children* 110 (1965): 382.

Chapter Eleven: Rabies

1. H. Koprowski, "Vaccines Against Rabies: Present and Future," *First International Conference on Vaccines Against Viral and Rickettsial Diseases of Man: Papers Presented and Discussions Held in Washington, D.C., November 7–11, 1966* (Washington, DC: Pan American Health Organization Scientific Publication No. 147, May 1967), 488.
2. Roger Vaughan, *Listen to the Music: The Life of Hilary Koprowski* (New York: Springer-Verlag, 2000), 67.
3. Deborah J. Briggs, "Human Rabies Vaccines," in *Rabies*, 2nd ed., ed. Alan C. Jackson and William H. Wunner (London: Academic Press, 2007), 506.
4. World Health Organization, WHO Expert Consultation on Rabies: Second Report, WHO Technical Report Series, No. 982 (Geneva: World Health Organization 2013), 8.
5. W. Suraweera et al., "Deaths from Symptomatically Identifiable Furious Rabies in India: A Nationally Representative Mortality Survey," *PLoS Neglected Tropical Diseases* 6, no. 10 (2012): e1847; M. K. Sudarshan, "Assessing Burden of Rabies in India: WHO-Sponsored National Multi-Centric Rabies Survey," *Association for the Prevention and Control of Rabies in India Journal* 6 (May 2004): 44–45.
6. George M. Baer, ed., *The Natural History of Rabies*, 2nd ed. (Boca Raton, FL: CRC, 1991): 523.
7. Ibid., 524.
8. Ibid.
9. Elisabeth Emerson, *Public Health Is People: A History of the Minnesota Department of Health from 1949 to 1999* (St. Paul: Minnesota Department of Health, 2002), 35, and 61–62, www.health.state.mn.us/library/publichealthispeople19491999.html.
10. Communicable Disease Center, "Annual Supplement: Summary 1965," *Morbidity and Mortality Weekly Report* 14, no. 53 (1966): 48.
11. U.S. Department of Health, Education and Welfare, National Office of Vital Statistics, "Annual Supplement: Reported Incidence of Notifiable Diseases in the United States, 1952," *Morbidity and Mortality Weekly Report* 1, no. 54 (1953): 7.
12. Louis Pasteur, "Méthode pour Prévenir la Rage Après Morsure," *Comptes Rendus de l'Academie des Sciences* 101: 765–74. For a concise English rendition of Pasteur's advance, see Hervé Bazin, "Pasteur and the Birth of Vaccines Made in the Laboratory," in *History of Vaccine Development*, ed. Stanley A. Plotkin (New York: Springer Science + Business Media, 2011), 39–41.
13. Madureira Pará, "An Outbreak of Post-vaccinal Rabies (Rage de Laboratoire) in Fortaleza, Brazil, in 1960: Residual Fixed Virus as the Etiological Agent," *Bulletin of the World Health Organization* 33 (1965): 177–82, www.ncbi.nlm.nih.gov/pubmed/5294589.
14. Ibid., 181.
15. "Franklin B. Peck Jr., Horace M. Powell, and Clyde G. Culbertson, "Duck Embryo Rabies Vaccine: Study of Fixed Virus Vaccine Grown in Embryonated Duck Eggs and Killed with Beta-Propiolactone (BPL)," *Journal of the American Medical Association* 162, no. 15 (1956): 1373; Hervé Bourhy, Annick Perrot, and Jean-Marc Cavaillon, "Rabies," in: Andrew W. Artenstein, ed., *Vaccines: A Biography* (New York: Springer Science + Business Media, 2010), 84.
16. Franklin B. Peck Jr., Horace M. Powell, and Clyde G. Culbertson, "New Antirabies Vaccine for Human Use," *Journal of Laboratory and Clinical Medicine* 45, no. 5 (1955): 679–83.
17. Peck, Powell, and Culbertson, "Duck Embryo Rabies Vaccine," 1373.

18. Koprowski, "Vaccines Against Rabies," 489.

19. Baer, *Natural History of Rabies*, 418.

20. Communicable Disease Center, "Annual Supplement: Summary 1965," *Morbidity and Mortality Weekly Report* 14, no. 53 (1966): 5.

21. Ibid., 49.

22. R. E. Kissling, "Growth of Rabies Virus in Non-nervous Tissue Culture," *Proceedings of the Society for Experimental Biology and Medicine* 98, no. 2 (June 1958): 223–25.

23. Tadeusz J. Wiktor, Stanley A. Plotkin, and Hilary Koprowski, "Development and Clinical Trials of the New Human Rabies Vaccine of Tissue Culture (Human Diploid Cell) Origin," *Developments in Biological Standardization* 40 (1978): 4.

24. Vaughan, *Listen to the Music*, 121.

25. T. J. Wiktor, M. V. Fernandes, and H. Koprowski, "Cultivation of Rabies Virus in Human Diploid Cell Strain WI-38," *Journal of Immunology* 93 (September 1964): 354–55.

26. T. J. Wiktor, M. V. Fernandes, and H. Koprowski, "Potential Use of Human Diploid Cell Strains for Rabies Vaccine," *Proceedings: Symposium on the Characterization and Uses of Human Diploid Cell Strains: Opatija 1963* (no location given: Permanent Section on Microbiological Standardization, International Association of Microbiological Societies, 1963): 354–56.

27. John A. Anderson, Frank T. Daly Jr., and Jack C. Kidd, "Human Rabies After Antiserum and Vaccine Postexposure Treatment: Case Report and Review," *Annals of Internal Medicine* 64, no. 6 (1966): 1297–1302.

28. Basil Rice, "Rabies Threat Worsens," *Kingsport* (TN) *Times-News*, May 9, 1964, 2.

29. Ibid.

30. Richard R. Leger, "Alarm over Rabies: Disease Infects Wildlife in More Areas, Posing Threat to Vacationers; Rabid Foxes Terrorize County in Tennessee; Convicts May Test New Type of Vaccine; Skunks and Bats Are Carriers," *Wall Street Journal*, May 19, 1965, 1.

31. Communicable Disease Center, "Rabies in Animals and Man—1964: Annual Surveillance Summary," *Morbidity and Mortality Weekly Report* 14, no. 31 (1965): 266.

32. Leger, "Alarm over Rabies."

33. Ibid.

34. Ibid.

35. Anderson, Daly, and Kidd, "Human Rabies After Antiserum," 1300–1301.

36. Communicable Disease Center, "Human Rabies: Minnesota," *Morbidity and Mortality Weekly Report* 13, no. 38 (1964): 330; Gary Sprick obituary, *Rochester* (MN) *Post-Bulletin*, September 2, 1964, www.findagrave.com/cgi-bin/fg.cgi?page=gr&GRid=114649224.

37. Communicable Disease Center, "Rabies in Animals and Man," 269.

38. Communicable Disease Center, "Epidemiologic Notes and Reports: Human Rabies Death—South Dakota," *Morbidity and Mortality Weekly Report* 15, no. 38 (1966): 3225–26.

39. Leger, "Alarm over Rabies"; Centers for Disease Control, "Recommendation of the Immunization Practices Advisory Committee (ACIP): Rabies Prevention," *Morbidity and Mortality Weekly Report* 29, no. 3 (1980): 267.

40. Communicable Disease Center, "Human Rabies Death: West Virginia," *Morbidity and Mortality Weekly Report* 14, no. 23 (1965): 195.

41. Wiktor, Fernandes, and Koprowski, "Cultivation of Rabies Virus," 353–60.

42. Wiktor, Plotkin, and Koprowski, "Development and Clinical Trials," 5.

43. T. J. Wiktor and H. Koprowski, "Successful Immunization of Primates with Rabies Vaccine Prepared in Human Diploid Cell Strain WI-38," *Proceedings of the Society for Experimental Biology and Medicine* 118 (1965): 1069–73.

44. Mario V. Fernandes, Hilary Koprowski, and Tadeusz J. Wiktor, "Method of Producing Rabies Vaccine," U.S. Patent 3,397,267, filed September 21, 1964, and issued August 13, 1968, http://www.google.com/patents/US3397267.

45. David Lansing, director of product acquisition and licensing, Research Corporation, to Thomas W. Norton, May 3, 1966, folder "SKF Correspondence 1968," Stanley Plotkin private papers, Doylestown, PA.

Chapter Twelve: Orphans and Ordinary People

1. Claude Bernard, *An Introduction to the Study of Experimental Medicine*, trans. Henry Copley Greene (New York: Dover, 1957), 101 (originally published in 1865).
2. Stanley Plotkin, "Protocol for Rubella Study," November 1, 1963, folder "St. Vincent's," Stanley Plotkin private papers, Doylestown, PA. Rosa Hoflacher, interview with the author, October 21, 2014; Jim Butler, "For Dependent Children, a 'Home' Is Not a Home," *Catholic Standard and Times* (Philadelphia), January 19, 1968, 8.
3. Mary Thérèse Hasson, telephone interview with the author, October 24, 2014; Hoflacher, interview with the author; Butler, "For Dependent Children," 8.
4. Ruth McClain, "St. Vincent's Hospital for Women and Children: Report of Inspection and Evaluation," Department of Public Welfare of the Commonwealth of Pennsylvania, April 1966, p. 5, Cardinal Krol Papers, Philadelphia Archdiocesan Historical Research Center.
5. Butler, "For Dependent Children," 8.
6. Hoflacher, interview with the author, October 21, 2014.
7. Mary Thérèse Hasson, telephone interview with the author, October 24, 2014.
8. Stanley Plotkin, interview with the author, December 18, 2012.
9. "Court's Abortion Rulings Termed 'Tragic' by Cardinal, Bishops, Pro-life Spokesmen," *Catholic Standard and Times* (Philadelphia), January 25, 1973, 1. See also John Cardinal Krol, "Statement on Abortion: A Statement Issued by the President of the National Conference of Catholic Bishops," January 22, 1973, www.priestsforlife.org/magisterium /bishops/73-01-22statementonabortionnccb.htm (accessed February 1, 2016).
10. Stanley A. Plotkin, David Cornfeld, and Theodore H. Ingalls, "Studies of Immunization with Living Rubella Virus: Trials in Children of a Strain Cultured from an Aborted Fetus," *American Journal of Diseases of Children* 110 (1965): 382.
11. Stanley A. Plotkin to Dr. Tepper (Division of Biologics Standards, National Institutes of Health), May 15, 1964, folder "DBS," Stanley Plotkin private papers.
12. Roderick Murray (director, Division of Biologics Standards) to Stanley Plotkin, May 21, 1964, p. 1, folder "DBS," Stanley Plotkin private papers.
13. Ibid., p. 2; "Continuously Cultured Tissue Cells and Viral Vaccines: Potential Advantages May Be Realized and Potential Hazards Obviated by Careful Planning and Monitoring: Report of a Committee on Tissue Culture Viruses and Vaccines," *Science* 139 (1963): 15–20.
14. Murray to Plotkin, May 21, 1964.
15. Stanley Plotkin, e-mail to the author, February 24, 2016.
16. Stanley Plotkin, interview with the author, May 25, 2015.
17. Stanley A. Plotkin to Theodore Ingalls, May 8, 1964, file folder "St. Vincent's," Stanley Plotkin private papers.
18. Y. Hiro and S. Tasaka, "Die röetheln sind eine Viruskrankheit," *Mschr Kinderheilk* 76 (1938): 328–32. (The article's title translates to "Rubella Is a Viral Disease.")
19. S. Krugman et al., "Studies on Rubella Immunization I: Demonstration of Rubella Without Rash," *Journal of the American Medical Association* 151, no. 4 (1953): 285–88.
20. John L. Sever et al., "Rubella Virus," *Journal of the American Medical Association* 162, no. 6 (1962): 663–71.
21. Stanley Plotkin, interview with the author, May 25, 2015.
22. Plotkin, Cornfeld, and Ingalls, "Studies of Immunization," 382.
23. Ibid.
24. Ibid.
25. Ibid.; Stanley A. Plotkin to George A. Jervis, November 6, 1964, file folder "Correspondence-out," Stanley Plotkin private papers.

26. Plotkin, Cornfeld, and Ingalls, "Studies of Immunization," 383–84.

27. Ibid., 387–88.

28. All of the information about the Wenzler family and their story was obtained from in-person and telephone interviews with Mary and Steve Wenzler on March 19, March 23, April 16, April 18, and May 22, 2015, and from follow-up e-mails.

29. J. C. McDonald, "Gamma-Globulin for Prevention of Rubella in Pregnancy," *British Medical Journal* 2, no. 5354 (1963): 416.

30. Ibid., 418.

31. Mary Wenzler, interview with the author, March 23, 2015.

32. Harold G. Scheie et al., "Congenital Rubella Cataracts: Surgical Results and Virus Recovery from Intraocular Tissue," *Archives of Ophthalmology* 77 (1967): 444.

33. John F. O'Neill, interview with the author, June 1, 2015.

34. John F. O'Neill, "The Ocular Manifestations of Congenital Infection: A Study of the Early Effect and Long-Term Outcome of Maternally Transmitted Rubella and Toxoplasmosis," *Transactions of the American Ophthalmological Society* 96 (1998): 839, 867.

35. Ibid., 867.

36. John F. O'Neill, interview with the author, June 1, 2015.

37. O'Neill, "Ocular Manifestations," 834; Scheie et al., "Congenital Rubella Cataracts," 442.

38. Norman McAlister Gregg, "Congenital Cataract Following German Measles in the Mother," *Transactions of the Ophthalmological Society of Australia* 3 (1941): 39.

39. M. E. Oster, T. Riehle-Colarusso, and A. Correa, "An Update on Cardiovascular Malformations in Congenital Rubella Syndrome," *Birth Defects Research Part A: Clinical and Molecular Teratology* 88, no. 1 (2010): 1–8.

40. S. Chess, P. Fernandez, and S. Korn, "Behavioral Consequences of Congenital Rubella," *Journal of Pediatrics* 93, no. 4 (1978): 699–703; William S. Webster, "Teratogen Update: Congenital Rubella," *Teratology* 58 (1998): 20; Stella Chess, "Autism in Children with Congenital Rubella," *Journal of Autism and Childhood Schizophrenia* 1, no. 1 (1971): 33.

41. C. J. Priebe Jr., J. A. Holahan, and P. R. Ziring, "Abnormalities of the Vas Deferens and Epidiymis in Cryptorchid Boys with Congenital Rubella," *Journal of Pediatric Surgery* 14, no. 6 (1979): 834–38.

Chapter Thirteen: The Devils We Know

1. Stanley A. Plotkin, "History of Rubella Vaccines and the Recent History of Cell Culture," in *Vaccinia, Vaccination and Vaccinology: Jenner, Pasteur and Their Successors*, S. Plotkin and B. Fantini, eds. (Paris: Elsevier, 1996), 275.

2. Eugene Buynak et al., "Live Attenuated Rubella Virus Vaccines Prepared in Duck Embryo Cell Culture," *Journal of the American Medical Association* 204, no. 3 (1968): 196.

3. David J. Rothman, *Strangers at the Bedside: A History of How Law and Bioethics Transformed Medical Decision Making* (New York: de Gruyter, 1991, 2003), 64.

4. Leslie J. Reagan, *Dangerous Pregnancies: Mothers, Disabilities, and Abortion in Modern America* (Berkeley and Los Angeles: University of California Press, 2010), 63–64.

5. Paul Parkman, interview with the author, April 7, 2015.

6. Ibid.

7. Ibid.; Reagan, *Dangerous Pregnancies*, 181.

8. Parkman, interview with the author.

9. Sarah Leavitt, "Dr. Paul Parkman Interview," June 7, 2005, p. 20, Office of NIH Oral History Program, National Institutes of Health.

10. Harry M. Meyer Jr., Paul D. Parkman, and Theodore C. Panos, "Attenuated Rubella Virus II: Production of an Experimental Live-Virus Vaccine and Clinical Trial," *New England Journal of Medicine* 275, no. 11 (1966): 575.

11. Leavitt, "Dr. Paul Parkman Interview," 19.

12. Rothman, *Strangers at the Bedside*, 56.

13. Leavitt, "Dr. Paul Parkman Interview," 21–22.
14. Meyer, Parkman, and Panos, "Attenuated Rubella Virus II," 575–80.
15. Mary Thérèse Hasson, telephone interview with the author, October 24, 2014.
16. "New Patients Arrive Monday," *Hamburg Item*, January 7, 1960; "Local T-B Hospital Changed by Law to Child Welfare," *Hamburg Item*, December 3, 1959, p. 1.
17. Francis Muller, interview with the author, May 21, 2015.
18. Jasper G. Chen See to Benjamin P. Clark, February 1, 1967, folder "Hamburg-I," Stanley Plotkin private papers, Doylestown, PA.
19. Jasper G. Chen See to Benjamin P. Clark, October 30, 1967, folder "Hamburg-I," Stanley Plotkin private papers, Doylestown, PA.
20. "Patients Move Around Like Robots: Retarded at Hamburg State School Are Kept Heavily Drugged According to Superintendent," *Observer-Reporter* (Washington, PA), April 20, 1973, p. A13.
21. Stanley Plotkin to Benjamin Clark, March 25, 1968, folder "Hamburg IX," Stanley Plotkin private papers; Francis Muller interview with the author, May 21, 2015.
22. Stanley A. Plotkin et al., "A New Attenuated Rubella Virus Grown In Human Fibroblasts: Evidence For Reduced Nasopharyngeal Excretion," *American Journal of Epidemiology* 86 (1967): 469.
23. Henry K. Beecher, "Ethics and Clinical Research," *New England Journal of Medicine* 274, no. 24 (1966): 1354–60, www.nejm.org/doi/pdf/10.1056/NEJM196606162742405.
24. Ibid., 1355.
25. Rothman, *Strangers at the Bedside*, 273.
26. Captain Robert Chamovitz et al., "Prevention of Rheumatic Fever by Treatment of Previous Streptococcal Infections I: Evaluation of Benzathine Penicillin G," *New England Journal of Medicine* 251, no. 12 (1954): 466–71.
27. Saul Krugman et al., "Infectious Hepatitis: Detection of Virus During the Incubation Period and in Clinically Inapparent Infection," *New England Journal of Medicine* 261 (1959): 729–34.
28. Elinor Langer, "Human Experimentation: Cancer Studies at Sloan-Kettering Stir Public Debate on Medical Ethics," *Science* 143 (1964): 552.
29. Rothman, *Strangers at the Bedside*, 87–90.
30. William H. Stewart, "Surgeon General's Directives on Human Experimentation," PPO #129 (Bethesda, MD: U.S. Public Health Service Division of Research Grants, revised July l, 1966).
31. Rothman, *Strangers at the Bedside*, 38; Dr. Ross, memo to Mr. Zernik, October 6, 1967, file folder "Vaccine Development Board," Stanley Plotkin private papers, Doylestown, PA.
32. Stanley Plotkin to Benjamin Clark, July 25, 1966, folder "Hamburg-I," Stanley Plotkin private papers.
33. Benjamin Clark to Stanley Plotkin, July 27, 1966, folder "Hamburg-I;" single sheet titled "1st group: 8–1 to 9–13," folder "Hamburg-I," Stanley Plotkin private papers.
34. Stanley Plotkin e-mail to the author, February 24, 2016.
35. Benjamin Clark to Stanley Plotkin, May 29, 1968, folder "Hamburg-I," Stanley Plotkin private papers.
36. Stanley A. Plotkin et al., "A New Attenuated Rubella Virus," 468–77.
37. James L. Bittle et al., "Results of Testing Production Lots of Oral Poliovirus Vaccine," *Journal of Infectious Diseases* 116, no. 2 (1966): 215–20.
38. Plotkin et al., "New Attenuated Rubella Virus," 473.
39. Werner Slenczka and Hans Dieter Klenk, "Forty Years of Marburg Virus," *Journal of Infectious Diseases* 196, supp. 2 (2007): S133.
40. Ibid.
41. Ibid., S131.
42. Ibid.

43. Ibid.
44. K. Todorovitch, M. Mocitch, and R. Klašnja, "Clinical Picture of Two Patients Infected by the Marburg Vervet Virus," in *Marburg Virus Disease*, Gustav Adolf Martini and Rudolf Siegert, eds. (Berlin and Heidelberg: Springer-Verlag, 1971), 19.
45. G. A. Martini, "Marburg Virus Disease: Clinical Syndrome," in Martini and Siegert, *Marburg Virus Disease*, 1.
46. Ibid., 2.
47. Ibid.
48. P. Gedigk, H. Bechtelsheimer, and G. Korb, "Pathologic Anatomy of Marburg Virus Disease," in Martini and Siegert, *Marburg Virus Disease*, 50.
49. Richard Preston, *The Hot Zone* (New York: Anchor Books, 1994), 38.
50. Slenczka and Klenk, "Forty Years of Marburg Virus," S131–32; Lawrence Corey, "Marburg Virus Disease," in chapter 207, "Rabies and Other Rhabdoviruses," in *Harrison's Principles of Internal Medicine Tenth Edition* (New York: McGraw-Hill, 1983), 1139.
51. C. E. Gordon Smith et al., "Fatal Human Disease from Vervet Monkeys," *Lancet* 290, no. 7526 (1967): 1119.
52. Richard Lyons, "Diseases Carried by Pets Increase," *New York Times*, October 26, 1967, p. 24; U.S. Department of Health, Education and Welfare Public Health Service, *National Cancer Institute Monograph December 29, 1968, Cell Cultures for Virus Vaccine Production*, 474; Leonard Hayflick, "Human Virus Vaccines: Why Monkey Cells?," *Science* 176 (1972): 813.
53. Preston, *Hot Zone*, 40–42.
54. Slenczka and Klenk, "Forty Years of Marburg Virus," S134.
55. Conference on Cell Cultures for Virus Vaccine Production, 474.
56. Ibid., 474–75.

Chapter Fourteen: Politics and Persuasion

1. Jacob Bronowski, "The Disestablishment of Science," *Encounter*, July 1971, 15.
2. "Wistar Institute Comparative Balance Sheets as of October 31, 1967," p. 4, UPT 50 R252, box 68, file folder 12, Isidor Schwaner Ravdin Collection, University Archives and Records Center, University of Pennsylvania.
3. Stanley Plotkin to Benjamin Clark, December 5, 1966, folder "Hamburg-I," Stanley Plotkin private papers, Doylestown, PA.
4. Earl Beck (scientist-administrator, Vaccine Development Branch, NIAID) to Stanley Plotkin, November 21, 1967, folder "DBS," Stanley Plotkin private papers.
5. Daniel Mullally to Hilary Koprowski, November 27, 1967, folder "Vaccine Development Board," Stanley Plotkin private papers.
6. Robert J. Ferlauto (director, Research-Microbiology, Smith, Kline & French Laboratories) to Stanley Plotkin, December 1, 1967, folder "SKF-Rubella," Stanley Plotkin private papers.
7. Plotkin to Clark.
8. Benjamin Clark to Stanley Plotkin, December 13, 1966, folder "Hamburg-I," Stanley Plotkin private papers.
9. Lois Colley, R.N. (director of nursing), memo to Dr. Clark (superintendent, Hamburg State School and Hospital), "Dr. Plotkin's Letter of June 15, 1967, Received Yesterday," June 16, 1967, folder "Rubella," Stanley Plotkin private papers.
10. Benjamin Clark to Stanley Plotkin, June 21, 1967, folder "Hamburg-VI," Stanley Plotkin private papers.
11. Stanley Plotkin to Miss Lois Colley, July 10, 1967, folder "Hamburg-VI," Stanley Plotkin private papers.
12. Bernard Frankel to Roderick Murray, January 10, 1968, folder "DBS," Stanley Plotkin private papers.

13. Stanley Plotkin to Theodore Ingalls (Epidemiologic Study Center, Framingham, Mass.), January 10, 1968, folder "DBS," Stanley Plotkin private papers.

14. Stanley Plotkin to Robert Dechert (Dechert, Price & Rhoads), January 17, 1968, folder "DBS," Stanley Plotkin private papers.

15. Paul D. Parkman et al., "Attenuated Rubella Virus I: Development and Laboratory Characterization," *New England Journal of Medicine* 275, no. 11 (1966): 569–74; Harry M. Meyer Jr., Paul D. Parkman, and Theodore C. Panos, "Attenuated Rubella Virus II: Production of an Experimental Live-Virus Vaccine and Clinical Trial," *New England Journal of Medicine* 275, no. 11 (1966): 575–80.

16. Harry M. Meyer Jr. et al., "Clinical Studies with Experimental Live Rubella Virus Vaccine (Strain HPV-77): Evaluation of Vaccine-Induced Immunity," *American Journal of Diseases of Children* 117 (1968): 648–54.

17. "Drs. Meyer, Parkman Win Joint Recognition for Rubella Research," *NIH Record* XIX, no. 22 (November 14, 1967): 3.

18. George L. Stewart et al., "Rubella-Virus Hemagglutination-Inhibition Test," *New England Journal of Medicine* 276, no. 10 (1967): 554–57; Sarah Leavitt, "Dr. Paul Parkman Interview," June 7, 2005, pp. 26–28, Office of NIH History Oral History Program, National Institutes of Health.

19. "Dr. Paul D. Parkman Named One of Ten Outstanding Young Men of the Year," *NIH Record XX*, no. 2 (January 23, 1968): 1 and 7.

20. Lyndon Baines Johnson to Paul D. Parkman, May 5, 1966, Name File P, file folder "Parkman, I–P," box 56, White House Central File, LBJ Presidential Library, Austin, TX.

21. Samuel J. Musser and Larry J. Hilsabeck, "Production of Rubella Virus Vaccine: Live, Attenuated, in Canine Renal Cell Cultures," *American Journal of Diseases of Children* 118 (1969): 356–57, 361.

22. Paul Offit, *Vaccinated: One Man's Quest to Defeat the World's Deadliest Diseases* (New York: HarperCollins, 2007), 76–78; Maurice Hilleman interview with Paul Offit, November 30, 2004. Audio file courtesy of Paul Offit.

23. Hilleman, interview with Offit.

24. Ibid.

25. Maurice R. Hilleman et al., "Live Attenuated Rubella Virus Vaccines: Experiences with Duck Embryo Cell Preparations," *American Journal of Diseases of Children* 118, no. 2 (1969): 166.

26. Eugene Buynak et al., "Live Attenuated Rubella Virus, Vaccines prepared in Duck Embryo Cell Culture," *Journal of the American Medical Association* 204, no. 3 (1968): 197 (table 2).

27. Hilleman, interview with Offit, 2004.

28. Robert E. Weibel et al., "Live Attenuated Rubella Virus Vaccines Prepared in Duck Embryo Cell Culture II: Clinical Tests in Families and in an Institution," *Journal of the American Medical Association* 205, no. 8 (1969): 558.

29. Ibid.

30. Hilleman, interview with Offit.

31. Ibid.

Chapter Fifteen: The Great Escape

1. John F. Morrison and William T. Keough, "Ex-Phila. Scientist Battles U.S. over Frozen Cells," *Philadelphia Evening Bulletin*, April 4, 1976.

2. International Association of Microbiological Societies, Permanent Section of Microbiological Standardization, "Minutes of the Fifth Meeting of the Committee on Cell Cultures," November 27, 1968, 21.

3. J. P. Jacobs and F. T. Perkins, "Supplying Cell Cultures Regularly to Distant Laboratories," *Bulletin of the World Health Organization* 40 (1969): 476–78.

4. J. P. Jacobs, C. M. Jones, and J. P. Baille, "Characteristics of a Human Diploid Cell Designated MRC-5," *Nature* 227 (1970): 168–70.
5. Harold M. Schmeck Jr., "Human Cells Given Role in Vaccines," *New York Times*, November 12, 1966, 36.
6. Jane Brody, "Cell Bank Is Suggested for Every Person at Birth," *New York Times*, April 3, 1967, 25.
7. "Wistar Institute Comparative Balance Sheet as of 4/30/66," fourth page: "Percentage Report as of 4/30/66—Grants," account number 60188, UPT 50 R252, box 68, folder 13 "Wistar Institute 1966", Isidor Schwaner Ravdin Papers, University Archives and Records Center, University of Pennsylvania; "Wistar Institute Comparative Balance Sheets as of 10/31/67," fourth page: "Percentage Report as of 10/31/67—Grants," account number 188, UPT 50 R252, box 68, folder 12 ("Wistar Institute 1966–67"), Isidor Schwaner Ravdin Papers, University Archives and Records Center, University of Pennsylvania.
8. Albert Sabin to Sidney Raffel, August 4, 1967, Correspondence, Individual (Graetz–Hayflick), series 1, box 11, folder "Hayflick, Leonard, 1964–81," Albert B. Sabin Collection, Henry R. Winkler Center for the History of the Health Professions, University of Cincinnati Libraries, Cincinnati, Ohio.
9. Nancy Pleibel, interview with the author, March 6, 2013.
10. Minutes, Wistar Institute Board of Managers, June 19, 1962; June 22, 1964; and June 15, 1965. All minutes were accessed in a bound volume of minutes of The Wistar Institute Board of Managers meetings, pp. 193, 152, and 213, respectively. Courtesy of The Wistar Institute of Anatomy and Biology, Philadelphia, PA.
11. Minutes, Wistar Institute Board of Managers, December 14, 1965. Accessed in a bound volume of minutes of the Wistar Institute Board of Managers meetings, p. 222. Courtesy of The Wistar Institute of Anatomy and Biology, Philadelphia, PA.
12. "Wistar Institute Comparative Balance Sheets as of 10/31/67," fourth page: "Percentage Report as of 10/31/67–Grants," box 68, file folder 12 ("Wistar Institute 1966–67"), Isidor Schwaner Ravdin Papers, UPT 50 R252, University Archives and Records Center, University of Pennsylvania, "Wistar Institute Comparative Balance Sheet as of 4/30/66," fourth page, Isidor Schwaner Ravdin Papers, UPT 50 R252, box 68, file folder 13 ("Wistar Institute 1966"); "Wistar Institute Comparative Balance Sheet as of 12/31/65," fourth page, University Archives and Records Center, University of Pennsylvania, Isidor Schwaner Ravdin Papers, UPT 50 R252, box 68, file folder 14 ("Wistar Institute 1966").
13. Factual Chronology, 7.
14. "Minutes of the Wistar Institute Board of Managers Meeting," December 16, 1966, p. 1, UPT 50 R252, box 68, folder 12 ("Wistar Institute 1966"), Isidor Schwaner Ravdin Papers, University Archives and Records Center, University of Pennsylvania.
15. Ibid., exhibit A; "Minutes of the Wistar Institute Board of Managers Meeting," February 24, 1967, p. 2, UPT 50 R252, box 68, folder 12 ("Wistar Institute, 1966–1967"), Isidor Schwaner Ravdin Papers, University Archives and Records Center, University of Pennsylvania.
16. Factual Chronology, 7.
17. John D. Ross, "Memorandum on Diploid Contract Conference, Minutes of Meeting," January 18, 1968, attachment B to Schriver Report, p 1.
18. Ibid.
19. Charles W. Boone to Hilary Koprowski, February 16, 1968, attachment C to Schriver Report, 2.
20. Leonard Hayflick, interview with the author, March 3, 2013.
21. Department of Health, Education, and Welfare, Public Health Service, National Institutes of Health, "Contract Number: PH43-62-157," February 6, 1962, Section 30: "Termination for the Convenience of the Government," part (g), p. HEW-315-6. Courtesy of Leonard Hayflick.

22. "Chronicle Burroughs Wellcome Proposed Agreement," folder "SKF Correspondence 1968," Stanley Plotkin private papers, Doylestown, PA.
23. Roger C. Egeberg (assistant secretary for health and scientific affairs) to Hilary Koprowski, August 13, 1970, Stanley Plotkin private papers.
24. A.C.C. Newman to Hilary Koprowski, October 16, 1968, file folder "Smith Kline French," Stanley Plotkin private papers.
25. Leonard Hayflick, interview with the author.
26. Hayflick Rebuttal to Schriver Report, 19; John Shannon to Leon Jacobs, May 7, 1976, investigations 9, file folder 1, Directors' Files, Office of the Director, National Institutes of Health.
27. Schriver Report, 4.

Chapter Sixteen: In the Bear Pit

1. Émile Roux is quoted in Stanley A. Plotkin, "Sang Froid in a Time of Trouble: Is a Vaccine Against HIV Possible?" *Journal of the International AIDS Society* 12, no. 2 (2009), http://www.ncbi.nlm.nih.gov/pmc/articles/PMC2647531/ doi:10.1186/1758-2652-12-2.
2. Maurice R. Hilleman et al., "Live Attenuated Rubella Virus Vaccines: Experiences with Duck Embryo Cell Preparations," *American Journal of Diseases of Children* 118 (1969): 171.
3. Samuel J. Musser and Larry J. Hilsabeck, "Production of Rubella Virus Vaccine: Live, Attenuated, in Canine Renal Cell Cultures," *American Journal of Diseases of Children* 118 (1969): 361.
4. George R. Thompson et al., "Intermittent Arthritis Following Rubella Vaccination," *American Journal of Diseases of Children* 125 (1973): 526.
5. Stanley A. Plotkin et al., "An Attenuated Rubella Virus Strain Adapted to Primary Rabbit Kidney," *American Journal of Epidemiology* 88 (1968): 97.
6. "Rubella: Vaccines May Be Licensed by Fall," *Science News* 95 (March 1, 1969): 209.
7. Stanley Plotkin to Sister Agape, April 4, 1968, folder "St. Vincent's," Stanley Plotkin private papers, Doylestown, PA.
8. Stanley Plotkin to John Cardinal Krol (archbishop of Philadelphia), April 5, 1968, folder "St. Vincent's," Stanley Plotkin private papers.
9. Archbishop of Philadelphia to Stanley Plotkin, April 11, 1968, Cardinal Krol papers, Philadelphia Archdiocesan Historical Research Center.
10. S. A. Plotkin et al., "Further Studies of an Attenuated Rubella Strain Grown in WI-38 Cells," *American Journal of Epidemiology* 39, no. 2 (1969): 236.
11. Stanley Plotkin, interview with the author, June 1, 2015.
12. Plotkin et al., "Further Studies," 232.
13. Ibid., 237.
14. Stanley Plotkin, "Status of Negotiations with SKF, Merieux and Wellcome," undated, file folder "SKF Correspondence 1968," Plotkin private papers. This paper is physically placed among papers dated autumn 1968.
15. Constant Huygelen, telegram to Robert Ferlauto, April 9, 1968, folder "SKF Correspondence 1968," Stanley Plotkin private papers.
16. Alan D. Lourie, memo to Ed Clay, January 8, 1968, file folder "SKF-Rubella," Stanley Plotkin private papers.
17. Stanley A. Plotkin et al., "A New Attenuated Rubella Virus Grown in Human Fibroblasts: Evidence for Reduced Nasopharyngeal Excretion," *American Journal of Epidemiology* 86, no. 2 (1967): 468–77.
18. Robert Ferlauto to Stanley Plotkin, August 28, 1968, folder "SKF Correspondence 1968," Stanley Plotkin private papers.
19. Robert Ferlauto to Hilary Koprowski, August 12, 1968, folder "SKF Correspondence 1968," Stanley Plotkin private papers.

20. R. Palmer Beasley et al., "Prevention of Rubella During an Epidemic on Taiwan," *American Journal of Diseases of Children* 118 (1969): 304.
21. Harold M. Schmeck Jr., "Test Finds Rubella Vaccine Effective," *New York Times*, October 17, 1968, 1, 27.
22. "Tests on Vaccines to Prevent Rubella Highly Effective," *NIH Record* 20, no. 22 (October 29, 1968): 1, 8, https://nihrecord.nih.gov/PDF_Archive/1968%20PDFs/19681029.pdf (accessed February 8, 2016).
23. Beasley et al., "Prevention of Rubella," 304.
24. Roderick Murray and Dorland J. Davis to Stanley Plotkin, September 30, 1968, folder "Washington, D.C.-1968," Stanley Plotkin private papers.
25. Plotkin et al., "Attenuated Rubella Virus Strain," 97–102.
26. Howard A. Rusk, "Rubella Vaccine Near: Likely to Be Available in 2 Months, After Production Guides Take Effect," *New York Times*, April 13, 1969.
27. Stanley A. Plotkin et al., "Attenuation of RA 27/3 Rubella Virus in WI-38 Human Diploid Cells," *American Journal of Diseases of Children* 118 (1969): 184.
28. Hilleman et al., "Live Attenuated Rubella Virus Vaccines," 167.
29. R. E. Weibel et al., "Live Rubella Vaccines in Adults and Children," *American Journal of Diseases of Children* 118, no. 2 (1969): 226–29.
30. Ibid.
31. Louis Z. Cooper et al., "Transient Arthritis After Rubella Vaccination," *American Journal of Diseases of Children* 118, no. 2 (1969): 218–25.
32. Weibel et al., "Live Rubella Vaccines," 229.
33. "Recommendation of the Public Health Service Advisory Committee on Immunization Practices: Prelicensing Statement on Rubella Virus Vaccine," *Morbidity and Mortality Weekly Report* 18, no. 15 (1969): 124–25.
34. "Leads from the MMWR: Rubella Vaccination During Pregnancy—United States, 1971–1988," *Journal of the American Medical Association* 261, no. 23 (1989): 3375.
35. Centers for Disease Control and Prevention, "Control and Prevention of Rubella: Evaluation and Management of Suspected Outbreaks, Rubella in Pregnant Women, and Surveillance for Congenital Rubella Syndrome," *Morbidity and Mortality Weekly Report* 50, no. RR-12 (2001): 16.
36. Ibid., 33.
37. J. E. Banatvala and D. W. G. Brown, "Seminar: Rubella," *Lancet* 363 (2004): 1128.
38. Plotkin et al., "Attenuation of RA 27/3 Rubella Virus," 184.
39. Roderick Murray, "Biologics Control of Virus Vaccines," *American Journal of Diseases of Children* 118 (1969): 336.
40. "Gamma Globulin Prophylaxis; Inactivated Rubella Virus; Production and Biologics Control of Live Attenuated Rubella Virus Vaccines: Discussion on Session V," *American Journal of Diseases of Children* 118 (1969): 377.
41. Ibid., 378.
42. Roger Vaughan, *Listen to the Music: The Life of Hilary Koprowski* (New York: Springer-Verlag, 2000), 54.
43. Edward Shorter, "The Health Century Oral History Collection," Bernice Eddy interview, December 4, 1986, p. 25, transcript at the National Library of Medicine, NIH.
44. "Gamma Globulin Prophylaxis," 378.
45. Ibid., 379.
46. Ibid., 379–80.
47. Stanley A. Plotkin, ed., *History of Vaccine Development* (New York: Springer Science + Business Media, 2011), 226.
48. Robert Q. Marston, director, National Institutes of Health, "Additional Standards; Rubella Virus Vaccine, Live," *Federal Register* 34, no. 109 (1969): 9072–75.

49. Louis Galambos with Jane Eliot Sewell, *Networks of Innovation: Vaccine Development at Merck, Sharp & Dohme, and Mulford, 1985–1995* (Cambridge: Cambridge University Press, 1995), 112.

50. Stanley A. Plotkin, "Rubella Vaccination," *Journal of the American Medical Association* 215, no. 9 (1971): 1492–93.

51. Thompson et al., "Intermittent Arthritis," 526.

52. William Schaffner et al., "Polyneuropathy Following Rubella Immunization: A Follow-up Study and Review of the Problem," *American Journal of Diseases of Children* 127 (1974): 684–88.

53. General Accounting Office, "Bid Protest-Negotiation-Specification Compliance Denial of Protest Against Rejection of Offer for Furnishing Live Rubella Vaccine to Veterans Administration on Basis That Vaccine Did Not Comply with the Requirements of the Amended Specifications," B-170817, September 25, 1970, www.gao.gov/products/429478#mt=e-report (accessed February 15, 2016).

54. Harold M. Schmeck Jr., "One of 3 Rubella Vaccine Producers Barred from Bidding for US Contract," *New York Times*, September 15, 1970, 13.

55. T. Norton, memo to Hilary Koprowski and Stanley Plotkin, January 15, 1970, folder "SKF-Rubella," Stanley Plotkin private papers.

56. Memorandum of a meeting held at Smith, Kline & French on September 22, 1970, folder "SKF-Rubella," Stanley Plotkin private papers.

57. Stanley Plotkin to Leonard Hayflick, October 2, 1970, folder "Correspondence-H," Stanley Plotkin private papers.

Chapter Seventeen: Cell Wars

1. Leonard Hayflick, "The Coming of Age of WI-38," *Advances in Cell Culture* 3 (1984): 303.

2. Robert Roosa, interview with the author, December 19, 2013.

3. Leonard Hayflick, telephone interview with the author, October 17, 2012.

4. Roger Vaughan, *Listen to the Music: The Life of Hilary Koprowski* (New York: Springer-Verlag, 2000), 112.

5. Stanley Plotkin, interview with the author, August 29, 2014.

6. Stanley Plotkin to Leonard Hayflick, July 9, 1968, file folder "Correspondence-H," Stanley Plotkin Rubella Papers.

7. International Association of Microbiological Societies, Permanent Section of Microbiological Standardization, "Minutes of the Fourth Meeting of the Committee on Cell Cultures," September 16, 1967, p. 63.

8. Factual Chronology, 7.

9. Schriver Report, 8–9.

10. Ibid., 9.

11. Hayflick Rebuttal to Schriver Report, 40–41.

12. International Association of Microbiological Societies, Permanent Section of Microbiological Standardization, "Minutes of the Fifth Meeting of the Committee on Cell Cultures," November 27, 1968, p. 20; Schriver Report, 8.

13. Schriver Report, 8.

14. Factual Chronology, 20.

15. Ibid.

16. Stanley Plotkin to Leonard Hayflick, December 8, 1969, file folder "Correspondence-H," Stanley Plotkin private papers, Doylestown, PA.

17. Factual Chronology, 32–33.

18. Stanley A. Plotkin to Leonard Hayflick, March 17, 1969, file folder "Correspondence-H," Stanley Plotkin private papers.

19. Stanley Plotkin to Leonard Hayflick, August 1, 1969, file folder "Correspondence-H," Stanley Plotkin private papers.

20. Leonard Hayflick to Stanley Plotkin, August 11, 1969, file folder "Correspondence-H," Stanley Plotkin private papers.

Chapter Eighteen: DBS Defeated

1. Testimony of Dr. Leonard Hayflick, Senate Subcommittee on Executive Reorganization and Government Research of the Committee on Government Operations, *Consumer Safety Act of 1972: Hearings on Titles I and II of S. 3419*, 92nd Cong., 2nd sess., April 20 and 21 and May 3 and 4, 1972, p. 34.

2. Debbie Bookchin and Jim Schumacher, *The Virus and the Vaccine: Contaminated Vaccine, Deadly Cancers and Government Neglect* (New York: St. Martin's, 2004), 127.

3. Jane E. Brody, "Vaccine Produced in Human Cells," *New York Times*, March 8, 1972, 18.

4. Nicholas Wade, "Division of Biologics Standards: The Boat That Never Rocked," *Science* 175 (1972): 1228.

5. Abraham Ribicoff, "Exhibit 55: Vaccine Safety," Senate Subcommittee on Executive Reorganization and Government Research, *Consumer Safety Act of 1972: Hearings*, 512–32.

6. Nicholas Wade, "DBS: Agency Contravenes Its Own Regulations," *Science* 175 (1972): 34.

7. Ribicoff, "Exhibit 55: Vaccine Safety," 527.

8. Testimony of Dr. Leonard Hayflick, 29–38.

9. Wade, "DBS: Agency Contravenes," 35.

10. *Consumer Safety Act of 1972: Hearings*, 36.

11. P. Stessel, memo to R. A. Schoellhorn, H. Perlmutter, J. Rose, G. J. Sella Jr., R. J. Vallancourt, P. J. Vasington (Lederle Laboratories), April 26, 1972, cited in Bookchin and Schumacher, *Virus and the Vaccine*, 127, 306.

12. Leonard Hayflick, "Human Virus Vaccines: Why Monkey Cells?" *Science* 176 (1972): 813–14.

13. Stanley A. Plotkin to Senator Abraham Ribicoff, *Consumer Safety Act of 1972: Hearings*, 419–20.

14. "Dr. Roderick Murray Named Special Assistant to the Director of NIAID," *NIH Record*, June 7, 1972, 5.

15. Bookchin and Schumacher, *Virus and the Vaccine*, 127.

16. S. Kops, "Oral Polio Vaccine and Human Cancer: A Reassessment of SV40 as a Contaminant, Based upon Legal Documents," *Anticancer Research* 20 (2000): 4746.

17. Bookchin and Schumacher, *Virus and the Vaccine*, 124.

18. David Oshinsky, "Polio," in: Andrew W. Artenstein, ed., *Vaccines: A Biography* (New York: Springer Science + Business Media, 2010), 219.

19. Nicoletta Previsani et al., "World Health Organization Guidelines for Containment of Poliovirus Following Type-Specific Polio Eradication: Worldwide, 2015," *Morbidity and Mortality Weekly Report* 64, no. 33 (2015): 913, www.cdc.gov/mmwr/preview/ mmwrhtml/mm6433a5.htm?s_cid=mm6433a5_w (accessed February 18, 2016).

20. World Health Organization Media Centre: "Government of Nigeria Reports 2 Wild Polio Cases, First Since July 2014: New Cases Come on the Two-Year Anniversary Since the Last Confirmed Case of Polio Was Reported in Africa" (news release), August 11, 2016, http://www.who.int/mediacentre/news/releases/2016/nigeria-polio/en/ (accessed September 8, 2016); Leslie Roberts, "Nigeria Outbreak Forces Rethink of Polio Strategies," Science Insider (online), September 6, 2016, http://www.sciencemag.org/news/2016/09 /nigeria-outbreak-forces-rethink-polio-strategies (accessed September 8, 2016).

Chapter Nineteen: Breakthrough

1. Rebecca Sheir, "Ebola Researcher Says Vaccinology Isn't Rocket Science—It's Harder," Metro Connection, WAMU Radio, October 23 2014. Transcript available here: http:// wamu.org/programs/metro_connection/14/10/23/ebola_researcher_says_vaccinology _isnt_rocket_science_its_harder_transcript (accessed September 1, 2016).

2. "Dr. Dorothy Horstmann dies—key in development of polio vaccine," *Yale Bulletin & Calendar* 29, no. 16 (2001), http://www.yale.edu/opa/arc-ybc/v29.n16/story18.html ; David M. Oshinsky, "Breaking the Back of Polio," *Yale Medicine* 40, no. 1 (2005), http://yalemedicine.yale.edu/autumn2005/features/feature/52012/; Daniel Wilson, unpublished interview with Dorothy Horstmann, 1990, Dorothy M. Horstmann Papers (MS 1700), box 12, Folder 257, Manuscripts and Archives, Yale University Library.

3. Jane E. Brody, "New Research on Rubella Challenges Effectiveness of Vaccination Program," *New York Times*, September 29, 1970, 8.

4. Dorothy M. Horstmann et al., "Rubella: Reinfection of Vaccinated and Naturally Immune Persons Exposed in an Epidemic," *New England Journal of Medicine* 283, no. 15 (1970): 771–78.

5. Scott B. Halstead et al., "Susceptibility to Rubella Among Adolescents and Adults in Hawaii," *Journal of the American Medical Association* 210 (10): 1881–83.

6. Te-Wen Chang, Suzanne DesRosiers, and Louis Weinstein, "Clinical and Serological Studies of an Outbreak of Rubella in a Vaccinated Population," *New England Journal of Medicine* 283, no. 5 (1970): 246–48; J. M. Forrest, M. A. Menser, and M. C. Honeyman, "Clinical Rubella Eleven Months After Vaccination," *Lancet* 300, no. 7774 (1972): 399–400.

7. Dorothy M. Horstmann et al., "Rubella: Reinfection of Vaccinated and Naturally Immune Persons Exposed in an Epidemic," *New England Journal of Medicine* 283, no. 15 (1970): 771–78.

8. Ibid., 775.

9. Chang, DesRosiers, and Weinstein, "Clinical and Serological Studies of an Outbreak of Rubella in a Vaccinated Population," *New England Journal of Medicine* 283, no. 5 (1970): 246–48.

10. Elias Abrutyn et al., "Rubella Vaccine Comparative Study: Nine-Month Follow-Up and Serologic Response to Natural Challenge," *American Journal of Diseases of Children* 120 (1970): 129–33; William J. Davis et al., "A Study of Rubella Immunity and Resistance to Infection," *Journal of the American Medical Association* 215, no. 4 (1971): 600–608; Jeanette Wilkins et al., "Reinfection with Rubella Virus Despite Live Vaccine-Induced Immunity," *American Journal of Diseases of Children* 118: (1969): 275–94; Harvey Liebhaber et al., "Vaccination with RA27/3 Rubella Vaccine: Persistence of Immunity and Resistance to Challenge After Two Years," *American Journal of Diseases of Children* 123 (1972): 134.

11. Horstmann et al., "Rubella: Reinfection," 777.

12. Dorothy Horstmann to Stanley Plotkin, January 30 1970.

13. Stanley Plotkin to Dorothy Horstmann, November 30, 1970, and April 16, 1971, folder "Horstmann," Stanley Plotkin private papers.

14. Dorothy Horstmann to Stanley Plotkin, March 12, 1970, file folder "Horstmann," Stanley Plotkin private papers.

15. Ann Schluederberg et al., "Neutralizing and Hemagglutination-Inhibiting Antibodies to Rubella Virus as Indicators of Protective Immunity in Vaccinees and Naturally Immune Individuals," *Journal of Infectious Diseases* 138, no. 6 (1978): 877–83.

16. Liebhaber et al., "Vaccination with RA 27/3," 133–36.

17. Chang, DesRosiers, and Weinstein, "Clinical and Serological Studies," 247; Wilkins et al., "Reinfection with Rubella Virus," 291; Horstmann et al., "Rubella: Reinfection," 775; Stanley A. Plotkin, John D. Farquhar, and Ogra L. Pearay, "Immunologic Properties of RA27/3 Rubella Virus Vaccine: A Comparison with Strains Presently Licensed in the United States," *Journal of the American Medical Association* 225, no. 6 (1973): 588.

18. Liebhaber et al., "Vaccination with RA 27/3," 134–35.

19. Ibid., 136.

20. Maurice Hilleman, interview with Paul Offit November 30, 2004. Audio file courtesy of Paul Offit.

21. Plotkin, Farquhar, and Pearay, "Immunologic Properties of RA 27/3," 585 and 589.

22. Stanley Plotkin to Leonard Hayflick, October 3, 1973, file folder "Correspondence-H," Stanley Plotkin private papers.
23. Hilleman, interview with Paul Offit.
24. Robert E. Weibel et al., "Clinical and Laboratory Studies of Live Attenuated RA 27/3 and HPV 77-DE Rubella Virus Vaccines," *Proceedings of the Society of Experimental Biology and Medicine* 165, no. 1 (1980): 44–49.
25. Ibid., 44.
26. Schluecderberg et al., "Neutralizing and Hemagglutination-Inhibiting Antibodies," 877–83.
27. Pamela Eisele (Merck spokesperson), e-mail to the author, August 31, 2015.

Chapter Twenty: Slaughtered Babies and Skylab

1. Leonard Hayflick, interview with the author, October 3, 2012.
2. Forrest Stevenson Jr., "Women, the Bible and Abortion," (Brighton, MI: Forrest Stevenson, 1972). Courtesy of Leonard Hayflick.
3. State of Michigan, Department of State, "Initiatives and Referendums Under the Constitution of the State of Michigan, 1963," December 5, 2008, 14.
4. James V. Siena to Forrest Stevenson, November 30, 1972. Courtesy of Leonard Hayflick.
5. Arthur F. Barkey to James V. Siena, December 7, 1972. Courtesy of Leonard Hayflick.
6. Forrest Stevenson to Dr. Hayflick, in "Letters to the Editor," *Gazette Times* (Heppner, OR), July 4, 1974.
7. Leonard Hayflick to *Life Line* (Passaic County Right to Life newsletter), July 20, 1973. Courtesy of Leonard Hayflick.
8. United Press International, "Human Cells to Be Orbited in Outer Space," *Los Angeles Times*, May 2, 1973.
9. James J. Ambrose to Joseph T. McGucken (archbishop of San Francisco), May 2, 1973. Courtesy of Leonard Hayflick.
10. Leonard Hayflick, telephone interview with the author, October 17, 2012.
11. P. O. Montgomery et al., "The Response of Single Human Cells to Zero Gravity," in Richard S. Johnston and Lawrence F. Dietlein, *Biomedical Results from Skylab, NASA SP-377* (Washington, DC: Scientific and Technical Information Office, National Aeronautical and Space Administration, 1977), 221–33.
12. Mrs. Raymond Somerville to Leonard Hayflick, June 14, 1973.

Chapter Twenty-one: Cells, Inc.

1. Leonard Hayflick, telephone interview with the author, October 16, 2012.
2. Hayflick Rebuttal to Schriver Report, 34.
3. Department of Health, Education, and Welfare, Public Health Service, National Institutes of Health, "Contract Number: PH43-62-157," February 6, 1962, Section 30: "Termination for the Convenience of the Government," part (g), p. HEW-315-6. Courtesy of Leonard Hayflick.
4. Schriver Report, attachment B.
5. Schriver Report, 9.
6. Leonard Hayflick, telephone interview with the author, October 11, 2012.
7. Schriver Report, 9; John E. Shannon and Marvin L. Macy, eds., *The American Type Culture Collection Registry of Animal Cell Lines*, 2nd ed. (Rockville, MD: American Type Culture Collection, 1972), 17; Robert F iy et al., eds., *The American Type Culture Collection: Catalogue of Strains II*, 2nd ed. (Rockville, MD: American Type Culture Collection, 1979), viii.
8. Schriver Report, 10.
9. Louis Rosenfeld, "Insulin: Discovery and Controversy," *Clinical Chemistry* 48, no. 12 (2002): 2280.
10. Jonas Salk, interview with Edward R. Murrow on *See It Now*, CBS, April 12, 1955. Quoted

in Elizabeth Popp Berman, *Creating the Market University: How Academic Science Became an Economic Engine* (Princeton, NJ: Princeton University Press, 2012), 5; and in Jane S. Smith, *Patenting the Sun: Polio and the Salk Vaccine* (New York: Morrow, 1990), 13.

11. Frederick J. Hammett, "Uncommitted Researchers," *Science* 117 (1953): 64.

12. Leonard Hayflick, telephone interview with the author, October 17, 2012.

13. Schriver Report, 9.

14. Ibid., 11; Schriver Rebuttal to Hayflick Rebuttal, 106.

15. Schriver Report, attachment A.

16. Ibid., 12.

17. Ibid., 6–7.

18. Ibid.; Schriver Rebuttal to Hayflick Rebuttal, 58–59.

19. Schriver Report, 12; Schriver Rebuttal to Hayflick Rebuttal, 106.

20. Factual Chronology, 38–39.

21. Ibid., 38.

22. Leonard Hayflick, interview with the author, March 4, 2013.

23. Schriver Report, 12.

24. Hayflick Rebuttal to Schriver Report, 14.

25. Stanley N. Cohen et al., "Construction of Biologically Functional Bacterial Plasmids *in Vitro,*" *Proceedings of the National Academy of Sciences* 70, no. 11 (1973): 3240–44.

26. Niels Reimers, "Stanford's Office of Technology Licensing and the Cohen/Boyer Cloning Patents," oral history conducted in 1997 by Sally Smith Hughes, p. 3, Regional Oral History Office, Bancroft Library, University of California at Berkeley, 1998.

27. Sally Smith Hughes, "Making Dollars Out of DNA: The First Major Patent in Biotechnology and the Commercialization of Molecular Biology, 1974–1980," *Isis* 92 (2001): 549; Berman, *Creating the Market University*, 64.

28. Rajendra K. Bera, "Commentary: The Story of the Cohen-Boyer Patents," *Current Science* 96, no. 6 (2009): 760.

29. Herbert W. Boyer, "Recombinant DNA Research at UCSF and Commercial Application at Genentech," oral history conducted in 1994 by Sally Smith Hughes, p. 98, Regional Oral History Office, Bancroft Library, University of California at Berkeley, www.oac.cdlib.org/view?docId=kt5d5nb0zs&brand=oac4&doc.view=entire_text (accessed February 26, 2016). Also cited in Berman, *Creating the Market University*, 66.

30. Stephen S. Hall, *Merchants of Immortality: Chasing the Dream of Human Life Extension* (Boston, New York: Houghton Mifflin Company, 2003), 37.

31. Ronald W. Lamont-Havers to Leonard Hayflick, October 10, 1974. Courtesy of Leonard Hayflick.

32. Hayflick Rebuttal to Schriver Report, 3–4; Leonard Hayflick, "Hayflick's Reply," *Science* 202 (1978): 129.

33. Schriver Rebuttal to Hayflick Rebuttal, 4–6.

34. Ibid., 6–7; Schriver Report, 1; Factual Chronology, 21.

35. Donald G. Murphy to Leonard Hayflick, January 31, 1975. Courtesy of Leonard Hayflick.

36. "Factual Chronology," 21.

37. Schriver Rebuttal to Hayflick Rebuttal, 6–7.

38. Leonard Hayflick, telephone interview with the author, October 16, 2012.

39. Schriver Rebuttal to Hayflick Rebuttal, 6.

40. "James Schriver Named Head of Newly Created OAM Audit Branch," *NIH Record* 25, no. 3 (February 12, 1963): 5; "James Schriver Retires After 17 Years at NIH," *NIH Record* 32, no. 7 (April 1, 1980): 4.

41. "James Schriver Named Head," 5.

42. Nicholas Wade, "Division of Biologics Standards: Scientific Management Questioned," *Science* 175 (1972): 966.

43. Philip M. Boffey, "The Fall and Rise of Leonard Hayflick, Biologist Whose Fight with US Seems Over," *New York Times*, January 19, 1982.

44. Richard Dugas, telephone interview with the author, April 27, 2013.

45. Nicholas Wade, telephone interview with the author, April 30, 2013.

46. All of Nancy Pleibel's recollections are drawn from interviews with the author on March 6 and 7, 2013.

47. Richard Dugas, telephone interview with the author, July 25, 2015.

48. Factual Chronology, 25, 28–29; Schriver Report, 5.

49. Nancy Pleibel, interview with the author, March 6, 2013.

50. Wistar Institute of Anatomy and Biology and Connaught Laboratories, License Agreement, January 1, 1973, folder "Connaught Correspondence," Stanley Plotkin private papers, Doylestown, PA.

51. Schriver Report, 7.

52. Ibid.

53. Factual Chronology, 26–27.

54. Ibid., 26.

55. Ibid., 27.

56. Ibid., 28.

57. Leonard Hayflick, telephone interview with the author, October 16, 2012.

58. Factual Chronology, 38.

59. Senate Subcommittee on Executive Reorganization and Government Research of the Committee on Government Operations, *Consumer Safety Act of 1972: Hearings on Titles I and II of S. 3419*, 92nd Cong., 2nd sess., April 20 and 21 and May 3 and 4, 1972, p. 35.

60. Factual Chronology, 38–39.

61. Ibid.

62. Leon Jacobs, memo to the record, "Re: Telephone Conversation with Mr. Don Brooks (Attorney), Merck and Company," March 31, 1976, investigations 9, file folder 1, Directors' Files, Office of the Director, National Institutes of Health.

63. M. F. Miller (Merck) "Memo for File: WI-38 Human Diploid Cells," July 11, 1974. Courtesy of Leonard Hayflick.

64. Riseberg Memo, 3.

65. Ibid.

66. Richard Thornburgh to Richard J. Riseberg, August 6, 1975, investigations 9, file folder 1, Directors' Files, Office of the Director, National Institutes of Health.

67. Factual Chronology, 33.

68. Donald G. Murphy, memos to the record, "PHS Working Group on WI-38," July 21 and 25, 1975, investigations 9, folder 1, Directors' Files, Office of the Director, National Institutes of Health.

69. J. E. Shannon, memo to Dr. R. Donvick, October 9, 1975, "Reference: Inventory of WI-38 Cells Delivered by NIH." Courtesy of Frank Simione, American Type Culture Collection, Manassas, VA.

Chapter Twenty-two: Rocky Passage

1. Nicholas Wade, "Hayflick's Tragedy: The Rise and Fall of a Human Cell Line," *Science* 192, no. 4235 (1976): 125.

2. Harold M. Schmeck Jr., "Investigator Says Scientist Sold Cell Specimens Owned by U.S.," *New York Times*, March 28, 1976, 1, 26.

3. Schriver Report, 1–14 and attachments A–C.

4. Factual Chronology, 51.

5. Hayflick Rebuttal to Schriver Report, 41, 45.

6. Ibid., 53; Nicholas Wade, "Vaccine Cells Found Mostly Contaminated," *Science* 194, no. 4260 (1976): 41.

7. Wade, "Vaccine Cells Found Mostly Contaminated," 41.

8. Leonard Hayflick, "Hayflick's Reply," *Science* 202 (1978): 131.

9. Leonard Hayflick, "Press Statement," Plotkin Rubella Papers, folder "Correspondence-H," Stanley Plotkin private papers.

10. Wade, "Hayflick's Tragedy," 125–27.

11. Nicholas Wade, telephone interview with the author, April 30, 2013.

12. Schmeck, "Investigator Says Scientist Sold," 26.

13. Schriver Report, 4; Factual Chronology, 13.

14. Factual Chronology, 34.

15. Wade, "Hayflick's Tragedy," 127.

16. Pan Demetrakakes, "Prof in Alleged Fund Misuse," *Stanford Daily* 169, no. 22 (March 3, 1976).

17. Factual Chronology, 43.

18. Ibid.

19. Clayton Rich, "Dean Rich Speaks on Hayflick Case," *Stanford University Campus Report* 8, no. 40 (July 21, 1976), investigations 9, file folder 1, Directors' Files, Office of the Director, National Institutes of Health.

20. William Fenwick, interview with the author, October 6, 2012.

21. Zhores A. Medvedev, "Letter: Hayflick's Tragedy," *Science* 192 (1976): 1182–84.

22. Hilleman is quoted in Stephen S. Hall, *Merchants of Immortality: Chasing the Dream of Human Life Extension* (Boston, New York: Houghton Mifflin Company, 2003), 39.

23. Albert Sabin to Bernard Strehler, December 1, 1981, box 11, file folder "Correspondence, Individual, Hayflick, Leonard, 1964–81," Correspondence-Individual (Graetz-Hayflick) series 1, Albert B. Sabin Collection, Henry R. Winkler Center for the History of the Health Professions, University of Cincinnati Libraries.

24. Fenwick, interview with the author.

25. *Cell Associates, Inc. and Leonard Hayflick v. National Institute* [sic] *of Health; Department of Health, Education and Welfare; and Donald Fredrickson, Director of National Institute* [sic] *of Health* (U.S. District Court for the Northern District of California, C76 601 RHS), March 25, 1976, p. 7, investigations 9, file folder 4, Directors' Files, Office of the Director, National Institutes of Health.

26. Herbert J. Kreitman to Leonard Hayflick, September 30, 1976, investigations 9, folder 2, Directors' Files, Office of the Director, NIH; Richard J. Riseberg, memo to Dr. Fredrickson and Mr. Schriver, "Subject: Cell Associates, Inc. v. NIH," November 2, 1978, investigations 9, folder 3, Directors' Files, Office of the Director, NIH; Nicholas Wade, "Despite the Length of Hayflick's Letter," *Science* 202 (1978): 136.

27. Leonard Hayflick, interview with the author, March 3, 2013.

28. Stanley A. Plotkin to Leonard Hayflick, May 27, 1976, file folder "Correspondence-H," Stanley Plotkin private papers.

29. Hayflick Rebuttal to Schriver Report, 1–65.

30. Schriver Rebuttal to Hayflick Rebuttal, 1–121.

31. Wade, "Vaccine Cells Found Mostly Contaminated," 41.

32. Schriver Report, 1.

33. *Cell Associates v. National Institute* [sic] *of Health*, 7.

34. "Human Cancer Cell Reconstruction and Transformation," 1 R01 CA18456-01, from 01/01/76 through 12/31/80, investigations 9, file folder 1, Directors' Files, Office of the Director, National Institutes of Health; Donald G. Murphy letter to Leonard Hayflick, July 30, 1976, Office of the Director, NIH, Directors' Files, investigations 9, folder 3.

35. Peter Libassi, memo to Thomas Morris and Donald Fredrickson, April 23, 1978, p. 2, Office of the Director, NIH, Directors' Files, investigations 9, folder 2.

36. Richard J. Riseberg, "Chronology," 4, attachment to Riseberg letter to Donald Fredrickson, January 6, 1978, Office of the Director, NIH, Directors' Files, investigations 9, folder 2.

37. Leonard Hayflick to Donald Fredrickson, November 8, 1978. Courtesy of Leonard Hayflick.
38. Richard J. Riseberg, memo to Robert N. Butler, "Subject: Application for Research Grant—Dr. Hayflick," 1–2, August 24, 1977, Office of the Director, NIH, Directors' Files, investigations 9, folder 2. The full Riseberg sentence reads: "It is my view, therefore, that there is no sound legal basis for rendering Dr. Hayflick categorically ineligible to serve as principal investigator on any NIH research grants."
39. Betty H. Pickett, memo to Donald Murphy, July 27, 1977. Courtesy of Leonard Hayflick.
40. Ibid.; Office of the Director, NIH, Directors' Files, investigations 9, folder 2.
41. Leon Jacobs, "Memorandum, Subject: Hayflick," December 8, 1977, Office of the Director, NIH, Directors' Files, investigations 9, folder 3.
42. Thomas D. Morris (inspector general), memo to the Hayflick File, "Subject: Meeting of January 9, 1978," Office of the Director, NIH, Directors' Files, investigations 9, folder 2.
43. Thomas D. Morris, memo to the Hayflick File, April 11, 1978. Courtesy of Leonard Hayflick.
44. Mary Miers, note to Dr. Malone, July 16, 1981, investigations 9, folder 2, Directors' Files, Office of the Director, National Institutes of Health.
45. Richard Riseberg, memo to Special Assistant to the Director, NIH, "Proposed Basis of Settlement in the Cell Associates Case," November 9, 1979, investigations 9, folder 2, Directors' Files, Office of the Director, National Institutes of Health.

Chapter Twenty-three: The Vaccine Race

1. Marcel Baltazard and Mehdi Ghodssi, "Prevention of Human Rabies: Treatment of Persons Bitten by Rabid Wolves in Iran," *Bulletin of the World Health Organization* 10, no. 5 (1954): 798.
2. Mahmoud Bahmanyar et al., "Successful Protection of Humans Exposed to Rabies Infection: Postexposure Treatment with the New Human Diploid Cell Rabies Vaccine and Antirabies Serum," *Journal of the American Medical Association* 236, no. 24 (1976): 2751–54.
3. T. J. Wiktor, S. A. Plotkin, and H. Koprowski, "Development and Clinical Trials of the New Human Rabies Vaccine of Tissue Culture (Human Diploid Cell) Origin," *Developments in Biological Standardization* 40 (1978): 3–9.
4. T. J. Wiktor et al., "Immunogenicity of Concentrated and Purified Rabies Vaccine of Tissue Culture Origin," *Proceedings of the Society for Experimental Biology and Medicine* 131 (1969): 799–805.
5. H. Koprowski, "In Vitro Production of Antirabies Virus Vaccine," in *International Symposium on Rabies, Talloires 1965*, Symposium Series on Immunobiological Standards, vol. 1 (Basel and New York: Karger, 1966), 363–64.
6. Centers for Disease Control, *Morbidity and Mortality Weekly Report* 34, nos. 2, 5, and 7 (1985), reprinted in *Journal of the American Medical Association* (March 15, 1985): 1540.
7. Ibid.; Wiktor, Plotkin, and Koprowski, "Development and Clinical Trials," 4.
8. Mario V. Fernandes, Hilary Koprowski, and Tadeusz J. Wiktor, "Method of Producing Rabies Vaccine," U.S. Patent 3,397,267, filed September 21, 1964, and issued August 13, 1968, www.google.com/patents/US3397267.
9. Wiktor, Plotkin, and Koprowski, "Development and Clinical Trials," 5; Tadeusz J. Wiktor, Stanley A. Plotkin, and Doris W. Grella, "Human Cell Culture Rabies Vaccine: Antibody Response in Man," *Journal of the American Medical Association* 224, no. 8 (1973): 1170–71.
10. Wiktor, Plotkin, and Grella, "Human Cell Culture Rabies Vaccine," 1170–71.
11. Wiktor, Plotkin, and Koprowski, "Development and Clinical Trials," 5, 7.
12. Bahmanyar et al., "Successful Protection of Humans," 2754.
13. Ibid.
14. Ibid.
15. Center for Disease Control, "Recommendation of the Immunization Practices Advisory Committee (ACIP): Rabies Prevention," *Morbidity and Mortality Weekly Report* 29, no. 3 (1980): 266.

16. Centers for Disease Control, "Rabies Postexposure Prophylaxis with Human Diploid Cell Rabies Vaccine: Lower Neutralizing Antibody Titers with Wyeth Vaccine," *Morbidity and Mortality Weekly Report* 34, no. 7 (1985): 90–92.

17. Ibid., 90–91.

18. United Press International, "Wyeth Laboratories Tuesday Recalled Its Wyvac Rabies Vaccine," February 19, 1985, http://www.upi.com/Archives/1985/02/19/Wyeth-Labora tories-Tuesday-recalled-its-Wyvac-rabies-vaccine-effective/5202477637200/.

19. Jeffrey P. Koplan and Stephen R. Preblud, "A Benefit-Cost Analysis of Mumps Vaccine," *American Journal of Diseases of Children* 136 (1982): 362; Kenneth B. Robbins, A. David Brandling-Bennett, and Alan R. Hinman, "Low Measles Incidence: Association with Enforce-ment of School Immunization Laws," *American Journal of Public Health* 71, no. 3 (1981): 270.

20. Michiaki Takahashi et al., "Live Vaccine Used to Prevent the Spread of Varicella in Chil-dren in Hospital," *Lancet* 2, no. 7892 (1974): 1288–90.

21. Michiaki Takahashi et al., "Development of Varicella Vaccine," *Journal of Infectious Diseases* 197 (2008): S41–44. See also Robert E. Weibel et al., "Live Attenuated Varicella Virus: Effi-cacy Trial in Healthy Children," *New England Journal of Medicine* 310, no. 22 (1984): 1409.

22. Beverly J. Neff et al., "Clinical and Laboratory Studies of KMcC Strain Live Attenuated Varicella Virus," *Proceedings of the Society for Experimental Biology and Medicine* 166, no. 3 (1981): 339–47.

23. Alan Shaw, interview with the author, March 16, 2014; Louis Galambos with Jane Eliot Sewell, *Networks of Innovation: Vaccine Development at Merck, Sharp & Dohme, and Mulford, 1985–1995* (Cambridge: Cambridge University Press, 1995), 231–32.

24. Neff et al., "Clinical and Laboratory Studies," 344.

25. Nicholas Wade, "Hayflick's Tragedy: The Rise and Fall of a Human Cell Line," *Science* 194, no. 4235 (1976): 125.

26. Philip Provost, interview with the author, December 18, 2012.

27. Weibel et al., "Live Attenuated Varicella Virus," 1409.

28. Wade, "Hayflick's Tragedy," 127.

29. E. L. Buescher, "Respiratory Disease and the Adenoviruses," *Medical Clinics of North America* 51 (1967): 773–74.

30. Centers for Disease Control, *Morbidity and Mortality Weekly Report: Recommendations and Reports* 39, no. RR-15 (1990): 1–18.

31. "1 Mil. Merck-Hayflick Contract for WI-38 Cells Revealed by NIH: Researcher Denies Wrongdoing, Sues Govt. for Defaming Character," *Blue Sheet: Drug Research Reports* 19, no. 13 (March 31, 1976): 3, investigations 9 (Human Diploid Cells Under Gvt. Owner-ship), file folder 1 (Jan–August 1976), Historical Files, Office of the Director, National Institutes of Health.

32. U.S. Food and Drug Administration, "WI-38 Cells Pre-tested for Vaccine Manufacture (ATCC 7/19/96)," response to Freedom of Information Act Request from Leonard Hay-flick, August 26, 1996. Courtesy of Leonard Hayflick.

33. John E. Shannon and Marvin Macy, eds., *The American Type Culture Collection Registry of Animal Cell Lines*, 2nd ed. (Rockville, MD: American Type Culture Collection, 1972), 17.

Chapter Twenty-four: Biology, Inc.

1. William Rutter, "The Department of Biochemistry and the Molecular Approach to Bio-medicine at the University of California, San Francisco," oral history conducted in 1992 by Sally Smith Hughes, p. 58, Regional Oral History Office, Bancroft Library, University of California at Berkeley, 1998.

2. Leonard Hayflick, interview with the author, March 5, 2013.

3. Ronald E. Cape, oral history conducted in 2003 by Sally Smith Hughes, Regional Oral History Office, Bancroft Library, University of California at Berkeley, http://digitalassets .lib.berkeley.edu/roho/ucb/text/cape_ron.pdf.

4. Robert Beyers, "Free Inquiry Must Be Rule in Research," *Campus Report* 13, no. 7 (1980): 1, 18. Found in Sally Smith Hughes, "Making Dollars Out of DNA: The First Major Patent in Biotechnology and the Commercialization of Molecular Biology, 1974–1980," *Isis* 92 (2001): p. 573.

5. Brook Byers, "Brook Byers: Biotechnology Venture Capitalist, 1970–2006," oral history conducted by Thomas D. Kiley, 2002–5, Regional Oral History Office, Bancroft Library, University of California, p. 19, http://digitalassets.lib.berkeley.edu/roho/ucb/text/byers_brook.pdf.

6. Wendy H. Schacht, "The Bayh-Dole Act: Selected Issues in Patent Policy and the Commercialization of Technology," *Congressional Research Service*, December 3, 2012, p. 2, https://www.fas.org/sgp/crs/misc/RL32076.pdf.

7. Elizabeth Popp Berman, *Creating the Market University: How Academic Science Became an Economic Engine* (Princeton, NJ: Princeton University Press, 2012), 106.

8. Ibid., 105. Readers interested in a more detailed history should read Berman's excellent account.

9. Ibid., 107–8.

10. Patent and Trademark Law Amendments Act, Public Law 96-517, U.S. Statutes at Large 94 (1980): 3015.

11. Berman, *Creating the Market University*, 108.

12. *Diamond v. Chakrabarty*, 447 U.S. 303 (1980), https://supreme.justia.com/cases/federal/us/447/303/case.html.

13. Ibid., 309.

14. The 2015 figures were provided by the Biotechnology Innovation Organization in Washington, DC.

15. Readers interested in close coverage of these converging forces should read Elizabeth Popp Berman's account in *Creating the Market University*, 69–79.

16. Berman, *Creating the Market University*, 83.

17. Hughes, "Making Dollars Out of DNA," : 569.

18. Maryann Feldman, Alessandra Colaianni, and Kang Liu, "Commercializing Cohen-Boyer, 1980–1997," *Druid* (Toronto: University of Toronto Rotman School of Management, 2005), 1.

19. Floyd Grolle (former licensing officer of the Cohen-Boyer patents), personal communication to Sally Smith Hughes, cited in Hughes, "Making Dollars Out of DNA," 570.

20. Michael Cleare (former executive director of Columbia's Science and Technology Ventures Office), personal communication to Richard R. Nelson (Henry R. Luce professor of international political economy at Columbia University) and Bhaven Sampat (assistant professor of health policy and management, Mailman School of Public Health, Columbia University), August 29, 2006, cited in Alessandra Colaianni and Robert Cook-Deegan, "Columbia University's Axel Patents: Technology Transfer and Implications for the Bayh-Dole Act," *Milibank Quarterly* 87, no. 3 (2009): 700, 711.

21. National Science Board, *Science and Engineering Indicators: 1993* (Washington, DC: National Science Foundation, 1993), 430; Association of University Technology Managers, *FY 2014 US Licensing Activity Survey* (Oakbrook Terrace, IL: Association of University Technology Managers, 2014), 23.

22. National Science Board, *Science and Engineering Indicators 2014* (Arlington, VA: National Science Foundation 2014): 5–54.

23. David Blumenthal et al., "University-Industry Research Relationships in Biotechnology: Implications for the University," *Science* 232, no. 4756 (1986): 1364.

24. Vincent B. Terlep Jr. to Thomas Malone, February 22, 1980, Directors' Files, NIH, Office of the Director, investigations 9, folder 2.

25. Richard J. Riseberg, note to Dr. Raub, July 24, 1989, investigations 9 (Human Diploid Cells Under Gvt. Ownership), file folder 2 (September 1976–July 1989), Historical Files,

Office of the Director, National Institutes of Health. This letter reads in part: "[Hayflick is] not aware that a principal consideration in causing the government to settle had nothing to do with the merits but just the fact that Jim Schriver had retired and it was proving very expensive to use him as a consultant." Riseberg was discussing the matter with William Raub, then the NIH's acting director, because Hayflick had goaded Raub by sending the top NIH official a new paper that Hayflick had written, giving a one-sided account of his conflict with the NIH and describing how some biologists had profited from the legal and policy changes of the previous decade. Hayflick's paper was titled "A New Technique for Transforming Purloined Human Cells into Acceptable Federal Policy."

26. Leonard Hayflick, Edmond C. Gregorian, Michael Hughes, and Vincent B. Terlep, "Settlement Agreement," September 15, 1981, 1–8, investigations 9, folder 1, Directors' Files, Office of the Director, National Institutes of Health.

27. Ibid., 2.

28. Bernard L. Strehler et al., "Hayflick-NIH Settlement," *Science* 215 (1982): 240, 242.

29. Leonard Hayflick, "Hayflick's Reply," *Science* 202 (1978): 128–36.

Chapter Twenty-five: Hayflick's Limit Explained

1. Leonard Hayflick and Paul S. Moorhead, "The Serial Cultivation of Human Diploid Cell Strains," *Experimental Cell Research* 25, no. 3 (1961): 585–621.

2. Leonard Hayflick, "The Limited *in Vitro* Lifetime of Human Diploid Cell Strains," *Experimental Cell Research* 37 (1965): 634.

3. P. L. Krohn, "Aging," *Science* 152 (1966): 392.

4. Ibid.; Sir Macfarlane Burnet, *Intrinsic Mutagenesis: A Genetic Approach to Ageing* (Lancaster, UK: Medical and Technical, 1974): 62; L. M. Franks, "Cellular Aspects of Aging," *Experimental Gerontology* 5 (1970): 281–89; R. L. Walford, *The Immunologic Theory of Aging* (Copenhagen: Munksgaard, 1969).

5. Burnet, *Intrinsic Mutagenesis*, 62.

6. Hayflick, "The Limited *in Vitro* Lifetime," 625.

7. Ibid.

8. G. M. Martin, C. A. Sprague, and C. J. Epstein, "Replicative Life-span of Cultivated Human Cells: Effects of Donor's Age, Tissue, and Genotype," *Laboratory Investigation: A Journal of Technical Methods and Pathology* 23, no. 1 (1970): 86–92; Y. Le Guilly et al., "Long-term Culture of Human Adult Liver Cells: Morphological Changes Related to in Vitro Senescence and Effect of Donor's Age on Growth Potential," *Gerontologia* 19, no. 5 (1973): 303–13; E. L. Bierman, "The Effect of Donor Age on the in Vitro Life Span of Cultured Human Arterial Smooth-Muscle Cells," *In Vitro* 14, no. 11 (1978): 951–55.

9. J. R. Smith and R. G. Whitney, "Intraclonal Variation in Proliferative Potential of Human Diploid Fibroblasts: Stochastic Mechanism for Cellular Aging," *Science* 207 (1980): 82–84.

10. Ibid., 82.

11. W. E. Wright and L. Hayflick, "Formation of Anucleate and Multinucleate Cells in Normal and SV40 Transformed WI-38 by Cytochalasin B," *Experimental Cell Research* 74 (1972): 187–94.

12. W. E. Wright and L. Hayflick, "Nuclear Control of Cellular Aging Demonstrated by Hybridization of Anucleate and Whole Cultured Normal Human Fibroblasts," *Experimental Cell Research* 96, no. 1 (1975): 113–21.

13. Leonard Hayflick, "Mortality and Immortality at the Cellular Level: A Review," *Biochemistry (Moscow)* 62 (1997): 1180–90.

14. Hermann J. Muller, "The Remaking of Chromosomes," *Collecting Net* 13 (1938): 181–98.

15. Barbara McClintock, "The Stability of Broken Ends of Chromosomes in *Zea mays*," *Genetics* 126 (1941): 234–82.

16. Alexey M. Olovnikov, "Telomeres, Telomerase, and Aging: Origin of the Theory," *Experimental Gerontology* 31, no. 4 (1996): 445.

17. A. M. Olovnikov ["Principles of Marginotomy in Template Synthesis of Polynucleoetides"], *Doklady Akademii Nauk SSSR* 201, no. 6 (1971): 1496–99 (in Russian).

18. A. M. Olovnikov, "A Theory of Marginotomy: The Incomplete Copying of Template Margin in Enzymic Synthesis of Polynucleotides and Biological Significance of the Phenomenon," *Journal of Theoretical Biology* 41 (1973): 186.

19. Ibid., 188.

20. Ibid., 181–90.

21. James D. Watson, "Origin of Concatameric T7 DNA," *Nature: New Biology* 239 (1972): 197–201.

22. Catherine Brady, *Elizabeth Blackburn and the Story of Telomeres: Deciphering the Ends of DNA* (Cambridge, MA, and London: MIT Press, 2007), 3.

23. Elizabeth H. Blackburn and Joseph G. Gall, "A Tandemly Repeated Sequence at the Termini of the Extrachromosomal Ribosomal RNA Genes in *Tetrahymena*," *Journal of Molecular Biology* (1978): 33–53.

24. Elizabeth H. Blackburn, Carol W. Greider, and Jack W. Szostak, "Telomeres and Telomerase: The Path from Maize, *Tetrahymena* and Yeast to Human Cancer and Aging," *Nature Medicine* 12, no. 10 (2006): 1134; Jack W. Szostak, "Biographical," Nobelprize.org, www.nobelprize.org/nobel_prizes/medicine/laureates/2009/szostak-bio.html.

25. Jack W. Szostak and Elizabeth H. Blackburn, "Cloning Yeast Telomeres on Linear Plasmid Vectors," *Cell* 29, no. 1 (1982): 245–55.

26. Janis Shampay, Jack W. Szostak, and Elizabeth H. Blackburn, "DNA Sequences of Telomeres Maintained in Yeast," *Nature* 310, no. 5973 (1984): 154–57.

27. Carol W. Greider, e-mail to the author, May 15, 2016.

28. Carol W. Greider and Elizabeth H. Blackburn, "Identification of a Specific Telomere Terminal Transferase Activity in Tetrahymena Extracts," *Cell* 43, no. 2, part 1 (1985): 405–13.

29. Guo-Liang Yu et al., "*In Vivo* Alteration of Telomere Sequences and Senescence Caused by Mutated *Tetrahymena* Telomerase RNAs," *Nature* 344 (1990): 126–32.

30. Carol W. Greider and Elizabeth H. Blackburn, "A Telomeric Sequence in the RNA of *Tetrahymena* Telomerase Required for Telomere Repeat Synthesis," *Nature* 337 (1989): 331–37.

31. Carol Greider, e-mail to the author.

32. Robert K. Moyzis et al., "A Highly Conserved Repetitive DNA Sequence (TTAGGG)n, Present at the Telomeres of Human Chromosomes," *Proceedings of the National Academy of Sciences* 85 (1988): 6622–26; Robin C. Allshire et al., "Telomeric Repeats from *T. thermophila* Cross Hybridize with Human Telomeres," *Nature* 332 (1988): 656–59.

33. Calvin B. Harley, A. Bruce Futcher, and Carol W. Greider, "Telomeres Shorten During Ageing of Human Fibroblasts," *Nature* 345 (1990): 459.

34. Stephen S. Hall, *Merchants of Immortality: Chasing the Dream of Human Life Extension* (Boston and New York: Houghton Mifflin, 2003), 58.

35. Andrea G. Bodnar et al., "Extension of Life-span by Introduction of Telomerase into Normal Human Cells," *Science* 279 (1998): 349–52.

36. Nicholas Wade, "Cells' Life Stretched in Lab," *New York Times*, January 14, 1998, p. A1, www.nytimes.com/1998/01/14/us/cells-life-stretched-in-lab.html; Nicholas Wade, "Cells Unlocked, Longevity's New Lease on Life," Week in Review, *New York Times*, January 18, 1998, www.nytimes.com/1998/01/18/weekinreview/ideas-trends-cells-unlocked-longevity-s-new-lease-on-life.html; Nicholas Wade, "Cell Rejuvenation May Yield Rush of Medical Advances," Science Times, *New York Times*, January 20, 1998, www.nytimes.com/1998/01/20/science/cell-rejuvenation-may-yield-rush-of-medical-advances.html.

37. Carl T. Hall, "Non-aging Human Cells Created in Lab; Bay Firm's Stock Soars on Hopes of Medical Advances," *San Francisco Chronicle*, January 14, 1998, p. 1.

38. David Stipp, "The Hunt for the Youth Pill: From Cell-Immortalizing Drugs to Cloned Organs, Biotech Finds New Ways to Fight Against Time's Toll," *Fortune*, October 11, 1999, 199, http://archive.fortune.com/magazines/fortune/fortune_archive/1999/10/11/267014/index.htm.

39. Mary Armanios and Elizabeth H. Blackburn, "The Telomere Syndromes," *Nature Reviews Genetics* 13 (2012): 693–704; Susan E. Stanley and Mary Armanios, "The Short and Long Telomere Syndromes: Paired Paradigms for Molecular Medicine," *Current Opinion in Genetics & Development* 33 (2015): 2–3.

40. Mary Armanios, "Telomeres and Age-Related Disease: How Telomere Biology Informs Clinical Paradigms," *Journal of Clinical Investigation* 123, no. 3 (2013): 996–1002.

41. Gil Atzmon et al., "Genetic Variation in Human Telomerase Is Associated with Telomere Length in Ashkenazi Centenarians," *Proceedings of the National Academy of Sciences* 107, supp. 1 (2010): 1710–17.

42. Nam W. Kim et al., "Specific Association of Human Telomerase Activity with Immortal Cells and Cancer," *Science* 266 (1994): 2011–15.

43. Blackburn, Greider, and Szostak, "Telomeres and Telomerase," 1137.

44. Stanley and Armanios, "Short and Long Telomere Syndromes," 3–6.

Chapter Twenty-six: Boot-Camp Bugs and Vatican Entreaties

1. Meredith Wadman, "Cell Division," *Nature* 498 (2013): 425.

2. Debi Vinnedge, telephone interview with the author, May 23, 2013.

3. Y. Takeuchi et al., ["Field Trial of Combined Measles and Rubella Live Attenuated Vaccine"], *Kansenshogaku Zasshi* [Japanese Journal of Infectious Diseases] 76, no. 1 (2002): 56–62.

4. The Japanese hepatitis A vaccine, Aimmugen, is made by the firm Kaketsuken using a line of cells from one African green monkey called GL37 cells. See David E. Anderson, "Compositions and Methods for Treating Viral Infections," U.S. Patent 20,140,356,399 A1, filed January 11, 2013, and issued December 4, 2014, section 0032.

5. Isabelle Claxton to Debra Vinnedge, November 1, 2000. Courtesy of Debi Vinnedge.

6. Bruce Ellis, assistant counsel, Merck & Co., to the Securities and Exchange Commission, December 20, 2002. Courtesy of Debi Vinnedge. Alex Shukhman, attorney adviser, Securities and Exchange Commission, "Response of the Office of Chief Counsel, Division of Corporate Finance, February 26, 2003, Re: Merck & Co., Inc., Incoming Letter Dated December 20, 2002." Courtesy of Debi Vinnedge.

7. Debi Vinnedge, e-mail to the author, April 17, 2016.

8. Debra L. Vinnedge to Cardinal Joseph Ratzinger, June 4, 2003. Courtesy of Debi Vinnedge.

9. Sarah Lueck, "Boot-Camp Bug Returns to the Barracks When Pentagon Pulls the Plug on Vaccine," *Wall Street Journal*, July 13, 2001.

10. Centers for Disease Control, "Two Fatal Cases of Adenovirus-Related Illness in Previously Healthy Young Adults—Illinois, 2000," *Morbidity and Mortality Weekly Report* 50, no. 26 (2001): 553–55.

11. Ibid., 555.

12. Ibid., 553; E. L. Buescher, "Respiratory Disease and the Adenoviruses," *Medical Clinics of North America* 51 (1967): 772, 778.

13. Lueck, "Boot-Camp Bug Returns."

14. Robert N. Potter et al., "Adenovirus-Associated Deaths in US Military During Postvaccination Period, 1999–2010," *Emerging Infectious Diseases* 18, no. 3 (2012): 507–9 (doi:10.3201/eid1803.111238); Project Director, "Adenovirus Vaccines Reinstated After Long Absence," The History of Vaccines Blog, April 17, 2012. www.historyofvaccines.org/content/blog/adenovirus-vaccines-reinstated-after-long-absence (accessed June 18, 2016).

15. Operational Infectious Diseases, Naval Health Research Center, "Febrile Respiratory Illness (FRI) Surveillance Update," 2015, week 38 (through September 26, 2015), 2, www .med.navy.mil/sites/nhrc/geis/Documents/FRIUpdate.pdf (accessed April 12, 2016).

16. The MacConnells' story is drawn from in-person interviews with Chip and Betsy MacConnell, June 27 and 28, 2015; from medical records provided by the MacConnells, and by Janet Gilsdorf; from several e-mails from Chip MacConnell to the author in 2014 and 2015; and from http://www.angelsforanna.com/.

17. Pontifical Academy for Life, "Moral Reflections on Vaccines Prepared from Cells Derived from Aborted Human Foetuses," Zenit, July 26, 2005, https://zenit.org/articles/on -vaccines-made-from-cells-of-aborted-fetuses/ (accessed April 17, 2016).

18. David Mitchell, "Merck Focusing on Combination Vaccine: Manufacturer Stops Sales of Monovalents for Measles, Mumps, Rubella," *American Academy of Family Physicians News* (online), December 24, 2008.

19. Centers for Disease Control and Prevention, "Measles, Mumps, and Rubella: Vaccine Use and Strategies for Elimination of Measles, Rubella, and Congenital Rubella Syndrome and Control of Mumps: Recommendations of the Advisory Committee on Immunization Practices (ACIP)," *Morbidity and Mortality Weekly Report* 47, no. RR-8 (1998): 1–57, www.cdc.gov/mmwr/preview/mmwrhtml/00053391.htm.

20. Mona Marin et al., "Use of Combination Measles, Mumps, Rubella, and Varicella Vaccine: Recommendations of the Advisory Committee on Immunization Practices (ACIP)," *Morbidity and Mortality Weekly Report Recommendations and Reports* 59, no. RR-3 (2010):1–12, www.ncbi.nlm.nih.gov/pubmed/20448530.

21. Merck & Co., Inc. to [name redacted], July 1, 2009, https://cogforlife.org/wp-content/ uploads/2012/05/merck2009Response.pdf (accessed April 18, 2016).

22. Children of God for Life, "Measles Outbreaks: Blame Merck—Not the Parents!" (press release), January 25, 2015, https://cogforlife.org/measlesPress.pdf (accessed April 16, 2016).

Chapter Twenty-seven: The Afterlife of a Cell

1. Alan Shaw, interview with the author, March 16, 2014.

2. Goberdhan P. Dimri et al., "A Biomarker That Identifies Senescent Human Cells in Culture and in Aging Skin *in Vivo*," *Science* 92 (1995): 9363–67.

3. Takeshi Shimi et al., "The Role of Nuclear Lamin B1 in Cell Proliferation and Senescence," *Genes and Development* 25, no. 24 (December 15, 2011): 2579–93.

4. Tal Ilani et al., "A Secreted Disulfide Catalyst Controls Extracellular Matrix Composition and Function," *Science* 341 (2013): 75–76.

5. Jessica Leung, Stephanie R. Bialek, and Mona Marin, "Trends in Varicella Mortality in the United States: Data from Vital Statistics and the National Surveillance System," *Human Vaccines & Immunotherapeutics* 11, no. 3 (2015): 662; M. Marin, J. X. Zhanag, and J. F. Seward, "Near Elimination of Varicella Deaths in the US After Implementation of the Vaccination Program," *Pediatrics* 128, no. 2 (2011): 214–20.

6. Centers for Disease Control and Prevention, "Notifiable Infectious Diseases and Conditions: United States, 2013," *Morbidity and Mortality Weekly Report* 62, no. 53 (2015): 27.

7. Ibid., 108; Leung, Bialek, and Marin, "Trends in Varicella Mortality," 663.

8. Centers for Disease Control and Prevention, "Prevention of Hepatitis A Through Active or Passive Immunization," *Morbidity and Mortality Weekly Report* 55, no. RR-7 (2006), www.cdc.gov/mmwr/pdf/rr/rr5507.pdf.

9. Division of Viral Hepatitis, Centers for Disease Control and Prevention, "Viral Hepatitis Surveillance: United States, 2013," 16, fig. 2.1, www.cdc.gov/hepatitis/statistics/2013 surveillance/pdfs/2013hepsurveillancerpt.pdf.

10. Alexia Harrist et al. "Human Rabies—Wyoming and Utah, 2015," *Morbidity and Mortality Weekly Report* 65, no. 21 (2016): 529–33. Available here: http://www.cdc.gov/mmwr/ volumes/65/wr/mm6521a1.htm.

11. L. Dyer et al., "Rabies Surveillance in the United States During 2013," *Journal of the American Veterinary Medical Association* 245, no. 10 (2014): 1111, http://avmajournals .avma.org/doi/pdf/10.2460/javma.245.10.1111.

12. Centers for Disease Control and Prevention, "Human Rabies: Human Rabies Surveillance," CDC.gov, September 21, 2015, www.cdc.gov/rabies/location/usa/surveillance/ human_rabies.html.

13. Harrist et al., "Human Rabies—Wyoming and Utah, 2015," 529–33.

14. P. D. Pratt et al., "Human Rabies: Missouri, 2014," *Morbidity and Mortality Weekly Report* 65, no. 10 (2016): 253–56, www.cdc.gov/mmwr/volumes/65/wr/mm6510a1.htm.

15. R. E. Willoughby et al., "Survival After Treatment of Rabies with Induction of Coma," *New England Journal of Medicine* 352, no. 24 (June 2005): 2508–14.

16. Ahmad Fayaz et al., "Antibody Persistence, 32 Years After Post-Exposure Prophylaxis with Human Diploid Cell Rabies Vaccine (HDCV)," *Vaccine* 29 (2011): 3742–45.

17. Centers for Disease Control, "Surveillance Summary: Rubella—United States," *Morbidity and Mortality Weekly Report* 19, no. 34 (1970): 336.

18. Dorothy M. Horstmann, "Maxwell Finland Lecture: Viral Vaccines and Their Ways," *Reviews of Infectious Diseases* 1, no. 3 (1979): 509.

19. Centers for Disease Control, "Rubella, United States, 1977–80," *Morbidity and Mortality Weekly Report* 29 (1980): 378.

20. Horstmann, "Maxwell Finland Lecture," 510.

21. Centers for Disease Control, "Annual Summary 1980: Reported Morbidity and Mortality in the United States," *Morbidity and Mortality Weekly Report* 29, no. 54 (1981): 12.

22. Centers for Disease Control and Prevention, "Summary of Notifiable Diseases: United States, 2001," *Morbidity and Mortality Weekly Report* 50, no. 53 (2003): 93.

23. Centers for Disease Control and Prevention, "Control and Prevention of Rubella: Evaluation and Management of Suspected Outbreaks, Rubella in Pregnant Women, and Surveillance for Congenital Rubella Syndrome," *Morbidity and Mortality Weekly Report* 50, no. RR-12 (2001): 1.

24. Pamela Eisele (Merck), e-mail to the author, August 31, 2015.

25. G. B. Grant et al., "Global Progress Toward Rubella and Congenital Rubella Syndrome Control and Elimination: 2000–2014," *Morbidity and Mortality Weekly Report* 64, no. 37 (2015): 1052–55.

26. Ryo Kinoshita and Hiroshi Nishiura, "Assessing Herd Immunity Against Rubella in Japan: A Retrospective Seroepidemiological Analysis of Age-Dependent Transmission Dynamics," *BMJ Open* 6, no. 1 (2016): 1.

27. Centers for Disease Control and Prevention, "Nationwide Rubella Epidemic: Japan, 2013," *Morbidity and Mortality Weekly Report* 62, no. 23 (2013): 457–62.

28. Kinoshita and Nishiura, "Assessing Herd Immunity," 1.

29. Centers for Disease Control and Prevention, "Three Cases of Congenital Rubella Syndrome in the Postelimination Era: Maryland, Alabama, and Illinois, 2012," *Morbidity and Mortality Weekly Report* 62, no. 12 2013): 226–29.

30. Nathaniel Lambert et al., "Seminar: Rubella," *Lancet* 385 (2015): 2297, 2300–2301.

31. Center for Biologics Evaluation and Research, Food and Drug Administration, *Guidance for Industry: Characterization and Qualification of Cell Substrates and Other Biological Materials Used in the Production of Viral Vaccines for Infectious Disease Indications* (Rockville, MD: U.S. Department of Health and Human Services, 2007), 36.

32. World Health Organization Expert Committee on Biological Standardization, *Fifty-sixth Report*, WHO Technical Report Series 941 (Geneva: World Health Organization, 2007), 103.

33. Alan Shaw, interview with the author, March 16, 2014.

34. Jean Wallace, "Turmoil Besets Wistar in Wake of Koprowski's Ouster," *Scientist* 6, no. 5 (March 2, 1992), www.the-scientist.com/?articles.view/articleNo/12194/title/Turmoil

-Besets-Wistar-In-Wake-Of-Koprowski-s-Ouster/; Maurice Hilleman, interview with Paul Offit, November 30, 2004. Hilleman said that the figure was $3.5 million.

35. Nancy Leone (communications director, Global Specialty Medicines, Teva), e-mail to the author, February 20, 2014.

36. Robert D. Truog, Aaron S. Kesselheim, and Steven Joffe, "Paying Patients for Their Tissue: The Legacy of Henrietta Lacks," *Science* 337 (2012): 37–38.

37. Meredith Wadman, "Cell Division," *Nature* 498 (2013): 426.

38. Steven Joffe, telephone interview with the author, February 23, 2013. In 2016 Joffe is a pediatric oncologist and bioethicist at the University of Pennsylvania Perelman School of Medicine.

39. Scott D. Kominers and Gary S. Becker, "Paying for Tissue: Net Benefits," *Science* 337 (2012): 1292–93.

40. E. Eiseman and S. B. Haga, *Handbook of Human Tissue Resources: A National Resource of Human Tissue Samples* (Santa Monica, CA: RAND Corporation, MR-954-OSTP, 1999).

41. "Federal Policy for the Protection of Human Subjects: Proposed Rules," *Federal Register* 80, no. 173 (September 8, 2015): 53936.

42. Jocelyn Kaiser, "Researchers Decry Consent Proposal," *Science* 352, no. 6288 (2016): 878–79; David Malakoff, "Panel Slams Plan for Human Research Rules," *Science* 353, no. 6295 (2016): 106–7; National Academies of Sciences, Engineering, and Medicine, *Optimizing the Nation's Investment in Academic Research: A New Regulatory Framework for the 21st Century* (Washington, DC: National Academies Press, 2016): 97–103.

43. *Code of Federal Regulations*, title 45, part 46, section 206: "Research Involving, After Delivery, the Placenta, the Dead Fetus or Fetal Material," October 1, 2014.

44. The following are the states that have *not* used an organ-donor law called the Uniform Anatomical Gift Act to require a woman's consent for the use of tissue from her aborted fetus in research: Florida, Iowa, Kentucky, Louisiana, Maryland, Michigan, Minnesota, Missouri, New Mexico, Nebraska, South Carolina, South Dakota, Washington, and Wisconsin. Michigan in 1978 enacted a separate law requiring such consent, *Michigan Compiled Laws Annotated,* section 333.2688.

45. Henry K. Beecher, "Ethics and Clinical Research," *New England Journal of Medicine* 274, no. 24 (1966): 1354–60, www.nejm.org/doi/pdf/10.1056/NEJM196606162742405; William H. Stewart, "Surgeon General's Directives on Human Experimentation," PPO #129 (Bethesda, MD: U.S. Public Health Service Division of Research Grants, revised July 1, 1966).

46. Jean Heller, "Syphilis Victims in US Study Went Untreated for Forty Years," Associated Press, July 24, 1972.

47. Allan M. Brandt, "Racism and Research: The Case of the Tuskegee Syphilis Study," *Hastings Center Report* 8, no. 6 (1978): 21–29.

48. Public Law 93-348, section 212, July 12, 1974.

49. W. J. Curran, "Government Regulation of the Use of Human Subjects in Medical Research: The Approaches of Two Federal Agencies," in *Experimentation with Human Subjects*, ed. P. A. Freund (New York: George Braziller, 1970), 443.

50. *Code of Federal Regulations*, title 45, part 46.

51. Meredith Wadman, "The Truth About Fetal Tissue Research," *Nature* 578 (2015): 178–81.

52. Molly Redden, "Vital Fetal Tissue Research Threatened by House of Representatives Subpoenas," the *Guardian*, April 1, 2016. https://www.theguardian.com/science/2016/apr/01/congress-subpoenas-fetal-tissue-research-abortion.

53. Denise Grady and Nicholas St. Fleur, "Fetal Tissue from Abortions for Research Is Traded in a Gray Zone," *New York Times*, July 27, 2015.

54. Redden, "Vital Fetal Tissue Research Threatened."

55. Bo Ma et al., "Characteristics and Viral Propagation Properties of a New Human Diploid Cell Line, Walvax-2, and Its Suitability as a Candidate Cell Substrate for Vaccine Production," *Human Vaccines & Immunotherapeutics* 11, no. 4 (2015): 998–1009.

56. Debi Vinnedge, "Scientists in China Create New Vaccines Using Body Parts from Nine Aborted Babies," LifeNews.com, September 9, 2015.

57. Pamela Eisele, e-mail to the author, July 1, 2016.

58. Leonard Hayflick, "Chain of Custody of Original Ampules of 8th Population Doubling Level WI-38 From June, 1962 Until July, 1996," undated. Courtesy of Leonard Hayflick.

59. Meredith Wadman, "Cell Division," *Nature* 498 (2013): 426, www.nature.com/news/medical-research-cell-division-1.13273.

Epilogue: Where They Are Now

1. Hermann Hesse, *The Glass Bead Game,* translated from the German by Richard and Clara Winston (New York: Holt, Rinehart and Winston, 1969, 1990), 169.

2. K. Stratton, D. A. Almario, and M. C. McCormick, eds., *Immunization Safety Review: SV40 Contamination of Polio Vaccine and Cancer* (Washington, DC: National Academies, 2002).

3. An archived version of the CDC fact sheet is available here: http://web.archive.org/web/20130522091608/http://www.cdc.gov/vaccinesafety/updates/archive/polio_and_cancer_factsheet.htm.

4. Leonard Hayflick, interview with the author, March 5, 2013.

5. Lara Marks, *"The Lock and Key of Medicine": Monoclonal Antibodies and the Transformation of Healthcare* (New Haven, CT, and London: Yale University Press, 2015), chapter 2.

6. Jean Wallace, "Turmoil Besets Wistar in Wake of Koprowski's Ouster," *Scientist*, March 2, 1992, www.the-scientist.com/?articles.view/articleNo/12194/title/Turmoil-Besets-Wistar-In-Wake-Of-Koprowski-s-Ouster/.

7. "Age Discrimination—Hilary Koprowski—Wistar Institute: Update: Koprowski, Wistar Settle Age Discrimination Suit—Pioneer in Immunotherapy Honored by Cancer Scientists," *Biotechnology Law Report* 12, no. 3 (1993): 261–62.

8. Silvana Regina Favoretto et al., "Experimental Infection of the Bat Tick *Carios fonsecai* (Acari: Ixodidae) with the Rabies Virus," *Revista da Socieda de Brasileira de Medicina Tropical* 46, no. 6 (2013): 788–90.

SELECTED BIBLIOGRAPHY

Articles and Books

Artenstein, Andrew W., ed. *Vaccines: A Biography.* New York: Springer Science + Business Media, 2010.

Baer, George M., ed. *The Natural History of Rabies.* 2nd edition. Boca Raton, FL: CRC, 1991.

Bahmanyar, Mahmoud, Ahmad Fayaz, Shokrollah Nour-Salehi, Manouchehr Mohammadi, and Hilary Koprowski. "Successful Protection of Humans Exposed to Rabies Infection: Postexposure Treatment with the New Human Diploid Cell Rabies Vaccine and Antirabies Serum." *Journal of the American Medical Association* 236, no. 24 (1976): 2751–54.

Banatvala, J. E., and D. W. G. Brown. "Seminar: Rubella." *Lancet* 363 (2004): 1127–37.

Barry, John M. *The Great Influenza: The Story of the Deadliest Pandemic in History.* New York: Viking Penguin, 2004.

Beecher, Henry K. "Ethics and Clinical Research." *New England Journal of Medicine* 274, no. 24 (1966): 1354–60. www.nejm.org/doi/pdf/10.1056/NEJM196606162742405.

Berman, Elizabeth Popp. *Creating the Market University: How Academic Science Became an Economic Engine.* Princeton, NJ: Princeton University Press, 2012.

Bodnar, Andrea G., Michel Ouellette, Maria Frolkis, Shawn E. Holt, Choy-Pik Chiu, Gregg B. Morin, Calvin B. Harley, Jerry W. Shay, Serge Lichtsteiner, and Woodring E. Wright. "Extension of Life-span by Introduction of Telomerase into Normal Human Cells." *Science* 279 (1998): 349–52.

Bookchin, Debbie, and Jim Schumacher. *The Virus and the Vaccine: Contaminated Vaccine, Deadly Cancers and Government Neglect.* New York: St. Martin's, 2004.

Brady, Catherine. *Elizabeth Blackburn and the Story of Telomeres: Deciphering the Ends of DNA.* Cambridge, MA, and London: MIT Press, 2007.

Burnet, Sir Macfarlane. *Intrinsic Mutagenesis: A Genetic Approach to Ageing.* Lancaster, UK: Medical and Technical, 1974.

Cohen, Adam. *Imbeciles: The Supreme Court, American Eugenics and the Sterilization of Carrie Buck.* New York: Penguin, 2016.

Cristofalo, Vincent J. "Profile in Gerontology: Leonard Hayflick, PhD." *Contemporary Gerontology* 9, no. 3 (2003): 83–86.

Davis, Allen F., and Mark H. Haller. *The Peoples of Philadelphia: A History of Ethnic Groups and Lower-Class Life, 1790–1940.* Philadelphia: Temple University Press, 1973.

Eddy, Bernice E., Gerald S. Borman, William H. Berkeley, and Ralph D. Young. "Tumors Induced in Hamsters by Injection of Rhesus Monkey Kidney Cell Extracts." *Proceedings of the Society for Experimental Biology and Medicine* 107, no. 1 (1961): 191–200.

Eddy, Bernice E., Gerald S. Borman, George E. Grubbs, and Ralph D. Young. "Identification of the Oncogenic Substance in Rhesus Monkey Kidney Cell Cultures as Simian Virus 40." *Virology* 17 (1962): 65–75.

Friedman, Meyer, and Gerald W. Friedland. *Medicine's 10 Greatest Discoveries.* New Haven, CT, and London: Yale University Press, 1998.

Galambos, Louis, with Jane Eliot Sewell. *Networks of Innovation: Vaccine Development at Merck, Sharp & Dohme, and Mulford, 1985–1995.* Cambridge: Cambridge University Press, 1995.

Gregg, Norman McAlister. "Congenital Cataract Following German Measles in the Mother." *Transactions of the Ophthalmological Society of Australia* 3 (1941): 35–46.

Greider, Carol W., and Elizabeth H. Blackburn. "Identification of a Specific Telomere Terminal Transferase Activity in *Tetrahymena* Extracts." *Cell* 43 (1985): 405–13.

Greider, Carol W., and Elizabeth H. Blackburn. "A Telomeric Sequence in the RNA of *Tetrahymena* Telomerase Required for Telomere Repeat Synthesis." *Nature* 337 (1989): 331–37.

Hall, Stephen S. *Merchants of Immortality: Chasing the Dream of Human Life Extension*. Boston and New York: Houghton Mifflin, 2003.

Harley, Calvin B., A. Bruce Futcher, and Carol W. Greider. "Telomeres Shorten During Ageing of Human Fibroblasts." *Nature* 345 (1990): 458–60.

Hawkes, Mary Q. *Excellent Effect: The Edna Mahan Story*. Arlington, VA: American Correctional Association, 1994.

Hayflick, Leonard. "Hayflick's Reply." *Science* 202 (1978): 128–36.

———. *How and Why We Age*. New York: Ballantine Books, 1994.

———. "The Limited *in Vitro* Lifetime of Human Diploid Cell Strains." *Experimental Cell Research* 37 (1965): 614–36.

Hayflick, Leonard, and Paul S. Moorhead. "The Serial Cultivation of Human Diploid Cell Strains." *Experimental Cell Research* 25, no. 3 (1961): 585–621.

Hayflick, Leonard, Stanley A. Plotkin, Thomas W. Norton, and Hilary Koprowski, "Preparation of Poliovirus Vaccines in a Human Fetal Diploid Cell Strain." *American Journal of Hygiene* 75 (1962): 240–58.

Hooper, Edward. *The River: A Journey to the Source of HIV and AIDS*. Boston, New York, and London: Little, Brown, 1999.

Hughes, Sally Smith. "Making Dollars Out of DNA: The First Major Patent in Biotechnology and the Commercialization of Molecular Biology, 1974–1980." *Isis* 92 (2001): 541–75.

Jackson, Alan C., and William H. Wunner, eds. *Rabies*. 2nd edition. London: Elsevier, 2007.

Jacobs, J. P., C. M. Jones, and J. P. Baille. "Characteristics of a Human Diploid Cell Designated MRC-5." *Nature* 227 (1970): 168–70.

Kim, Nam W., Mieczyslaw A. Piatyszek, et al. "Specific Association of Human Telomerase Activity with Immortal Cells and Cancer." *Science* 266 (1994): 2011–15.

Kirkwood, Tom. *Time of Our Lives: The Science of Human Aging*. New York: Oxford University Press, 1999.

Koprowska, Irena. *A Woman Wanders Through Life and Science*. Albany: State University of New York Press, 1997.

Korsman, Stephen N. J., Gert U. van Zyl, Louise Nutt, Monique I. Andersson, and Wolfgang Preiser, eds. *Virology: An Illustrated Colour Text*. Edinburgh: Churchill Livingstone Elsevier, 2012.

Lambert, Nathaniel, Peter Strebel, Walter Orenstein, Joseph Icenogle, and Gregory A. Poland. "Seminar: Rubella," *Lancet* 385 (2015): 2297–2307.

Levenstein, Lisa. *A Movement Without Marches: African American Women and the Politics of Poverty in Postwar Philadelphia*. Chapel Hill: University of North Carolina Press, 2009.

Miller, Patricia G. *The Worst of Times: Illegal Abortion—Survivors, Practitioners, Coroners, Cops, and Children of Women Who Died Talk About Its Horrors*. New York: HarperCollins, 1993.

Mukherjee, Siddhartha. *The Emperor of All Maladies: A Biography of Cancer*. New York: Scribner, 2010.

Mulford, Robert D. "Experimentation on Human Beings." *Stanford Law Review* 20 (1967): 99–117.

National Communicable Disease Center. *Rubella Surveillance Report No. 1*. Washington, DC: U.S. Department of Health, Education, and Welfare, Public Health Service, 1969.

Norrby, Erling. *Nobel Prizes and Life Sciences*. Singapore: World Scientific, 2010.

Nossif, Rosemary. *Before Roe: Abortion Policy in the States*. Philadelphia: Temple University Press, 2001.

O'Donnell, Donna Gentile. *Provider of Last Resort: The Story of the Closure of the Philadelphia General Hospital*. Philadelphia: Camino, 2005.

O'Neill, John F. "The Ocular Manifestations of Congenital Infection: A Study of the Early Effect and Long-Term Outcome of Maternally Transmitted Rubella and Toxoplasmosis." *Transactions of the American Ophthalmological Society* 96 (1998): 813–79.

Offit, Paul A. *The Cutter Incident: How America's First Polio Vaccine Led to the Growing Vaccine Crisis*. New Haven, CT, and London: Yale University Press, 2005.

———. *Vaccinated: One Man's Quest to Defeat the World's Deadliest Diseases*. New York: Harper-Collins, 2007.

Oshinsky, David M. *Polio: An American Story*. New York: Oxford University Press, 2005.

Plotkin, Stanley A. ed. *History of Vaccine Development*. Springer Science + Business Media, 2011.

———. "Virologic Assistance in the Management of German Measles in Pregnancy." *Journal of the American Medical Association* 190 (1964): 265–68.

Plotkin, Stanley A., André Boué, and Joelle Boué. "The *in Vitro* Growth of Rubella Virus in Human Embryo Cells." *American Journal of Epidemiology* 61 (1965): 71–85.

Plotkin, Stanley A., David Cornfeld, and Theodore H. Ingalls. "Studies of Immunization with Living Rubella Virus: Trials in Children of a Strain Cultured from an Aborted Fetus." *American Journal of Diseases of Children* 110 (1965): 381–89.

Plotkin, Stanley A., and Antti Vaheri. "Human Fibroblasts Infected with Rubella Produce a Growth Inhibitor." *Science* 156 (1967): 659–61.

Plotkin, Stanley A., John D. Farquhar, Michael Katz, and Fritz Buser. "Attenuation of RA 27/3 Rubella Virus in WI-38 Human Diploid Cells." *American Journal of Diseases of Children* 118 (1969): 178–85.

Plotkin, Stanley A., and Pearay L. Ogra. "Immunologic Properties of RA27/3 Rubella Virus Vaccine: A Comparison with Strains Presently Licensed in the United States." *Journal of the American Medical Association* 225, no. 6 (1973): 585–90.

Plotkin, Stanley A., Michael Katz, and Theodore H. Ingalls. "A New Attenuated Rubella Virus Grown in Human Fibroblasts: Evidence for Reduced Nasopharyngeal Excretion." *American Journal of Epidemiology* 86 (1967): 468–77.

Preston, Richard. *The Hot Zone*. New York: Anchor Books, Doubleday, 1995. First published 1994 by Random House.

Reagan, Leslie J. *Dangerous Pregnancies: Mothers, Disabilities, and Abortion in Modern America*. Berkeley and Los Angeles: University of California Press: 2010.

———. *When Abortion Was a Crime: Women, Medicine and Law in the United States, 1867–1973*. Berkeley and Los Angeles: University of California Press, 1997.

Rothman, David J. *Strangers at the Bedside: A History of How Law and Bioethics Transformed Medical Decision Making*. New York: Aldine de Gruyter, 1991, 2003.

Rybicki, Edward P. *A Short History of the Discovery of Viruses*. Buglet Press e-book, 2015. This book is available for purchase at https://itunes.apple.com/us/book/short-history-discovery -viruses/id1001627125?mt=13. The various parts of the e-book are also freely available online:

   part 1: https://rybicki.wordpress.com/2012/02/06/
      a-short-history-of-the-discovery-of-viruses-part-1/
   part 2: https://rybicki.wordpress.com/2012/02/07/
      a-short-history-of-the-discovery-of-viruses-part-2/
   part 3: https://rybicki.wordpress.com/2015/01/29/
      a-short-history-of-the-discovery-of-viruses-part-3/
   part 4: https://rybicki.wordpress.com/2015/02/11/
      a-short-history-of-the-discovery-of-viruses-part-4/

Shah, Keerti, and Neal Nathanson. "Human Exposure to SV40: Review and Comment." *American Journal of Epidemiology* 103, no. 1 (1976): 1–12.

Shay, Jerry W., and Woodring E. Wright. "Hayflick, His Limit, and Cellular Ageing." *Nature Reviews Molecular Cell Biology* 1 (2000): 72–76.

Shorter, Edward. *The Health Century*. New York: Doubleday, 1987.

Skloot, Rebecca. *The Immortal Life of Henrietta Lacks*. New York: Crown, 2010.

Slenczka, Werner, and Hans Dieter Klenk. "Forty Years of Marburg Virus." *Journal of Infectious Diseases* 196 (2007): S131–35.

Szostak, Jack W., and Elizabeth H. Blackburn. "Cloning Yeast Telomeres on Linear Plasmid Vectors." *Cell* 20, no. 1 (1982): 245–55.

Vaughan, Roger. *Listen to the Music: The Life of Hilary Koprowski.* New York: Springer-Verlag, 2000.

Wade, Nicholas. "Hayflick's Tragedy: The Rise and Fall of a Human Cell Line." *Science* 192, no. 4235 (1976): 125–27.

Wadman, Meredith. "Cell Division." *Nature* 498 (2013): 422–26. http://www.nature.com/news/medical-research-cell-division-1.13273.

———. "The Truth About Fetal Tissue Research." *Nature* 528 (2015): 178–81. www.nature.com/news/the-truth-about-fetal-tissue-research-1.18960.

Warren, Leonard. "A History of the Wistar Institute of Anatomy and Biology" (unpublished manuscript). March 25, 2014.

Webster, William S. "Teratogen Update: Congenital Rubella." *Teratology* 58 (1998): 13–23. http://teratology.org/updates/58pg13.pdf.

Wiktor, Tadeusz J., Mario Fernandes, and Hilary Koprowski. "Cultivation of Rabies Virus in Human Diploid Cell Strain WI-38." *Journal of Immunology* 93 (1964): 353–66.

Wiktor, Tadeusz J., Stanley A. Plotkin, and Doris W. Grella. "Human Cell Culture Rabies Vaccine: Antibody Response in Man." *Journal of the American Medical Association* 224, no. 8 (1973): 1170–71.

Wistar Institute of Anatomy and Biology. *Biennial Report, 1958–1959.* Courtesy of the Wistar Institute.

———. *Biennial Report, 1960–1961.* Courtesy of the Wistar Institute.

———. *Biennial Research Report 1962–1963.* Courtesy of the Wistar Institute.

Wright, W. E., and L. Hayflick. "Nuclear Control of Cellular Aging Demonstrated by Hybridization of Anucleate and Whole Cultured Normal Human Fibroblasts." *Experimental Cell Research* 96 (1975): 113–21.

Congressional Testimony

U.S. Congress. Senate. Subcommittee on Executive Reorganization and Government Research of the Committee on Government Operations. *Consumer Safety Act of 1972: Hearings Before the Subcommittee on Executive Reorganization and Government Research of the Committee on Government Operations on Titles I and II of S. 3419.* 92nd Cong., 2nd sess., April 20 and 21 and May 3 and 4, 1972.

Documents Obtained Under the Freedom of Information Act

Division of Management Survey and Review. "Factual Chronology." Investigations 9, folder 1. Directors' Files, Office of the Director, National Institutes of Health. (Abbreviated in notes as "Factual Chronology." This document is appended to the Riseberg Memo, detailed below.)

Hayflick, Leonard. "Requested Amendments and Changes to the Records Pertaining to Doctor Leonard Hayflick." April 13, 1976. Investigations 9, folder 1. Directors' Files, Office of the Director, National Institutes of Health. (Abbreviated in notes as "Hayflick Rebuttal to Schriver Report.")

———, Edmond C. Gregorian, Michael Hughes, and Vincent B. Terlep. "Settlement Agreement." September 15, 1981. Investigations 9, folder 1. Directors' Files, Office of the Director, National Institutes of Health.

Riseberg, Richard J. (legal adviser, NIH). Memo to Mary Goggin (chief, Administrative Law Branch, Business & Administrative Law Division, U.S. Department of Health, Education and Welfare, Office of the General Counsel). "Subject: Cell Associates, Inc., et al, v. National Institutes of Health, et al.," May 6, 1976. (Abbreviated in notes as "Riseberg Memo.")

Schriver, James W. "Investigation of Activities Relating to the Storage, Distribution and Sale of Human Diploid Cell Strains WI-38 and WI-26." P-75-211. January 30, 1976. Investigations 9, folder 1. Directors' Files, Office of the Director, National Institutes of Health. (Abbreviated in notes as "Schriver Report.")

———. (director, Division of Management Survey and Review, ODA). "Comments to Dr. Hayflick's Request for Amendments and Changes to DMSR Reports." September 1, 1976. Investigations 9, folder 1. Directors' Files, Office of the Director, National Institutes of Health. (Abbreviated in notes as "Schriver Rebuttal to Hayflick Rebuttal.")

Interviews, Oral Histories, and Private Collections

"Interviewee: Bernice Eddy, December 4, 1986." Edward Shorter. *The Health Century* Oral History Collection. National Library of Medicine, History of Medicine Division, Bethesda, MD.

"Leonard Hayflick Interviews." Meredith Wadman. October 3*, 4*, 10, 11, 12, 16, 17, 18, 23, and 24, 2012; November 7 and 26, 2012; February 22, 2013; March 3*, 4*, and 5*, 2013; May 13, 30, and 31, 2013; June 2, 2013; July 1, 2013; March 23 and 31, 2014; May 19, 20, and 28, 2014; June 9, 12, 17, 18, and 23, 2014; July 3, 10, and 11, 2014; November 19, 2014*; and July 29, 2015. (Interviews marked with an asterisk were conducted in person. The others were conducted by telephone.)

"Leonard Hayflick Interviews." *Web of Stories*. www.webofstories.com/people/leonard.hayflick.

"Maurice Hilleman Interview," Paul Offit. November 30, 2004. Courtesy of Paul Offit.

"Interviewee: Maurice Hilleman, February 6, 1987." Edward Shorter. *The Health Century* Oral History Collection. National Library of Medicine, History of Medicine Division, Bethesda, MD.

"Ruth L. Kirschstein Oral History Interview, Part 2." Victoria Harden and Caroline Hannaway. October 29, 1998. Office of NIH History, Oral History Archive, Bethesda, MD. https://history.nih.gov/archives/oral_histories.html.

"Dr. Paul Parkman Interview." Sarah Leavitt. June 7, 2005. Office of NIH History, Oral History Archive, Bethesda, MD. https://history.nih.gov/archives/oral_histories.html.

"Stanley Plotkin Interviews." Meredith Wadman. December 18, 2012; March 19, August 29, September 20, October 20 and 21, 2014; May 25, June 1, and September 8, 2015.

Stanley Plotkin private papers, Doylestown, PA.

Niels Reimers. "Stanford's Office of Technology Licensing and the Cohen/Boyer Cloning Patents." Oral history conducted in 1997 by Sally Smith Hughes. Regional Oral History Office, Bancroft Library, University of California, Berkeley, 1998.

INDEX